STREETCARS AND THE SHIFTING GEOGRAPHIES OF TORONTO

To David Cape:

Many thanks for buying our book. We hope you enjoy it!

BRIAN DOUCET AND MICHAEL DOUCET

STREETCARS
AND THE SHIFTING GEOGRAPHIES OF TORONTO

A Visual Analysis of Change

UNIVERSITY OF TORONTO PRESS
Toronto Buffalo London

© University of Toronto Press 2022
Toronto Buffalo London
utorontopress.com
Printed in Canada

ISBN 978-1-4875-0010-8 (paper)
ISBN 978-1-4875-1019-0 (EPUB)
ISBN 978-1-4875-1018-3 (PDF)

Every effort has been made to contact copyright holders; in the event of an error or omission, please notify the publisher.

Library and Archives Canada Cataloguing in Publication

Title: Streetcars and the shifting geographies of Toronto : a visual analysis of change / Brian Doucet and Michael Doucet.
Names: Doucet, Brian, 1980– author. | Doucet, Michael J., author.
Description: Includes bibliographical references and index.
Identifiers: Canadiana (print) 20210369310 | Canadiana (ebook) 20210369396 | ISBN 9781487500108 (paper) | ISBN 9781487510190 (EPUB) | ISBN 9781487510183 (PDF)
Subjects: LCSH: Cable cars (Streetcars) – Ontario – Toronto – History – 20th century. | LCSH: Cable cars (Streetcars) – Ontario – Toronto – History—20th century – Pictorial works. | LCSH: Repeat photography – Ontario – Toronto. | LCSH: Documentary photography – Ontario – Toronto. | LCSH: Toronto (Ont.) – Historical geography. | LCSH: Toronto (Ont.) – History – 20th century.
Classification: LCC TF727.T6 D68 2022 | DDC 388.4/609713541 – dc23

We wish to acknowledge the land on which the University of Toronto Press operates. This land is the traditional territory of the Wendat, the Anishnaabeg, the Haudenosaunee, the Métis, and the Mississaugas of the Credit First Nation.

University of Toronto Press acknowledges the financial support of the Government of Canada, the Canada Council for the Arts, and the Ontario Arts Council, an agency of the Government of Ontario, for its publishing activities.

This book is dedicated to Helen, Natalie, Christine, Elodie, Hugo, Felix, and Emmett.

CONTENTS

List of Figures and Tables ... ix

Acknowledgments and Preface ... xiii

Introduction: Streetcar Photography and the Changing City ... 1

Chapter 1: The Changing Geography of Toronto ... 7

Chapter 2: Toronto in a Global Context ... 35

Chapter 3: Neighbourhood Change ... 51

Chapter 4: Visual Methodologies and Repeat Photography ... 85

Chapter 5: Photographing Streetcars; Picturing Toronto ... 97

Chapter 6: A Short History of Toronto's Streetcars ... 123

Photo Portfolios ... 135

Portfolio 1: Downtown ... 137

Portfolio 2: (De)industrialization ... 165

Portfolio 3: Neighbourhoods ... 189

Chapter 7: Interpreting Visual Change in a Divided City ... 231

Chapter 8: Neighbourhood Change, Mobility, and Socially Just Solutions ... 247

Notes ... 265

Index ... 289

FIGURES AND TABLES

Figures

I.1.	Brian and Michael Doucet, 1982 and 2019	4
1.1.	City of Toronto population, 1911–2016	9
1.2.	A Sunday stop on Roncesvalles Avenue, 2005	10
1.3.	TTC ridership, 1960–2020	10
1.4.	Aerial view of apartments in St. James Town and inner suburbs from Commerce Court, 1975	14
1.5.	CLRVs on King Street during the December 2004 ice storm	17
1.6.	A shuttle bus at Yonge and Manor in January 1999	18
1.7.	Streetcars out of service at Russell carhouse during the blackout of August 2003	19
1.8.	Looking east from the observation deck on the 57th floor of Commerce Court, May 1975	23
1.9.	Average Toronto house prices, 1965–2019	23
1.10.	Three views from a helicopter above Toronto, 1972	26
1.11.	Views from the CN Tower, 1985/6 and 2020	28
2.1.	Yonge-Dundas Square	46
3.1.	Broadview Avenue near Riverdale Avenue, 1971 and 2020	53
3.2.	Queen Street East at Lee Avenue, 1968 and 2020	54
3.3.	Queen Street East at Greenwood, 1969 and 2020	55
3.4.	Gladstone Avenue between Dundas and College Streets, 2019	56
3.5.	Dundas Street West, near Roncesvalles, 1967 and 2016	57
3.6.	The Streetcar City and the maximum extent of Toronto's streetcar network (1928)	61
3.7.	The Streetcar City and the current streetcar and subway network (2021)	61
3.8.	Jane Jacobs speaking at a rally in Toronto	65
3.9.	Retail gentrification in Little India (Gerrard Street East), 2011 and 2014	72
3.10.	A bistro opposite a coin laundry, Gerrard Street East at Greenwood Avenue, 2019	73
3.11.	The "condofication" or "condomiumization" of Toronto. Bathurst Street at Fort York Boulevard, 2004 and 2019	76
4.1.	Repeat photograph at Bathurst and Fleet Streets in 2004, 2015, and 2019	90
4.2.	The Honest Ed's site and CLRV streetcars, 1984 and 2019	91
4.3.	Two images from the Rocky Mountain Legacy Project, 1927 and 2009	92
4.4.	Four images of south view along Bathurst Street at Queen Street West: 1965, 2005, 2015, and 2019	94
5.1.	A typical three-quarters streetcar image. Roncesvalles Avenue at Grenadier Road, 23 July 2016	100
5.2.	A less conventional type of streetcar-subject photography. Bathurst and Fort York Boulevard, 6 December 2019	101

5.3.	CLRV 4187 in its new home on the family farm of Alex Glista, near Priceville, Ontario, 17 August 2020	101
5.4.	Johnstown, Pennsylvania, the last small city trolley system in America	104
5.5.	Downtown. Queen Street West and York Street, 1967 and 2020	105
5.6.	The intersection of Queen, King, and Roncesvalles, 1964, 2020, 1964, 2020	106
5.7.	Neville Park at the end of the 501 Queen line, 1968 and 2014	108
5.8.	An example of "walking the streetcar line," in which a photographer stops to take a picture when a streetcar approaches. Westbound streetcars, King Street West, between Roncesvalles and Dufferin	109
5.9.	Two streetcar-subject photography images by Roberta Hill. Queen and Roncesvalles, 1967, and Pape and Danforth, 1965	110
5.10.	Two streetcar images of cities which are not major tourist draws: Philadelphia and Duisburg, Germany	112
5.11.	Paris and Rome as seen through their tram networks	113
5.12.	The late John F. Bromley	119
5.13.	A scan of a John F. Bromley slide-mount, showing details of what information was recorded	119
5.14.	The late Robert D. McMann	119
5.15.	Queen Street East and Coxwell Avenue, 1954 and 2019	120
6.1.	A Toronto Peter Witt streetcar	124
6.2.	An Air-Electric PCC streetcar	125
6.3.	A Cincinnati PCC streetcar that would later be sold to Toronto	126
6.4.	An ex-TTC, ex-Cincinnati Street Railway PCC running in its third home, Tampico, Mexico	127
6.5.	An ex-TTC, ex-Birmingham, Alabama, PCC running in its third home, Philadelphia, 1980	128
6.6.	A prototype CLRV depicted new in 1978	128
6.7.	PCCs awaiting scrapping at Wychwood yard in the 1980s	129
6.8.	Former TTC PCCs at the Halton County Radial Railway Museum in Milton, Ontario	129
6.9.	A Bombardier Flexity LRT in Brussels, Belgium	130
6.10.	The final night of CLRV operation in Toronto. Bathurst and Harbord Streets, 28 December 2019	131
6.11.	A farewell parade of CLRVs after 40 years of service in Toronto, 29 December 2019	131
6.12	Three generations of Toronto streetcars, 2016	133
7.1.	Two different views of Toronto's streetcars	232
7.2.	Kim's Convenience and an eastbound 501 streetcar	232
7.3.	Development proposal sign for 2 Bloor Street West	235
7.4.	Contrasting images of the pandemic city	237
7.5.	Rogers Road, one of the last frontiers of gentrification in the Streetcar City	243
8.1.	Two images of Amsterdam showing distinct changes to the built environment to shift focus away from cars, and towards pedestrians, bikes, and transit	249
8.2.	Waterloo Region's new ION LRT	250
8.3.	The transformation of King Street as a result of the King Street Pilot	251
8.4.	Dedicated bus lane in the Eglinton East Corridor	255

Tables

1.1.	Selected census variables, 1961	9
1.2.	Rapid transit and LRT development since 1954	11

1.3.	Selected census variables, 1981	13
1.4.	Toronto's employment structure, 1983, 2003, and 2019	14
1.5.	Ten largest employers in Toronto, 1983 and 2019	15
1.6.	Selected census variables, 2016	20
1.7.	Milestones in the development of Toronto's skyline	24
2.1.	Canadian cities in GaWC classification 2000 and 2018	42
3.1	Characteristics of the Streetcar City and the automobile city, 2016	62
3.2.	Characteristics of the Streetcar City and the automobile city, 1971	63

ACKNOWLEDGMENTS AND PREFACE

We acknowledge the land we live and work on is the traditional territory of many nations including the Mississaugas of the Credit, the Anishnabeg, the Chippewa, the Haudenosaunee, and the Wendat peoples, and is now home to many diverse First Nations, Inuit, and Métis peoples. We also acknowledge that Toronto is covered by Treaty 13 with the Mississaugas of the Credit.

The University of Waterloo and the City of Kitchener are situated on the Haldimand Tract, land that was promised to the Haudenosaunee of the Six Nations of the Grand River, and is within the traditional territory of the Neutral, Anishinaabeg, and Haudenosaunee peoples.

This project has been many years in the making. Much has changed in our own lives in the time it has taken to go from concept to published book. When we began to think about using streetcar photography as a method to interpret and analyse urban change, Michael had recently retired from his position as a Professor of Geography at Ryerson University. Brian had recently completed his PhD in Geography at Utrecht University and was starting his academic career in the Netherlands. In the intervening years, Michael has become a grandfather four-times over and Brian has returned to Canada with his family to take up a position as a Canada Research Chair in the School of Planning at the University of Waterloo.

Despite these changes, a constant throughout most of our lives has been an interest in streetcars and transit. Michael rode the opening day of the Yonge subway with his father in 1954. Brian enjoyed many opening day rides together with Michael. Our mutual interest in transit was supported by living on a street with its own pedestrian bridge over the Yonge subway line.

Both of us have taken pictures of streetcars for decades – Michael beginning in the early 1970s and Brian in the late 1990s. To start, our approach was similar to many other streetcar enthusiasts: we focused on the different vehicles, different routes, enthusiast charters, the appearances of historic streetcars and on deviations from day-to-day operations of the system. Like other enthusiasts, we also collected historic slides of streetcars in Toronto and other cities, largely dating from the 1950s, 1960s, and 1970s. To begin with, our interest was in the streetcars themselves.

Over time, however, we began to realize that these images were about more than streetcars. When examining everything within them, it became clear that these photographs and slides presented us with an entirely new and unique lens from which to explore the changing geographies of the city. Consequently, our approach to photography shifted. We rode the streetcars and walked the lines with these historic images in hand, rephotographing them throughout the city. This photography was made easier once Brian returned to Southern Ontario; however, during regular visits to Toronto while Brian was still living in the Netherlands, it was not uncommon for us to ride the entire streetcar network in the span of a few weeks' visit. In our growing collection of historic streetcar slides, we also sought images that showed ordinary views of the city and its streets, rather than slides where the streetcar filled the majority of the frame.

A book of this sort is, by its very nature, slow science. Rather than trying to publish as many articles in the shortest possible time span, a project of this type requires an approach that is contemplative and reflective, both qualities that require sufficient time. Our repeat photography utilizes images that are more than fifty years old. A lot has changed in Toronto since PCCs ran along Bloor Street or Rogers Road. Likewise, a lot has changed since we first started thinking about this project around 2013.

As a result, some of our "second views" have been further updated with third, or even fourth views in order to depict a city that is constantly changing. We even added some images of Toronto during the first waves of the COVID-19 pandemic to illustrate changing social geographies, even if the buildings themselves have remained unaltered.

Because this has been such a large project, there are many people to thank and acknowledge for their support, advice, and guidance. To start, we thank the editors and staff at the University of Toronto Press, Jodi Lewchuk, who took over as the acquisitions editor midway through this project, and Doug Hildebrand, who originally supported it. Both have been extremely enthusiastic and supportive of this research. Leah Connor helped with production, and the copy-editing was done by Anne Laughlin. We also wish to thank the anonymous reviewers of our manuscript for their helpful suggestions and encouragement.

We have used images from a number of streetcar photographers who were active in the 1960s, 1970s, and 1980s. We thank Ted Wickson and John F. Bromley for allowing us to use their images in this book. Sadly, John F. Bromley passed away in late 2019 and was unable to see this final book. In addition to kindly allowing us to use his excellent photos, he also helped identify the correct photographer of some of the images we use in this book, including several by the late Robert D. McMann. John F. Bromley's widow, Margaret, kindly provided some additional information about John, and provided a photograph (figure 5.12). Jane Naylor also provided some background on the life of the late Harvey Naylor. This information, combined with what other enthusiasts have told us, has allowed us to go beyond simply stating a photographer's name; it has also enabled us to provide insightful information about the people behind the camera.

Many original slides in our collection were taken by the late Robert D. McMann (see figure 5.14). McMann passed away in the mid-1990s, never married, and had no children. He began taking slides of Toronto streetcars in 1961 and was an active photographer until shortly before his death. His vast collection of slides was sold after his death; many were purchased by John F. Bromley, who resold them to other collectors and enthusiasts. Since then, they have appeared on many different websites. The original slides of images attributed to Robert D. McMann are all in our personal collection, unless otherwise noted.

We have made every attempt to credit all the images in this book to the photographers who took them and have made all reasonable attempts to contact the photographers, or their descendants or relatives. If we remain uncertain as to the photographer, we have labelled the source as "unknown."

Other streetcar enthusiasts have assisted us in a variety of ways. In addition to providing insights about the photographers, they have provided us with information about the streetcars, the city, and several images from their own personal collections that have helped to illustrate key places, trends, and processes. Thanks to Rob Hutchinson, Robert Lubinski, John Knight, and James Bow.

We purchased many images from Hampton Wayt, who sold the slide collection of his late father on eBay and also assisted us with some additional information about the images we were acquiring. Ken Josephson provided copies of a large number of images taken by Roberta Hill, as well as some details about her life and photography. Conrad Kickert was helpful in identifying the location of the Cincinnati image (figure 6.3).

Our academic colleagues at several institutions provided feedback and support. At Utrecht University, they included Jan van Weesep, Jan Prillwitz, Matthieu Permentier, Rianne van Melik, Pieter Hooijmeier, and the late Ronald van Kempen. Utrecht University's Department of Human Geography and Planning also provided funding for the acquisition of some of the images used in this book. At Erasmus University, colleagues included Roy Kemmers and Ward Vloeberghs. At the University of Waterloo, Pierre Filion provided helpful insights and critical feedback, particularly in our concluding chapters, and the book has benefited from conversations with Markus Moos, Martine August, and Jennifer Dean. We would also like to thank Andrew Trant for supplying images from the Rocky Mountain Legacy Project (figure 4.3).

Other important people to acknowledge include Larry Bourne, Donna Williams, and Chris Livett (who encouraged us early on in this project to look beyond the streetcars). Sean Marshall produced the maps featured throughout this book. Giuseppe Tolfo analysed the census data on the Streetcar City. The indexing was done by Cheryl Lemmens. Joe Mihevc graciously supplied an image of the reconstruction of Wychwood, and Lindsey Kwan supplied a photo of Wychwood Artscape. Alex Glista welcomed Brian and his children to his family's farm in Priceville, Ontario, where CLRV 4187 now resides. Rebecca Ng at the City of Toronto Planning Department assisted with the data from the City's Employment Survey. Library staff at the Ryerson Library, the Robarts Library, and the Toronto Reference Library all provided help with data acquisition. While we thank and acknowledge these individuals and organizations, all errors, omissions or mistakes are our own.

This book was supported with a grant provided by the office of the Dean of Arts, Ryerson University. Additional funding for it was provided through the Canada Research Chairs Program, under the award number 950-213821. Parts of this book, including some text from chapter 4 and some of the photo sets, were published in Brian Doucet, "Repeat Photography and Urban Change: Streetcar Photographs of Toronto since the 1960s," *City* 23, nos. 4–5 (2019): 411–38.

Brian would like to thank Helen Hamilton-Doucet for her love and support. Michael would like to thank Natalie Doucet for her encouragement and love.

A note about the treatment of the original slides and digital images: The original views and many of the figures were originally taken on slide film, often Kodachrome. These images have been scanned using a high-resolution Nikon slide scanner. Because of their age, it has been necessary for these images to undergo post-production editing in Lightroom. Our approach to post-production has been to do as little editing as possible. Kodachrome generally keeps its subtle colours over time, even after more than fifty years; however, it has been necessary to correct white balance, exposure, contrast, highlights, and shadows on many images. When editing these scanned images, we have tried to restore the original colour, white balance, and tone of the slide. We have also used cloning tools to eliminate dust marks and scratches on the slides.

Recent images have all been taken on digital cameras. Digital images were lightly edited in Lightroom for colour, light, and exposure. When cameras are tilted upwards, vertical lines have a tendency to converge towards the top. We have used post-production tools in Lightroom to correct this as much as possible in both the slides and digital images. Where needed, the angle of both digital and scanned images has been levelled to avoid any tilt on the horizon line. In some cases, a minimal amount of cropping was done. Apart from this, these images are unmanipulated.

On the new Bombardier streetcars, and many other modern LRVs and buses, the route and destination signs are digital. These signs flicker, although this is not visible to the naked eye. In bright light, or on images taken with a shutter speed faster than 1/100 of a second, this flickering makes the digital display less readable. Rather than either trying to recreate the text on these digital displays in post-production or blanking them out entirely, in the interests of manipulating the photography as little as possible, we have opted to leave these signs as they were photographed. Where possible, we have attempted to photograph stationary LRVs at lower shutter speeds in order to avoid this issue.

INTRODUCTION:
STREETCAR PHOTOGRAPHY AND THE CHANGING CITY

Toronto is constantly changing. It is now the fourth largest city in North America, behind Mexico City, New York, and Los Angeles, and the Greater Toronto Area (GTA) has ranked as one of North America's fastest-growing regions for decades. Toronto, for many years English Canada's primary manufacturing city, has over the past five decades become one of the world's major financial hubs. It has moved from being a rather insignificant provincial city to Canada's largest and most important metropolis, and has entered the upper echelons of world cities. Waves of immigration have transformed Toronto, once predominantly white and British, into one of the most multicultural cities on earth. Shifting ideas about planning, design, and policy have influenced how communities, neighbourhoods, and urban space have changed. Just as the city's economy, demography, and policies have changed dramatically, so too has its social geography. Fifty years ago, Toronto was a predominantly middle-class, middle-income city, with most poverty concentrated within its urban core; today, the city is highly polarized and unequal, with both growing affluence and poverty, while those same inner-city neighbourhoods have become gentrified and expensive.

This book is one of many that has attempted to document, interpret, and analyse these long-term trends. However, unlike most scholarly studies, ours does not rely primarily on statistics, policy analysis, or interviews. Instead, our research centres on the visual: what these changes look like in ordinary, everyday spaces of the city and what visual changes tell us about social, spatial, and economic shifts. Our enquiry begins with the question of how we can visualize these major transformations as well as the forces of change that shape them.

To search for visual images to do this, we could delve into the city's official archives, or look through old newspapers. However, these pictures were often tinted with particular social or political objectives and do not necessarily cover all aspects of the city.[1] Instead, we have turned to an unlikely, albeit quintessentially Toronto resource: photographs taken by streetcar enthusiasts. At first glance, photos of old trolleys may seem of little use to academic researchers interested in topics such as deindustrialization, gentrification, and world city formation; the men (and the vast majority of streetcar photographers were, and are, men) who took, and continue to take, photographs of streetcars, are more focused on the vehicles themselves, rather than the wider geographies around them. Their main aim has always been to photograph the streetcars, rather than to visually document the complex economic, social, and political changes taking place in the cities in which they ran. Streetcars are inherently part of the Toronto's landscape the way they are in no other North American city. As Toronto Star columnist Christopher Hume has stated, "Streetcars in Toronto aren't just a means of getting around; they are part of the city's heritage, its self-image. In that most over-rated word, streetcars are iconic. No representation of Toronto is complete without one."[2]

However, these images contain much more than just the streetcars themselves. Since the 1960s – the starting point of our long-term visual analysis – Toronto has had the largest streetcar network in North America. This makes it an ideal place to use this unique visual approach. The photos taken by streetcar enthusiasts offer us a rich and unique perspective of the ordinary, day-to-day spaces in Toronto. They provide windows into the everyday city that help us to understand changing urban form and morphologies, land-use, spatial patterns, planning ideas, economic trends, and social relations. When contextualized within a deeper theoretical and conceptual understanding of the major economic, social, and political forces of change that have shaped Toronto, these photographs from the 1960s and 1970s offer new insights for a robust analysis and interpretation of the city.

On their own these images present us with a historical story of what Toronto used to be like. However, when these old photographs are "brought into a current dialogue [that] extend[s] a conversation about place over time,"[3] they become highly relevant for current planning, policy, and political debates. To do this, we use a practice called repeat photography (sometimes known as rephotography), which involves rephotographing images at different moments in time.[4] Repeat photography has its origins in the natural sciences, where it has been used to study, among other things, glacial, geological, or vegetation changes over many decades, but it is an approach that is also used by a growing number of social scientists to examine urban change.[5] To paint a clearer picture of the ways in which major economic, political, and social forces of change have shaped the visual landscapes of Toronto, we have travelled throughout the city with copies of these old images at hand, gone to the same places where streetcar enthusiasts took their photos over fifty years ago, and rephotographed the same scenes they did. These "before and after" views then act as the starting point, or "data set," for a visual analysis of change.

Simply looking at these before-and-after pictures, it can be easy to conclude that change is a natural progression, part of the organic evolution and development of cities. However, as critical urban scholars, we argue that there is nothing natural or inevitable about the changes that have taken place in Toronto over the past decades. The visual manifestations of changes (or lack thereof) in the urban landscape are the products of economic shifts, political choices, social transformations, and planning decisions. In other words, they are the outcomes of the wider political economy of the city. Additionally, differences in urban form, morphology, and the built environment, specifically revolving around differences in the city that developed with the streetcar as the dominant transport technology (pre-1945) and the city dominated by the automobile (post-1945), are important to take into account. Therefore, some explanation as to these urban structural changes and the political, planning, and policy decisions that shape urban space is necessary. This is why this book is not merely a collection of before-and-after photographs. Before delving into the visual, it is important to ground these views in both urban theories and the shifting political economy of Toronto. The photos and the text are a two-way street: an understanding of urban change helps us to interpret these images while the images themselves also contribute to an enhanced analysis of the changing geographies of Toronto.

While today almost everyone has a camera on their smartphone, in the 1960s photography was much less common. Most photos from this era depicted either major family, community, or civic events (such as family gatherings or a Stanley Cup parade) or tourist landmarks. Urban photography was largely the realm of professional photographers such as official archivists or journalists. Some pioneering amateur photographers did seek to document particular urban conditions, but what makes streetcar photographers unique is the accidental, or unintentional, nature by which they photographed urban space; their primary gaze was on the vehicles themselves. For us, it is everything around these streetcars that acts as our starting point of visual analysis.

There is a growing interest in both visual methodologies as a tool for interpreting the complexities of the city and the need to better understand long-term changes that shape urban space. This book combines these two approaches to provide a critical visual analysis of Toronto and the uneven nature of its growth, development, and change. Here photography is more than an appendage to illustrate a main point or finding: it plays the central role in the analysis.[6]

Aims of the Book

Photography can play a pivotal role in critical urban scholarship through its ability to show fine-grained detail and physical manifestations of change, and to make visible what is often rendered invisible. Because of this, visual analysis can work towards enhancing existing conversations about the city, while also starting new ones. Urban planning professor Peter Marcuse has outlined three pillars of critical urban planning: expose injustices or inequities, politicize key urban processes, and propose alternatives.[7] As we discuss in detail in chapter 4, photography is ideally placed to contribute to all three pillars of this critical analysis of the city.

There are two main aims of this book. The first is to analyse how the major forces of change that have shaped Toronto since the 1960s are visually manifested in uneven geographies across the city. While we are confined to the boundaries of the Streetcar City – the parts of Toronto that developed before World War II with the streetcar as the dominant transport technology – throughout the book, we also reflect on how what we see here relates to the rest of the city and the wider region beyond. There are several key questions that have guided this book. Does a visual analysis of long-term change reinforce, challenge, or contradict analysis done using other methods? What patterns or trends are made visible through a photographic analysis? What new insights do we gain from studying long-term visual changes and how can we use this knowledge to address key policy and planning challenges, particularly around inequality, housing, and mobility?

These questions relate to the second aim of the book: to bridge the divide between academic research and wider policy, political, planning, and public debates. There is a remarkable curiosity and thirst for knowledge among Torontonians about their city. This is not just a matter of academic enquiry; ordinary residents who have spent their entire lives in the city as well as relative newcomers are eager to know more about Toronto, how it has changed, and why. Our book responds to a desire for genuine knowledge and explanation, rather than just descriptive reminiscing about days gone by. In this sense, our choice to focus on the visuals of the streetcars is about more than just nostalgia or the availability of a large supply of historic images: we recognize that the city's streetcars are themselves political. To some Torontonians, they are an iconic mode of daily travel and part of a much-sought-after urban authenticity; to others, they are either irrelevant, an impediment to mobility, or a symbol of a downtown "elite."

The issues rendered visible in the photos we present also spark new questions about current (and historic) fault lines within the city. Therefore, we have attempted to merge academic rigour with an accessible style that will be of value to scholars, students, professionals, and the wider public. The images and analysis presented in this book are intended to inspire and spark curiosity, as well as challenge all types of readers to look at the city with an enhanced, critical perspective. We want to move beyond siloed debates about cities; it is increasingly evident that discussions about transit, mobility, land-use, housing, economic development, the environment, racism, and equity need to be part of the same conversation. Visual analysis is well-positioned to play a leading role in these debates.

We are both academics who believe that it is important to share our knowledge with a wide audience. Both of us have strong and long-standing roots in Toronto.[*] Our teaching methods focus on visual representations of change and we encourage our students to critically examine the world around them by observing and exploring. With this book, we want to employ these same techniques, approaches, and principles for a much wider audience than just our academic peers and our own students. We hope that the photographs we have presented in this book challenge you to look at Toronto from new perspectives, make you question why things are the way they are and where they are, and imagine solutions for a better, and more socially just, city.

Outline and Structure of the Book

While our book emphasizes the visual, the text is designed to help the reader contextualize and analyse the images that are presented. The first chapter discusses the changing nature of Toronto. Relying on a combination of statistics from the Census of Canada and other sources, it outlines the major changes in Toronto's urban, economic, and social geography since the 1960s. This discussion is interlaced with our

---

[*] Michael Doucet has spent his entire life living in Toronto. He is an expert in the retail and immigration geographies of the city and, while working in the Department of Geography at Ryerson University, developed and taught a course specifically about Toronto. His son, Brian Doucet, was born and raised in Toronto and studied geography and history at the University of Toronto. Brian lived in the Netherlands from 2004 to 2017 and regularly brought his Dutch university students to Toronto on field trips. He is now the Canada Research Chair in Urban Change and Social Inclusion at the University of Waterloo.

Figure I.1 | Brian and Michael Doucet in 1982 and 2019.

own personal stories and experiences growing up in different periods in Toronto's history.

Chapter 2 places Toronto within a wider theoretical context that is rooted in economic shifts, deindustrialization, world city formation, and neoliberalism. Chapter 3 builds on this and examines how Toronto's emergence as a major global city impacts neighbourhood change. In this chapter, we will introduce the "Streetcar City," and outline its distinct urban form, morphology, land use, and density patterns, and compare these to the city and suburbs constructed after World War II. We will also discuss gentrification – a dominant trend in the Streetcar City – and situate it within an increasingly unequal and polarized city.

Chapter 4 begins our discussion of visual methodologies by focusing on a critical and constructive approach to interpreting visual imagery. This is followed by an introduction to repeat photography. In chapter 5, we explore the genre of streetcar- and railway-subject photography, including its geographies, and we critically examine the types of images it produces. We provide a brief overview of some important Toronto photographs, before turning to a discussion of the specific methods we have used for this book.

Chapter 6 provides a concise history of Toronto's streetcars within the context of wider urban developments in the city. While this book is not explicitly about Toronto's streetcars from a historical or technical perspective, an overview of the city's transit history helps to place the images used in this book within the wider trajectories of transit in Toronto specifically and in North America more broadly.

While we anticipate many readers will look at the photos first, we encourage you to read the chapters that precede them and then return to the photo sets in the middle of the book with this new information at hand and think about the active ways in which a variety of stakeholders (politicians, planners, developers, businesses, architects, households, community groups, activists) have influenced the type of economic, social, and architectural changes that are depicted in this book.

In total, we have selected almost seventy locations across the streetcar network; each features at least two views, spanning several decades. The first views generally date from the 1960s, although there are a few more recent photographs as well. Paired with these are updated views, taken between 2014 and 2020. In some instances, there are multiple views to reflect more recent changes, as well as a few views that depict the city during the COVID-19 pandemic. We have divided the photographs into three portfolios: the downtown core, (former) industrial districts, and neighbourhoods. Rather than limiting ourselves to today's network, we have included images taken throughout the streetcar network, as it existed in the 1960s, including thoroughfares such as Bloor Street and Rogers Road.

Two final chapters provide further analysis to help explain what we have seen and bring these photographs into conversation with current planning and policy debates. In chapter 7, we return to the idea of Toronto being a divided city. We examine issues of transportation, politics, and housing, and assess the role the fault line between the city built by the streetcars and the city built by the automobile plays in these divisions. Finally, in chapter 8, we propose some alternatives for a more socially just city and use our visual analysis to question some long-held assumptions about how Toronto can become a city for all who want to call it home.

1

THE CHANGING GEOGRAPHY OF TORONTO

In 1914, Charles Trick Currelly, the ROM's founding director, opened his brand-new institution with a very important personal imperative intact. He insisted that the museum be on a streetcar line, so it could be reached by anyone and everyone, regardless of their social or economic status. (Murray White, 2016)[1]

Anyone who lives in a major city knows that for transit to be successful, it must be reliable, frequent, fast and affordable. It must offer people a viable alternative to driving a car. This is especially true in downtown Toronto, where many of our iconic streetcar routes run along narrow streets with general car traffic.… King St. isn't working as well as it could, especially when it comes to providing reliable and efficient transit service … Streetcars are often crowded and forced to crawl through heavy traffic congestion … King St. needs to change. (Jennifer Keesmaat, former chief planner, City of Toronto, 2016)[2]

If the rise of the suburbs after the Second World War created a new type of city, then what is changing our city today is the rebirth of the downtown. Toronto's core is being transformed into a high-density, high-rise residential community and the largest centre of employment in the [Greater Toronto and Hamilton Area]. (Bill Freeman, 2015)[3]

Michael's Toronto

The older of the two authors of this book, I was born in the Toronto Western Hospital at the northeast corner of Dundas and Bathurst Streets, where two streetcar lines intersected, in February of 1948. Dealing with transit and urban change soon would become part of my experience. My first two years were spent in a sixth-floor apartment located at 1 Harbord Street, a building on the southwest corner of that street and St. George Street, at the then-western edge of the campus of the University of Toronto. Dating from 1899, and known as St. George Mansions, this six-storey building was the first purpose-built apartment building in the City of Toronto.[4] By 1965, it had been demolished and the site became the university's Ramsay Wright Zoological Laboratories. My parents did not purchase their first automobile until 1958, so transit use and riding streetcars were part of my experience from my earliest days. The Harbord Street apartment was two blocks east of the streetcar tracks on Spadina Avenue; trips could be made to the downtown area via the Harbord route.

In 1950, my family moved to a semi-detached house, on the south side of Eglinton Avenue East, that was four blocks east of Yonge Street. At that time, streetcar tracks extended along this portion of Eglinton Avenue between Yonge Street and Mount Pleasant Road to facilitate the movement of streetcars between these two important thoroughfares. Transit service along Eglinton, however, was provided by buses and trolleybuses. In late March of 1954, Canada's first subway line opened along Yonge Street from Eglinton to Union Station. My father took me to ride the subway on its first day of operation, and opening day rides would become a part of my experience whenever the streetcar or subway networks were expanded. The original Yonge subway line included 12 stations and was 7.4 kilometres in length. It was built at a cost of $50.5 million ($488.52 million in 2019 dollars), including rolling stock.[5] Trips downtown in my youth nearly always involved an ice cream treat, either a soft ice cream cone at the Eaton's Annex store or a dish of clown ice cream after lunch at the Round Room in Eaton's College Street store.

In 1956, our Eglinton Avenue house, along with all the others in the block between Lillian Street and Redpath Avenue, was purchased by a developer to make way for a commercial redevelopment project that became a 10-storey office structure known as the Union Carbide Building. Opened in 1958, that building would last until 1999, when it was torn down to make way for a 17-storey condominium structure known as 123 Eglinton Avenue East.

Following the sale of our home on Eglinton Avenue East, my family moved a little more than a kilometre east to a detached home on Mann Avenue, a street that ran south from Eglinton for one block to Soudan Avenue and that lay one block west of Bayview Avenue. In those days, the City of Toronto stopped at Bayview Avenue, where the Town of Leaside began. For me, the nearest Catholic school was St. Anselm's on Millwood Road in Leaside. It was there that I would complete elementary school.

I married my wife, Natalie, in July 1972. Our first two years together were spent in an apartment in a low-rise building on the east side of Bathurst Street two blocks south of St. Clair Avenue. Both thoroughfares were home to streetcar tracks. In 1974, we moved to our first house which was north and east of the intersection of St. Clair and Old Weston Road. The house was midway between the St. Clair and Rogers Road streetcar lines, which provided our links to downtown. In 1978, we moved to our current home, which was built in 1923. It is located south and slightly west of the intersection of Yonge Street and Eglinton Avenue. It was here that our two children, Brian and Christine, were raised, some distance from streetcar tracks but very close to ones used for Toronto's original subway line.

Since the earliest photos we intend to use in this book date from the early 1960s, we thought 1961, a census year in Canada, would make a good starting point for our discussion of change in Toronto. Coincidentally, in September 1961, I started high school at St. Michael's College School, located at St. Clair Avenue West and Bathurst Street, some six kilometres from my home. At least one-third of the daily journey to and from high school was completed via streetcar. So would begin a daily commuting pattern via public transportation to and from an educational institution, either as a student or faculty member, that would last for almost half a century until my retirement from Ryerson University[*] in August of 2010.

Toronto in 1961

Toronto in 1961 was a provincial and much less diverse place than it is today. It was a growing city, but one of little global significance, and still very much in the shadow of Canada's largest city, Montreal. Compared to contemporary Toronto, in 1961 it was truncated in terms of both skyline and spatial extent. Excluding industrial chimneys, only four buildings in Toronto, all in the downtown area, stood taller than 100 metres – the Bank of Commerce (1931, 145 metres and 34 storeys), the Royal York Hotel (1929, 124 metres and 28 storeys), the Bank of Nova Scotia (1951, 115 metres and 27 storeys), and the clock tower of Old City Hall (1899, 104 metres). Just 21 buildings exceeded 50 metres in height, including three church steeples, and the city was home to only 44 buildings that contained at least 12 storeys. The tallest residential structure in Toronto in 1961 was a 21-storey apartment building on Logan Avenue known as Ray McCleary Towers, which had been built in 1958 to house senior citizens. It was the 16th tallest building in the city at that time.[6]

According to the 1961 Canadian census, the then–City of Toronto was home to 672,407 people (figure 1.1). Clearly, suburbanization was well under way as the City of Toronto housed just 41.5 per cent of the residents of the municipality of Metropolitan Toronto, which had been created by the provincial government in 1954 and, at the time when the census was taken, encompassed Toronto and 12 other municipalities, each of which still had its own fire department. The Metropolitan Toronto Police Force dated from just 1957, and its officers patrolled Metro Toronto in yellow police cars, serving a population of 1,618, 800.[7]

For the 1961 census, the Toronto Census Metropolitan Area (CMA) attempted to capture Toronto's commuter shed, and had been expanded outward to include Oakville and what is now Mississauga to the west, Vaughan,

---

[*] On 26 August 2021, Ryerson's board of directors voted to change the name of the university due to Egerton Ryerson's role in establishing the Indian residential school system in Canada.

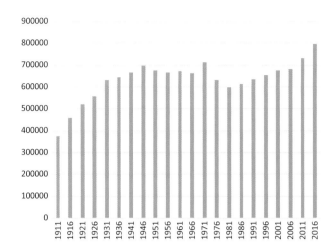

Figure 1.1 | City of Toronto population, 1911–2016. Source: Census of Canada, various years.

Table 1.1. Selected census variables, 1961.

Census variable	City of Toronto	Toronto CMA
Born outside Canada	41.9%	33.3%
British Ethnicity	51.8%	60.7%
Roman Catholic	34.7%	26.2%
Protestant	51.7%	63.2%
Jewish	2.8%	4.9%
Occupied dwellings	172,864	482,540
Single-detached	28.5%	55.7%
Apartments	36.5%	26.7%
Others types	35.0%	17.6%
Owner-occupied	56.5%	67.4%
Built before 1920	54.6%	24.0%
Built after 1945	13.5%	51.6%
Employment	**Males – CMA**	**Females – CMA**
Craftsmen, production process & related	32.8%	14.9%
Managerial	14.2%	2.6%
Professional & technical	11.3%	11.6%
Clerical	11.1%	41.8%
Services & recreation	8.6%	18.5%
Sales	8.1%	7.5%
Transportation & communication	7.4%	2.1%
Labourers	5.5%	1.5%
Primary	1.6%	0.2%

Richmond Hill and Markham to the north, and Pickering and Ajax to the east. The CMA contained 1,824,481 people in 1961 and circumscribed an area of some 1,286.4 square kilometres, roughly double the size of Metro Toronto (626 square kilometres) and 13 times the size of the City of Toronto (97.6 square kilometres). Settlement in the areas beyond the Metro Toronto boundaries was sparse, with 206,713 people living there in 1961. Much of the land in these areas remained in agricultural use, and the same was true for the northern portions of Etobicoke, North York, and Scarborough. The City of Toronto was far denser than the other 12 Metro municipalities, at 6,889.4 people per square kilometre, compared with only 1,789.1 for the CMA. Density was extremely low in the rest of the CMA beyond Metro's boundaries at 313.0 people per square kilometre.[8] Not surprisingly, these density differences were reflected in the streetscapes to be seen in various parts of the CMA. Here, as some of our photographs will show, housing types and periods of construction played a role in the look of the city.

In terms of demographic characteristics, in 1961 33.3 per cent of CMA residents and 41.9 per cent of City residents had been born outside of Canada. The 1961 census did not record visible minorities, but it did present figures on ethnic groups and religious affiliations, which point to an overwhelmingly white and Christian population. For example, fully 60.7 per cent of the CMA population self-identified with the British Isles, with the comparable figure for the City at 51.8 per cent (table 1.1). Toronto's Christian character was reflected in its transit system. Beginning in the 1920s, distinctive yellow Sunday stops were placed outside churches that were considered to be too far from existing stops (figure 1.2). These remained in use until mid-2015 along streetcar routes and mid-2016 along bus lines, with 41 served by streetcars and 28 by buses. They were removed both as part of a stop rationalization process and as acknowledgment of the city's changing religious landscape.[9]

In 1961, the Toronto CMA contained 482,540 occupied dwellings. At this spatial scale, a majority of the units were classified as single-detached. Fully 67.4 per cent of the dwellings in the CMA were owner-occupied, with the rest tenant-occupied. The median value of owner-occupied dwellings in 1961 was $17,301 ($150,309 in 2019 dollars), with an average monthly contract rent in tenant-occupied dwellings of $101 ($878 in 2019 dollars). Most had been erected since 1945, clear signs of growth and suburbanization in the post–World War II period.

Within the City of Toronto in 1961, no housing type predominated, but apartment units were the most

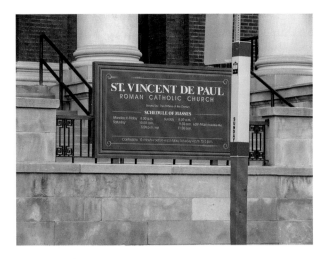

Figure 1.2 | A Sunday stop on Roncesvalles, 2005.

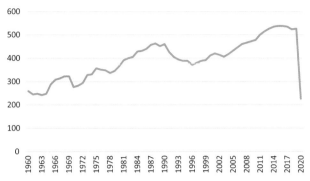

Figure 1.3 | TTC ridership, 1960–2020.

numerous. A small majority, 56.3 per cent, of dwellings in the City of Toronto were owner-occupied, with a median value of such dwellings of $17,523 ($152,238 in 2019 dollars), just above the CMA figure. Within the City, the average monthly contract rent in tenant-occupied dwellings was $99 ($860 in 2019 dollars). A majority of the dwellings in the City in 1961 had been built before 1920 (table 1.1).

On the employment front, Toronto in 1961 was very different from the place we find today. Just 3.1 per cent of males and 2.1 per cent of females were categorized as looking for work at the time the census was taken. Using the employment categories found in the census of that year, in the Toronto CMA, the largest proportion of males could be found in the craftsmen, production process, and related workers category (table 1.1). The average wage and salary income for males in the Toronto CMA in 1961 was $4,330 ($37,619 in 2019 dollars). For females in the CMA, the largest proportion could be found in the clerical category. The average wage and salary income for females in the Toronto CMA at that time was $2,338 ($20,312 in 2019 dollars).

Three final observations about the state of Toronto in 1961 are worth sharing. In terms of retailing, the main focus was on the downtown core; there Canada's department store giants, Eaton's and Simpsons, sat opposite each other on the west side of Yonge Street at Queen Street. Other retail activity could be found along major streets, with a handful of small shopping centres beginning to dot the landscape. Toronto's first indoor mall, Thorncliffe (now East York Town Centre), opened in 1961, but the big regional malls would not appear for a few years. In 1953, fully 73.2 per cent of all retail floor space in the Toronto region was to be found within the City of Toronto; by 1966, that figure had fallen to just 45.8 per cent, a trend that would continue.[10] No condominiums existed in Toronto in 1961 as Ontario's first Condominium Act was several years away. Toronto, however, was about to witness an apartment construction boom, both within the City and along arterial roads in suburban districts. Finally, work had begun on Highway 401 and other expressways. At the time, people referred to the 401 as the Toronto Bypass Highway, as it was constructed at the fringes of the built-up area of the region.

In 1961, vehicles operated by the Toronto Transit Commission (TTC) carried 267.6 million passengers (figure 1.3). Its fleet consisted of 140 subway cars, 145 trolley coaches, 574 buses, and 836 streetcars (744 PCCs acquired between 1938 and 1957 and 92 Peter Witts, which dated from the early 1920s). The streetcar network, most of which was located within the old City of Toronto, comprised almost 319 kilometres of surface track. At that time, the TTC still divided Metropolitan Toronto into two zones. Basic fares within each zone were 20 cents cash ($1.74 in 2019 dollars), but tickets or tokens could be purchased at a cost of 7 for $1.00 (7 for $8.69 in 2019 dollars). Children could ride the system for 10 cents cash (87 cents in 2019 dollars), with tickets being sold at the rate of 4 for 25 cents (4 for $2.17 in 2019 dollars). Net earnings for the TTC in 1961 amounted to $375,422 ($3,261,688 in 2019 dollars). At that time, the TTC had some 5,600 employees.[11]

Table 1.2. Rapid transit and LRT development since 1954.

Link or node name	Line[a]	Opened	Length[b]	Total[b]	Stations	Cost[c]	Cost/km[c]
Eglinton-Union	YUS	30/03/54	7.4	7.4	12	$50.5	$6.8
Union-St. George	YUS	28/02/63	3.9	11.3	6	$45.0	$11.5
Keele-Woodbine	B-D	26/02/66	12.9	24.1	18	$160.0	$12.4
Keele-Islington	B-D	11/05/68	5.5	29.6	6		
Woodbine-Warden	B-D	11/05/68	4.3	34.0	3	$77.7	$18.1
Eglinton-York Mills	YUS	31/03/73	4.2	38.1	2		
York Mills-Finch	YUS	30/03/74	4.3	42.6	2	$140.0	$32.6
St. George-Wilson	YUS	28/01/78	10.0	52.6	8	$220.0	$22
Warden-Kennedy	B-D	22/11/80	2.9	55.5	1	$71.4	$24.6
Islington-Kipling	B-D	22/11/80	1.4	57.0	1	$38.6	$27.6
Kennedy-McCowan	SRT	22/03/85	7.2	64.2	5	$200.0	$27.8
North York Centre	YUS	18/06/87	-	-	1	$19.9	n.a.
Harbourfront LRT		22/06/90	2.1		2[d]	$58.3	$27.8
Wilson-Downsview	YUS	31/03/96	1.6	65.8	1	$150.0	$93.8
Spadina LRT		27/07/97	3.7		1[e]	$140.9	$38.1
Harbourfront LRT Extension		21/07/00	0.8		N/A	$13.3	$16.6
Yonge/Sheppard-Don Mills	SHP	24/11/02	6.4	72.3	5	$945.0	$147.7
Downsview[f]-Vaughan Metropolitan Centre	YUS	17/12/17	8.6	80.9	6	$3,180.0	$369.8

[a] YUS = Yonge/University/Spadina; B-D = Bloor/Danforth; SRT = Scarborough RT; SHP = Sheppard
[b] in kilometres for subway and SRT lines
[c] in millions of dollars
[d] route includes 2 underground stations and 4 surface-level stops
[e] 2 underground stations and ~12 surface stops
[f] Downsview was renamed Sheppard West in 2017
Source: Compiled by M. Doucet from various TTC publications

Toronto as a Model for Urban Transportation

It may be hard to imagine today, but in the decades after World War II, Toronto was seen as a model city for urban transit. The public transit system largely kept pace with demand and growth. It was frequently cited as an example for other, particularly American, cities to follow for building subways and investing in mass transit. The subway network expanded in the 1960s and a grid of frequent bus routes in suburban areas provided reliable feeder routes directly to outlying subway stations. During the 1970s, Toronto's rapid transit system was expanded by more than 50 per cent. Ridership on the TTC increased by 25 per cent over the decade, with more than a million riders every weekday. Only New York topped Toronto in the use of public transit among cities in Canada and the United States, and not even New York could approach Toronto's transit dependence if figures were expressed per capita instead of in their raw form.[12]

The interval between 1963 and 1980 remains as the most active period for subway construction in Toronto's history. In all, Toronto's subway network grew by 49.6 kilometres and 47 stations in that time frame (table 1.2). Since that time, just 16.8 kilometres of line and 12 stations have been added to the network. In other words, just more than seven-tenths (71.6 per cent) of the current TTC rapid transit network had been completed by 1980. Moreover, the provincial government established a regional rail network known as GO Transit in 1967, with lines radiating out from Union Station. By 1982, the system was well developed, with all seven of its current rail lines in at least partial operation.

Brian's Toronto

The younger of the two authors of this book, I was born in Toronto's Women's College Hospital, a stone's throw from the Carlton streetcar line, in late December 1980. While growing up, I never lived on a street with streetcar tracks, but my parents' home was 50 metres from the open-cut section of the Yonge subway line that exists to this day between Eglinton and Davisville stations. I could watch the passing trains from my bedroom window. Our street had a pedestrian bridge over the subway tracks and, as kids, my friends and I used to watch the subways passing by, waving at the drivers and hoping for a wave back, or, if we were lucky, a friendly honk of the horn! This practice continues today among a new generation of children on the same street, and no doubt in other areas around the city.

Given my father's interest in public transportation, first-day rides would become part of my Toronto experience, beginning with the opening of the Scarborough Rapid Transit line in March 1985. I also recall attending the opening day of the North York Centre station as well as the Harbourfront LRT in 1990. But, as we noted above, the greatest period of subway expansion ended shortly before I was born. While the original red subway cars and the red and cream PCC streetcars are familiar vehicles for readers of my father's generation, for those of us born in the 1980s, these exist only in brief and now distant childhood memories. But given the interest in transit in our household, I distinctly remember riding these vehicles, and trying (often successfully) to convince my parents, eager to get home, to wait for just "one more train" in hope that one would arrive.

Toronto in 1981

By 1981, Toronto had exploded upwards and outwards. The Toronto region's population surpassed that of Montreal in 1976. Its skyline was much more dramatic, and already boasted Canada's two tallest buildings: the CN Tower (1976, 553 metres) and First Canadian Place (1976, 298 metres, 72 storeys). Four buildings contained at least 50 storeys, with another six standing at least 40 storeys tall. The city's tallest building in 1961, the Bank of Commerce (now called Commerce Court North), had fallen to 10th place by 1981. In all, 45 buildings were at least 100 metres in height, with 353 at least 50 metres tall. The city's tallest residential structure was the ManuLife Centre (1972, 160 metres, 51 storeys).

According to the 1981 Canadian census, the City of Toronto was home to 599,217 people, a decline of 10.9 per cent compared with the 1961 figure, and the lowest total since 1926 (figure 1.1).[13] While still the largest of the then six municipalities that comprised Metropolitan Toronto, the old city now housed just 28 per cent of the 2,137,395 residents of Metro Toronto. If anything, suburbanization of the population had accelerated in the 20 years since 1961. For example, the Toronto Census Metropolitan Area in 1981 stood at 3,742.94 square kilometres, almost three times the size of the 1961 CMA. It was home to 2,998,947 people. Of the total CMA population in 1981, just 20 per cent lived in the old City of Toronto, with 51.3 per cent in the other five Metro municipalities, and 28.7 per cent in the areas beyond the Metro Toronto borders. Even with its reduced population, the old city, at 6,139.5 people per square kilometre, remained the most densely settled part of the CMA. The density figure for Metro Toronto minus the old city was 2,911.0 people per square kilometre, and for the parts of the CMA that lay beyond the Metro borders it was just 276.4 people per square kilometre, even lower than it had been in 1961, reflecting a growing boundary that included many still-rural areas.[14]

Toronto in 1981 was a city of immigrants, with 37.8 per cent of the CMA's population and 43.0 per cent of the City's population having been born outside of Canada. Both of these figures were higher than those for 1961. As in 1961, the 1981 Canadian census did not record the visible minority status of the population, but it did contain figures on ethnic origin. These point to a still overwhelmingly white population, but one that was less British than in 1961. For the CMA, 46.7 per cent self-identified as British, with the comparable figure for the City at 39.4 per cent. In terms of religious affiliation, Toronto remained a primarily Christian place in 1981 (table 1.3).

In 1981, the Toronto CMA contained 1,040,340 occupied dwellings, more than twice the total in 1961. The largest number were classified as single-detached and more than half of the dwellings in the CMA, 56.4 per cent, were owner-occupied. The average value of the CMA's owner-occupied dwellings in 1981 was $114,284 ($300,933 in 2019 dollars), up considerably from the median value of $17,301 for 1961. The average gross monthly rent for the CMA in 1981 was $364 ($959 in 2019 dollars), more than 3.5 times higher than in 1961. More than a quarter of the dwellings in the CMA had been built between 1971 and 1981.

The City of Toronto in 1981 contained 190,750 occupied dwellings, only 10.3 per cent more than in

Table 1.3. Selected census variables, 1981.

Census variable	City of Toronto	Toronto CMA
Born outside Canada	43.0%	37.8%
British ethnicity	39.4%	46.7%
Roman Catholic	39.4%	35.8%
Protestant	34.4%	43.3%
Jewish	4.1%	4.1%
Occupied dwellings	**190,720**	**1,040,340**
Single-detached	20.6%	40.3%
Apartments with 5+ storeys	32.7%	29.5%
Other types	46.7%	30.2%
Owner-occupied	40.7%	56.4%
Built before 1946	53.9%	19.7%
Built 1971–81	13.7%	28.9%

1961. Of these, the largest number were categorized as other housing forms, such as semi-detached, row/town houses, and low-rise apartments. Only 40.7 per cent of the dwellings in the City of Toronto were owner-occupied in 1981, down from the figure in 1961. The changes in both the dwelling type and tenure patterns between 1961 and 1981 were the result of a noticeable boom in the construction of high-rise and low-rise apartment buildings in the City, through both urban renewal, and in the growing suburbs, particularly along wide arterial roads that delineate neighbourhoods of single-family homes (figure 1.4).[15] The average value of Toronto's owner-occupied dwellings was $130,397 ($343,362 in 2019 dollars), about 14 per cent higher than the average for the CMA. As in the CMA, this was considerably higher than in 1961. For tenant-occupied dwellings in 1981, the average gross monthly rent was $356 ($937 in 2019 dollars), almost identical to the figure for the CMA, and more than 3.5 times the figure for 1961. A majority of the dwellings in the City in 1981, 53.9 per cent, had been constructed prior to 1946, with just 13.7 per cent built between 1971 and 1981.

To demonstrate employment trends in Toronto in the early 1980s, we utilize data from both the Canadian census and Toronto's annual employment survey, which began in 1983. The focus in this section will be on the CMA and the Municipality of Metropolitan Toronto. Looking first at the census data, the unemployment rate for males in the CMA was 3.4 per cent, and 4.6 per cent for females. For Metro Toronto, the figures were 3.8 per cent for males and 4.6 per cent for females. The average income for males in the CMA was $18,936 ($49,862 in 2019 dollars), and for females it was $9,831 ($25,887 in 2019 dollars). For Metro Toronto, the comparable figures were $18,140 ($47,766 in 2019 dollars) for males and $10,009 ($26,565 in 2019 dollars) for females.

At the time of the first Toronto Employment Survey in 1983, Metropolitan Toronto exhibited an employment structure that reflected the era prior to both the signing of free-trade agreements with the United States, and later Mexico, and the globalization of manufacturing (table 1.4). Just more than one-fifth of the city's 1.1 million jobs could be found in manufacturing and warehousing, with another third found in retail, service, and institutional positions. The largest concentration of jobs, just more than two-fifths, was found in the office sector. Only three organizations employed more than ten thousand workers in Metropolitan Toronto in 1983, and the list of the ten largest employers included the three levels of government, two utilities, two banks, a university, a major retailer, and what we would now call an IT company (table 1.5).

To conclude this discussion of Toronto in the early 1980s, several observations are worth noting. While Eaton's and Simpsons remained in operation on Yonge Street, the former had moved north to Dundas in the mid-1970s to anchor the new Toronto Eaton Centre. Downtown, however, no longer dominated the retail scene. By 1981, all of the regional malls now in operation in what then was known as Metropolitan Toronto had opened: Yorkdale (1964), Fairview Mall (1970), Sherway Gardens (1971), and Scarborough Town Centre (1972). In 1983, the City of Toronto accounted for 51.4 per cent of the retail space within Metropolitan Toronto, but just 30.5 per cent within the Toronto region.[16] Condominiums had begun to appear, in both townhouse and apartment-building complexes. The GO Transit regional commuter rail system was well established, with seven rail lines radiating outwards from Union Station. Following on the decision by then Ontario premier William Davis to halt construction of the Spadina Expressway at Eglinton Avenue in 1971, the era of expressway construction within Toronto more or less came to an end.

In 1981, the TTC carried 392 million passengers, 46.5 per cent more than in 1961 (figure 1.3). Its fleet consisted of 632 subway cars, 151 trolley coaches, 1,394 buses, and 446 streetcars (2 Peter Witts, 256 PCCs, and 188

Figure 1.4 | Aerial view of apartments in St. James Town and inner suburbs from Commerce Court. Photographer: Michael Doucet.

Table 1.4. Toronto's employment structure, 1983, 2003, 2019.

Category	1983	2003	2019
% Manufacturing/warehouse	21.9	13.6	8.7
% Retail	12.1	11.4	9.8
% Service	10.7	11.4	12.6
% Office	41.8	45.7	48.0
% Institutional	11.8	15.2	17.4
% Other	1.7	2.6	3.6
Total employment	1,100,000	1,251,300	1,569,800

Source: City of Toronto, *Profile Toronto: Toronto Employment Survey 2003* (Toronto: Toronto Urban Development Services Policy and Research, 2004) and *Profile Toronto: Toronto Employment Survey 2019* (Toronto: City Planning Division, Strategic Initiatives, Policy and Analysis, Research and Information, 2020).

new Canadian Light Rail Vehicles (CLRVs). Whereas streetcars accounted for 41.5 per cent of the "route miles" operated by the TTC in 1961, that figure had fallen to just 8.6 per cent in 1981. Subway expansion had meant the elimination of several streetcar lines. By 1981, only the King, Queen, Dundas, Carlton, Kingston Road/Downtowner, St. Clair, Long Branch, and Bathurst lines remained. Indeed, the entire system had been threatened with closure in the early 1970s. A group of citizens, led by University of Toronto psychology professor Andrew Biemiller, founded Streetcars for Toronto in 1972 and lobbied successfully for the retention of electric traction in the city. By 1981, a new fleet of Canadian Light Rail Vehicles (CLRVs) was in service. They served Toronto for more than 40 years, before being retired at the end of 2019.[17] Basic fares for adults were 65 cents cash ($1.71 in 2019 dollars), with tickets or tokens sold at the rate of 7 for $4 (7 for $10.53 in 2019 dollars). For children, the cash fare was 20 cents (53 cents in 2019 dollars), with tickets costing 90 cents for a strip of five ($2.37 in 2019 dollars). Tickets for students and seniors sold at a cost of $2 for seven ($5.27 in 2019 dollars). The unlimited-use monthly Metropass was introduced in May of 1980 at a price of $26.00 ($78.23 in 2019 dollars). In 1981, the TTC employed 8,906 people.[18]

Table 1.5. Ten largest employers in Toronto, 1983 and 2019.

Organization, 1983	Total employment
Public Works Canada	21,608
Government Services Ministry (Government of Ontario)	19,912
Bell Canada	10,884
Ontario Hydro	9,711
Municipality of Metropolitan Toronto	9,320
Robert Simpson Company, Ltd.	8,609
Canadian Imperial Bank of Commerce	8,217
York University	7,215
Bank of Nova Scotia	6,632
IBM Canada, Ltd.	6,443

Organization, 2019	Total employment
Toronto District School Board	37,950
City of Toronto	30,300
TD Canada Trust	23,700
Canadian Imperial Bank of Commerce	23,560
Royal Bank of Canada	22,780
Government of Ontario	21,780
University of Toronto	19,750
Scotiabank	17,190
Toronto Transit Commission	16,190
Bank of Montreal	14,030

Source: Toronto Employment Survey – 1983 and 2019. Prepared by Research & Information, Strategic Initiatives, Policy & Analysis, City Planning, City of Toronto, 25 January 2020.

Toronto Today (Felix and Emmett's Toronto)[19]

We use data from a variety of time points in this section.[20] To complicate matters, effective 1 January 1998, the Ontario government, under the leadership of then premier Mike Harris, forced the amalgamation of the six municipalities within Metropolitan Toronto, forming the new City of Toronto.[21]

The TTC's reputation has suffered during this most recent period; no longer do outsiders come to study it as a model for their own cities to emulate. No significant additions were made to Toronto's subway system during the period between 1981 and 1995, though the lower-capacity Scarborough Rapid Transit line (SRT) did open in 1985. In a five-part series published during March of 1990, *Toronto Star* columnist Peter Howell described the 1980s as "The Ten Lost Years" for the TTC. Toronto, Canada's first subway city, in fact, fell behind Montreal during the 1980s as Canada's rapid transit leader. In the fall of 1990 CBC television aired a program about the TTC that was aptly entitled "Spinning Its Wheels."[22] Even the TTC's vaunted safety record was challenged. A combination of equipment failure and human error produced a subway collision between the St. Clair West and Dupont stations that killed three passengers on 11 August 1995.[23] Ridership on the TTC tumbled during the recession of the 1990s, and did not return to its 1987 peak until 2009 (figure 1.3). Furthermore, over the past decade, much time has been wasted at City Hall in debates about which mode of rapid transit to build – subways or LRTs – with precious little to show for it, save for the 8.6-kilometre extension to Vaughan Metropolitan Centre that opened in 2017.

In 2018, the TTC carried 521.4 million passengers, 33.0 per cent more than in 1981, and 94.8 per cent more than in 1961. Its fleet consisted of 848 subway cars, 28 Scarborough RT cars, 2,010 buses, and 245 streetcars (113 CLRVs, 15 Articulated Light Rail Vehicles (ALRV), and 117 low-floor, 30.2-metre-long Bombardier Flexity articulated streetcars).[24] Overall, streetcars carried 36 million passengers, or 6.9 per cent of the TTC's total ridership in 2018. Unlike the situation between 1961 and 1981, when streetcar lines were eliminated, the period after 1981 saw an expansion of the network with the completion of the Harbourfront LRT (22 June 1990), the Spadina LRT (27 July 1997), the Harbourfront LRT extension (21 July 2000), and the introduction of the 514 Cherry route[25] (18 June 2016), adding 8.5 kilometres to the system (table 1.2).[26] In addition, after a very prolonged process, the 512 St. Clair line was converted to a private right-of-way that was fully operational on 30 July 2010.[27] By 2016, the TTC's 11 streetcar lines operated over 323.8 kilometres of track. Six of the TTC's ten busiest surface routes were streetcar lines (504 King, 510 Spadina, 501 Queen, 506 Carlton, 512 St. Clair, and 505 Dundas). The basic adult ticket fare in 2021 was $3.20, with the cash fare set at $3.25, more than 4.5 times higher than in 1981. Monthly charges for a Metropass were $156.00 for adults and $128.15 for students and seniors in 2021.[28] Sadly, the TTC stands as the most poorly funded transit agency in North America, with about two-thirds of its annual revenues coming from passenger fares. As a consequence, fares have risen at a much faster rate than inflation for many decades. For example, the monthly Metropass has gone up in price at twice the rate of inflation ($156.00 versus an inflation value of $82.41). The average number of TTC employees in 2018 stood at 14,812, 164.5 per cent more than in 1961 and 66.3 per cent higher than in 1981.[29]

While ridership on the TTC has shown a significant increase over the years under study, the numbers have displayed some degree of fluctuation due to several factors. These include economic conditions, labour disputes, weather conditions, and threats to human safety. In terms of impact, poor economic conditions, often of a global nature, periodically have caused grief for the heavily fare-dependent TTC. Recessions in 1960–1, 1974–5, 1981–2, 1990–2, and 2008–9 all resulted in either flat lines or dips in the ridership graph (figure 1.3).[30]

Labour disruptions have had some short-lived impacts on ridership. Since 1960, TTC workers have staged seven strikes (1970, 1974, 1978, 1991, 1999, 2006, and 2008) and one 41-day work-to-rule campaign (1989). The longest strike, in 1974, lasted 23 days, resulting in almost no ridership growth over 1973. Following the 2008 strike, the then-Liberal provincial government passed legislation in 2011 that made the TTC an essential service, thus removing the right to strike.[31]

Certain types of weather events can have a short-term impact on transit operations and ridership, with electrically powered vehicles especially susceptible to some conditions. Snow and ice often create problems for streetcar lines and the above-ground portions of subway routes. Freezing rain in sufficient amounts will coat the overhead cables that supply electrical power to streetcars, causing service disruptions. This certainly was the case on 23 December 2004 (figure 1.5). Normally, it takes a day or two to clear the ice from the overhead, so service disruptions of this sort do not last long. Blizzards, on the other hand, may be more disruptive. Early in January of 1999, a series of storms dumped more than 110 centimetres of snow on Toronto, forcing the TTC to shut down subway and streetcar operations for several days so that snow could be removed from the tracks. The then mayor, Mel Lastman, asked the federal government to send in army troops to help clear the snow. Buses replaced subway service during the period (figure 1.6).

Less common are events that pose threats to human health and safety. In mid-August of 2003, Toronto's streetcars and subways were shut down by a massive power failure that struck much of northeastern North America. The blackout also posed a safety threat to Torontonians as it rendered street and traffic lights inoperable. It took several days to fully restore electrical power within the affected area, during which time Toronto's electrically powered transit vehicles were forced to remain in their storage yards (figure 1.7). According to the then chair of the TTC board, Howard Moscoe, "ridership dropped by more than 2.5 million trips, as subway and streetcar service was knocked out for several days."[32]

The year 2003 also witnessed a significant threat to public health with the arrival of cases of Severe Acute Respiratory Syndrome (SARS) in February. Originating in southern China, Toronto was one of the first places beyond Asia affected by the disease. The World Health Organization issued an advisory against non-essential travel to Toronto in late April; it was finally rescinded in early July. Toronto suffered through a four-month, two-stage SARS outbreak that was the largest outside of Asia, recording almost 250 cases and 39 deaths, with some 27,000 people quarantined during that period. Most cases were travel-related and confined to hospitals, with little community spread of the disease.[33] Nevertheless, both tourism and the TTC suffered as a result of the SARS outbreak. According to Howard Moscoe:

> The SARS crisis caused a significant dip in ridership. The Commission estimated a loss of about 3.5 million riders as a result of the outbreak, which lasted several months. A further 1 million rides were lost due to the extended economic slowdown. However, the excellent TTC service proved to be a big hit with many of the Rolling Stones fans attending a special SARS benefit concert at Downsview Park on July 30.[34]

As bad as SARS had been, nothing prepared Toronto and the world for a viral infection that began in Wuhan, China, and quickly spread around the globe early in 2020. On 11 March, the World Health Organization declared a global pandemic for a novel coronavirus that has come to be known as COVID-19. Within days, most Canadian provinces, including Ontario, had declared states of emergency. In what amounted to a lockdown, citizens were asked to stay at home unless they were

Figure 1.5 | CLRVs on King St. during the December 2004 ice storm. Photographer: Michael Doucet.

classified as essential workers, except for trips to attend medical appointments or to get essentials, such as groceries. International travel was restricted, and the border between Canada and the United States was closed to non-essential trips. Many businesses closed their premises, and those employees who could began to work from home. Schools and religious buildings also closed, with classes and services going online. People were asked to wash their hands frequently, stay at least two metres apart from people with whom they were not living, and to avoid non-essential trips from home. Later, citizens were ordered to wear masks or a facial covering in many situations. By early August 2020, more than 15,000 COVID-19 cases had been recorded in the City of Toronto, and 1,159 Torontonians had died from the disease. Unlike with SARS, community spread was a factor with COVID-19.[35]

The impact on the TTC was immediate and profound. Ridership plummeted by more than 80 per cent, and TTC revenue fell by $21 million per week, forcing the TTC to cut service by 15 per cent and lay off some 1,200 employees. Beginning on 2 July, masks were made mandatory for anyone wishing to use the TTC. By the end of November, vending machines for the sale of items of personal protective equipment (PPE), such as masks and gloves, had been installed in ten subway stations. The revenue shortfall by the end of 2020 was projected to be $700 million. It remained unclear at the time of writing how this shortfall would be covered. By the end of June 2020, ridership was making a slow recovery, but remained at almost 78 per cent below projections made prior to the pandemic. Overall, TTC ridership totalled just 225 million in 2020, the lowest figure in 80 years, and

Figure 1.6 | A shuttle bus at Yonge and Manor in January 1999. The subway was shut down because of the blizzard. Photographer: Michael Doucet.

few imagined that it would be much better in the short-term future.

Later in this book, we discuss the growing economic, racial, and housing disparities in contemporary Toronto. The arrival of COVID-19 served to underscore the nature of these inequalities. For example, to the extent that there was a ridership recovery by mid-2020, it tended to be associated with suburban areas that were served by bus routes linking low-income residents, often living in high-rise apartments and holding jobs that could not be done from home, to industrial and warehousing jobs in Etobicoke, North York, and Scarborough. At certain times of the day, many of these bus routes were overcrowded, making physical distancing difficult for riders.[36]

Toronto continued to soar upwards and outwards after 1981. At the time of writing, Toronto had 12 buildings with at least 60 storeys (10 of which had been built since 2011, with another 8 well under construction and expected to open by 2023, with 3 of them to be at least 85 storeys), 30 buildings of between 50 and 59 storeys, and another 73 buildings that stood at between 40 and 49 floors. The addition of tall buildings to the city's skyline has been explosive of late. Of the 100 tallest buildings in Toronto in early 2020, 44 had been completed since 2015. According to the Council on Tall Buildings and Urban Habitat, Toronto had the third-largest number of skyscrapers in North America in early 2020, after New York and Chicago, and was poised to pass Chicago in the near future. While the two tallest structures in 1981 (the CN Tower and First Canadian Place) retained their positions, the 1961 champion (Commerce Court North) had fallen to 91st place in the rankings. All told, 305 Toronto buildings stood at least 100 metres tall, with 1,406

Figure 1.7 | Streetcars out of service at Russell carhouse during the blackout of August 2003. Photographer: Michael Doucet.

at least 50 metres tall. The tallest residential building in 2020, the Aura condominium tower, soared to 272 metres (78 storeys and 985 dwelling units) within the College Park complex near Yonge and Gerrard. Aura has been described by one architecture critic as "highly visible and titanically bad."[37]

According to the 2016 Canadian census, the area that had comprised the old City of Toronto was home to a record-high 797,642 people, 33.1 per cent more than in 1981, and 17.1 per cent more than in 2006, with many of the additional citizens housed in high-rise condominium towers.[38] This is where we will focus most of our visual analysis; this growth, however, has not been evenly distributed throughout the old City of Toronto, with many residential neighbourhoods experiencing declines in population during this time. In spite of this growth, the old City was home to just 29.2 per cent of the 2,731,571 residents of the new City of Toronto, about the same percentage as in 1981. Suburbanization, however, had continued apace in the Toronto region. By 2016, the Toronto Census Metropolitan Area had been expanded to include an area of 5,905.84 square kilometres, and was home to 5,928,040 people. This meant that the old City of Toronto was home to just 13.5 per cent of CMA residents, with the remainder of the new City of Toronto housing another 32.6 per cent.

The so-called 905 portion of the CMA now housed 53.9 per cent of all residents.[39] Density patterns continued to vary throughout the CMA, with the highest, 8,172.6 per square kilometre in the old City, 3,660.0 per square kilometre in the rest of the new City, and 605.4 per square kilometre in the remainder of the CMA, low, but more than twice what it had been in 1981.[40]

Toronto in 2016 remained a city of immigrants, with 46.1 per cent of the CMA's population, 47.0 per cent of those living in the new City, but just 33.1 per cent of residents of the old City having been born outside of Canada, reflecting its gradual transition from an immigrant settlement area to a patchwork of many gentrified, and expensive, neighbourhoods. As a result, the percentage of immigrants living in the old City in 2016 was lower than it had been in both 1961 and 1981, and, for the first time, it was lower than that for the CMA. Toronto's Indigenous population is small, especially when compared with that of other cities in Canada. In the city of Toronto, 23,065 indicated Aboriginal identity in the 2016 census, representing 0.9 per cent of the total population; in the CMA, that percentage was slightly lower, at 0.8 per cent. Across Canada, 1,673,785 people, or 4.9 per cent of the population indicated Aboriginal identity. Saskatoon (11.3 per cent) and Winnipeg (12.2 per cent) have far higher Indigenous populations than Toronto; even Hamilton's figure of 12,135 constitutes 2.3 per cent of that city's population, more than double Toronto's percentage.

By the time of the 2016 census, data were no longer gathered concerning religious affiliation, so it was no longer possible to monitor trends in that area. On the positive side, Statistics Canada had begun to collect information about visible minorities. These data pointed to an increasingly non-white Toronto, a clear reflection of the countries of origin for recent immigrants. For the CMA, fully 51.4 per cent of the population self-identified as belonging to a visible minority group. The comparable figures for the new City and the old City were 51.5 per cent and 33.5 per cent, respectively. Clearly, the face of the Toronto CMA's population had changed dramatically between 1961 and 2016. This was less true of the old City of Toronto, where European categories still accounted for 66.5 per cent of the population in 2016, compared with only 47 per cent in the CMA (table 1.6).

In 2016, the Toronto CMA contained 2,135,910 occupied dwellings, 105.3 per cent more than in 1981. Of these, the most numerous were classified as single-detached and almost two-thirds were owner-occupied, with those units having an average value of $734,924, or about 6.4 times more than in 1981. The average monthly shelter cost for rental dwellings in 2016 was $1,264, or 3.5 times the 1981 figure, 1.4 times if corrected for inflation. The 2016 census also reported that 20.9 per cent of the CMA's dwelling units were classified as condominiums.

The old City of Toronto contained 386,580 occupied dwellings in 2016, an increase of 102.7 per cent over the 1981 total. Of these, a slight majority were found in buildings (apartments or condos) having more than five storeys. Just 43.5 per cent of the old City's dwelling units were owner-occupied at that time. While 40.2 per cent of the old City's housing units had been erected before 1961, almost one of every five units had been constructed between 2006 and 2016, a reflection of the boom in condominium construction in central Toronto that continues to the present day.[41] According to the 2016 census, 32.0 per cent of all dwelling units in the old City could be categorized as condominiums.

As we alluded to above, new development has tended to be concentrated in specific parts of the

Table 1.6. Selected census variables, 2016

Census variable	Old City of Toronto	Toronto CMA
Born outside Canada	33.1%	46.1%
Visible minority	33.5%	51.4%
British ethnicity	36.8%	22.0%
Occupied dwellings	**386,580**	**2,135,910**
Single-detached	11.0%	39.6%
Apartments with 5+ storeys	50.1%	29.4%
Other types	38.7%	31.0%
Owner-occupied	43.5%	66.5%
Built before 1961	40.2%	20.4%
Built 2006–16	19.8%	17.0%

old City, often resulting in dramatic changes to the demography and landscape of neighbourhoods. For example, in the region planners refer to as the central area (roughly south from Davenport to the Toronto Harbour and extending east from Bathurst to the Don Valley), growth has been spectacular. The population in this part of the old City grew from 107,066 in 1981 to 255,763 in 2016, an increase of 138.9 per cent. Dwelling unit growth in this area was even greater, rising from 52,090 units in 1981 to 146,630 in 2016, an increase of 181.5 per cent, and a reflection of the small size of many of the new condominium units, which are designed for one- or two-person households. Fully 59.4 per cent of the dwelling units in this area were classified as condominiums in the 2016 census. Within this central area, the formerly industrial Harbourfront (south of Front Street between Bathurst and the Don River), saw population rise from 2,814 in 1981 to 47,774 in 2016, with dwelling units increasing from 1,480 to 28,595 over the same period, with 89.4 per cent of all units classified as condominiums. In the previously industrial area extending west along both sides of King Street from Bathurst to Jameson, population rose from 17,223 in 1981 to 28,159 in 2016; housing units increased from 7,095 to 16,284 during the same time frame, with 36.7 per cent of all units in 2016 categorized as condominiums. Little wonder that the 504 King streetcar route is the busiest surface line in all of Toronto.

The form of the new housing in these areas is unmistakeable, part of what Dutch architect Hans Ibelings has called the "condominiumization of Toronto." Clearly, the look of Toronto has changed as a result, but not always to universal approval. As Ibelings observed:

> The rise and sprawl of the condo tower is undeniably [Toronto's] most remarkable development in recent years. The number of towers, their size, mass, volume, height, and the speed with which they are built is astounding. The only thing that isn't remarkable about the rise and sprawl of Toronto's condo towers is their architecture ... One doesn't have to be an architectural connoisseur to see that Toronto's condo boom hasn't produced much, if any, outstanding architecture ... bland, interchangeable, and forgettable ... they are fascinating as a cultural phenomenon, as architecture minus the art of architecture.[42]

Yet beside these towers are many neighbourhoods that have lost population since the 1970s. Often referred to as "stable residential neighbourhoods" they have, in fact, seen population losses as high as 200 people per hectare,[43] as gentrification has led to the conversion of apartments and rooming houses back to single-family homes for more affluent households. This, combined with planning restrictions that have effectively prohibited any increases in density throughout much of the city means that, even near transit-rich streets such as Bloor, Danforth, or College, very little new housing has actually been built. Growth and the demand for urban living are therefore shoehorned into small pieces of the city, most of which were formerly industrial sites.

According to the 2016 census, the unemployment rate for males was 7.4 per cent in the CMA and 8.0 per cent in the new City of Toronto. For females, the corresponding figures were 8.1 per cent and 8.5 per cent, respectively. The average income for males at this time was $60,343 in the CMA and $62,667 in the new City. For females, the comparable figures were $41,343 and $42,807. While income levels had increased for both males and females, the unemployment rates were considerably higher than in either 1961 or 1981, largely owing to Toronto's altered economic structure.

For 2019, the Toronto Employment Survey revealed a vastly different economic composition than had been present in the early 1980s (see table 1.4). While total employment had increased by 32.8 per cent over this period, growth was not evenly spread over the reported employment categories. Not surprisingly, the big loser was manufacturing, which saw its share of employment fall from 21.9 per cent to 8.7 per cent. Retail employment also fell, though far less dramatically. The biggest winners were office employment, which had risen to 48.0 per cent of all jobs by 2019 and the institutional sector (17.4 per cent), fuelled by the expansion of Toronto's universities, community colleges, and medical centres during this period. In terms of Toronto's largest employers in 2019, the Toronto District School Board and the City of Toronto captured the top two spots. The remaining eight positions on the list belong to the Government of Ontario, five major banks, the University of Toronto, and the Toronto Transit Commission (table 1.5).

Toronto 1961–2016: Concluding Reflections

Without question, Toronto is a much different place today than it was in the early 1960s. This book visually chronicles the ascendancy of Toronto during a period in which it became both more global and more divided. It is bigger in every urban sense of the term, having grown to be an important political, cultural, and economic centre in Ontario, Canada, and beyond. Furthermore, it is a much more diverse place than it was in 1961, and, as a consequence, is far less white, Christian, and British than the city of Michael's youth. At the same time, it has also become a much more unequal city, with stark fault lines around race, employment, politics, and geography. As David Hulchanski has argued (see chapter 3), there now are three distinct cities within Toronto, with an expansion of both wealthy and poor neighbourhoods and a sharp decline of those categorized as middle income. Much of the wealth is concentrating in the old City of Toronto and along the central north-south spine of the city, areas well served by transit.[44] If the old City receives immigrants today, they are less likely to arrive in an impoverished state. The old City is both more densely settled and more affluent than ever before.

In 1981, Toronto's main department stores were Eaton's at Yonge and Dundas and Simpsons at Yonge and Queen. Those stores catered to all tastes and incomes. Symbolic of the demographic changes occurring in the old City, those same buildings are now occupied by much higher-end retailers – Saks Fifth Avenue at Yonge and Queen, and Nordstrom's at Yonge and Dundas. Unlike the case in many US cities, Toronto's downtown remains a desirable location for retailers. According to one major study, downtown was home to 8.9 per cent of retail businesses and 9.2 per cent of all retail space in the Greater Toronto Area in 2010.[45] The region's largest department stores remain in the downtown area. This stands in sharp contrast to several other Great Lakes cities, such as Cleveland and Detroit, which no longer have any major downtown department stores. Furthermore, along many of Toronto's retail strips, chain stores have begun to supplant independent retailers, and in the 1990s and 2000s in particular, the arrival of a Starbucks was a sure sign of gentrification along Toronto's streetcar streets. Along those thoroughfares, store signage has changed from neon and Vitrolite to back-lit and front-lit forms. Many such changes will be found in the photos found later in this book.

In 1966, 64 per cent of the 23.23 million square feet of office space in Metropolitan Toronto could be found in the downtown area. By 2013, Toronto's downtown area still captured 40 per cent of the then-169,509,118 million square feet of office space in the Greater Toronto Area. Just 26.4 per cent of the GTA's 1,351 office buildings, however, could be found in the downtown area, a reflection of their large size. Almost no office space was added to downtown Toronto during the 1990s and early 2000s. Of the almost 68 million square feet of office space in the downtown area in 2013, 4.3 million square feet were added between 2009 and 2013, with another 5.3 million square feet added by 2016, much of it in the so-called South Core Financial District between Front Street and Lake Ontario. Symbolic of the health of Toronto's core area, the office vacancy rate in the downtown area was just 2.6 per cent in late 2017, compared with a figure of 6.6 per cent for the entire GTA.[46] It is too early to tell the longer-term impacts that working from home during the COVID-19 pandemic will have on the demand for office space.

The past half century or so has brought about considerable changes to the look of Toronto. Some of the changes have been subtle, such as the removal of television antennae from residential rooftops, or the replacement of an independent retailer by a national or international chain. Along the streetcar lines, the area between the tracks has changed from cobblestones and granite setts to surfaces of poured concrete. But other changes have been dramatic, especially in areas that have seen both gentrification and deindustrialization. Especially in the downtown area, surface parking lots, once plentiful, have become an endangered species (figure 1.8).[47]

Old buildings have found new life with different uses – factories have been transformed into lofts, and some churches have been converted to residential condominiums. Another feature of central Toronto's recent makeover has been the preservation of historic façades, a process that has not been universally welcomed.[48] Some have referred to this process as "façadism"; Robert Allsopp has dubbed it "urban taxidermy."[49] While many buildings in the inner city have been renovated, it is in the social changes, rather than the structural ones, where the alterations to central Toronto's neighbourhoods have been the most dramatic.

In Toronto, the matter of house prices is a major topic of conversation. The Toronto housing market is an important element of the change that has occurred in the city over the past several decades. Without question, the average price for a single-family home has risen substantially over the past half century (figure 1.9). In raw terms, the average home price has grown from $18,883 in 1965 to $819,319 in 2019, a 43-fold increase that was anything but evenly spread over the period. Prices rose steadily from 1965 to 1985, soared from 1985 to 1989, fell from 1989 to 1996, rose steadily from 1996 to 2009, and soared again from 2009 to 2017, with a slight dip in 2018. If we control for inflation, a slightly more volatile housing market is revealed. Such an analysis shows not one, but two, periods of price decline – 1976 to 1984 and 1989 to 1996. Indeed, in constant 2019 dollars, the peak of 1989 was not matched until 2002. Even controlling for inflation, the periods from 1985 to 1989 and 2009 to 2017 stand out as times when average prices rose very rapidly. In December of 2019, average sale prices by type in the amalgamated City of Toronto were as follows: detached – $1,363,357; semi-detached – $1,004,477; townhouse – $717,369; and condominium – $656,233. The figures for the surrounding 905 region were $956,792, $706,651, $657,577, and $508,173, respectively.[50] While Toronto is not the most expensive housing market in Canada, the rise in house prices in the city has been a significant factor in its growing levels of inequality.

While light rail/streetcar systems have been undergoing a revival in North America, streetcars have

Figure 1.8 | Looking east from the observation deck on the 57th floor of Commerce Court, May 1975. Photographer: Michael Doucet.

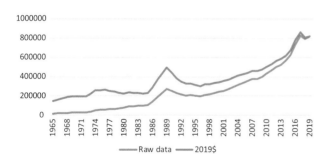

Figure 1.9 | Average Toronto house prices, 1965–2019. Source: Toronto Real Estate Board, *Market Watch*, various years.

been a continuous feature of Toronto since 1861, and in their electrically powered form since 1892.[51] A 2007 National Geographic publication entitled *Journeys of a Lifetime* rated Toronto's 501 Queen line as one of the world's top ten trolley rides.[52] In spite of proposals in the 1970s, and more recently from former mayor Rob Ford, to get rid of streetcars, they are unlikely to disappear from Toronto's streets. This is especially true given the recent purchase of 204 new, and accessible, Flexity Outlook Light Rail Vehicles (LRVs) from Bombardier, and the construction of a new maintenance and storage facility for them, the Leslie Barns. In 2021, the TTC placed a $568 million order for 60 additional streetcars from Alstom, a French-based manufacturer that had purchased Bombardier's rail division in early 2021.[53] A century ago, it was a similar investment in rolling stock and infrastructure during the decade following the creation of the TTC in 1921 that prevented Toronto from following the lead of most American cities in abandoning their streetcar systems in the 1930s and 1940s. Moreover, the Toronto system has expanded in recent years. A new line, the 514, opened in mid-2016, adding a few hundred metres of track to the network. It ran from the Dufferin loop at the west end of the Liberty Village neighbourhood, then along King Street to Sumach

Table 1.7. Milestones in the development of Toronto's skyline.

	Year	Name of building	Height of building Feet	Metres	Floors
Spires/Chimneys/Towers	1850	Cupola of St. Lawrence Hall (151 King St. E.)	120	36.6	n.a.
	1866	Spire of St. Michael's Cathedral (57 Bond St.)	260	79.2	n.a.
	1874	Spire of St. James Cathedral (106 King St. E.)	306	93.3	n.a.
	1899	Clock tower of Old City Hall (60 Queen St. W.)	285	86.9	n.a.
	1971	Hearn Generating Station chimney (Portlands – 440 Unwin Ave.)	715	218.0	n.a.
	1976	CN Tower	1,815	553.0	n.a.
Office	1895	Temple Building (62 Richmond St. W. – demolished 1970)	121	36.9	11
	1906	Traders Bank Building (61 Yonge St.)	188	57.3	15
	1913	Canadian Pacific Building (1 King St. E.)	245	74.7	15
	1914	Royal Bank Building (2 King St. E.)	300	91.4	20
	1929	Toronto Star Building (80 King St. W. – demolished 1972)	285	86.9	29
	1931	Canadian Bank of Commerce Building (25 King St. W.)	476	145.1	34
	1967	Toronto-Dominion Tower (55 King St. W.)	758	231.0	56
	1972	Commerce Court West (243 Bay St.)	784	239.0	57
	1979	First Canadian Place (100 King St. W.)	970	295.7	72
Hotels	1903	King Edward Hotel (37 King St. E.)	100	30.0	8
	1929	Royal York Hotel (100 Front St. W.)	439	133.8	26
	1972	Sheraton Centre Hotel (123 Queen St. W.)	443	135.0	43
	2012	St. Regis Hotel and Condo (325 Bay St. – formerly Trump Hotel)	909	277.0	58
Residential	1904	St. George Mansions (1 Harbord St. – demolished)	60	18.3	6
	1905	The Alexandra (184 University Ave. – demolished)	70	21.3	7
	1929	Park Plaza Apartment Hotel (4 Avenue Rd.)	207	63.0	18
	1955	City Park Co-Operative (3 towers on Wood St.)	135	41.0	14
	1958	Ray McCleary Towers (444 Logan Ave.)	217	66.0	21
	1967	The Halifax (280 Wellesley St. – St. Jamestown)	308	94.0	32
	1970	Leaside Towers Apartments (85 Thorncliffe Park Dr. – twin towers)	423	128.9	44
	1975	ManuLife Centre Apartments (44 Charles St. W.)	545	166.1	51
	2014	Aura at College Park (386 Yonge St.)	892	272.0	78
	2023	YSL Residences (383 Yonge St.)	981	299.0	85
	2023	The One (1 Bloor St. W.)	1,109	338.0	94

Sources: Lydia Dotto, "Toronto's Temple Building, Once Empire's Tallest, Comes Down," *Toronto Star*, Saturday, 1 August 1970. Data on building heights were extracted from Patricia McHugh, *Toronto Architecture: A City Guide*, 2nd. ed. (Toronto: McClelland & Stewart, 1989), 95–8; and William Dendy and William Kilbourn, *Toronto Observed: Its Architecture, Patrons, and History* (Toronto: Oxford University Press, 1986), 152–223. Other notable studies of Toronto's downtown skyline include Gunter Gad and Deryk Holdsworth, "Building for City, Region, and Nation: Office Development in Toronto 1834–1984," in Victor L. Russell (ed.), *Forging a Consensus: Historical Essays on Toronto* (Toronto: University of Toronto Press, 1984), 272–319; and Gunter Gad, "Toronto's Financial District," *The Canadian Geographer* 35 (1991): 203–7. Heights for the Traders' Bank, Canadian Pacific, and Royal Bank buildings were provided to us by Dr. Gunter Gad of the Department of Geography, University of Toronto. Gad is the unquestioned authority on changes to Toronto's skyline. Data on building heights were extracted from *The 1991 Canadian World Almanac*, 481 and 539, McHugh, *Toronto Architecture*, 95–8, and the database found at https://www.emporis.com/search/Toronto. On the Toronto-Dominion Centre see Collier, *Contemporary Cathedrals*; and John Bentley Mays, "'The Coffins' Leave Them Cold," *Globe and Mail*, Wednesday, 30 October 1991, C1; and "Embodying Lofty Virtues," *Globe and Mail*, Wednesday, 6 November 1991, C1. See also McHugh, *Toronto Architecture*, 95–8; and Dendy and Kilbourn, *Toronto Observed*, 152–223. On the Toronto Star building see Michael Kluckner, *Toronto The Way It Was* (Toronto: Whitecap Books, 1988), 53. On the construction of First Canadian Place see Peter Foster, *The Master Builders: How the Reichmanns Reached for an Empire* (Toronto: Key Porter Books, 1986), 23–33. On Toronto's early apartment buildings see Richard Dennis, *Toronto's First Apartment House Boom: An Historical Geography, 1900–1920*, Research Paper No. 177 (Toronto: Centre for Urban and Community Studies, University of Toronto, 1989); and "Apartment Housing in Canadian Cities, 1900–1940," *Urban History Review* 26 no. 2 (March 1998): 17–31. On the City Park Co-Operative Apartments see Chris Bateman, "The First Modern Apartment Complex in Toronto," http://spacing.ca/toronto/2017/08/26/first-modern-apartment-complex-toronto/.

Street and then south to Cherry Street, to serve both the residential and commercial developments in the Distillery District and a condominium development at the site of the former Athletes' Village, which was built for the 2015 Pan American Games. In order to help solve congestion along the 504 King route, this line was short-lived and has now been incorporated into different branches of the King streetcar, which helps to provide more frequent service along the central portion of the line.

To improve service along King Street, the so-called King Street Pilot project was introduced in November 2017. Under its mandate, car traffic between Jarvis and Bathurst Streets was restricted. Ridership on the streetcar quickly increased from around 65,000 to 84,000 riders each weekday and reliability and scheduling improved.[54] In 2019, City Council voted to make these changes permanent. As Ken Greenberg has observed:

> The natural companions of the legacy streetcar were the many continuous main streets, lined with neighbourhood shopping, readily accessible on the journey home. But with greater intensity in the city, legacy streetcars in mixed traffic, in our narrow twenty-metre rights-of-way like King Street, had become immobilized in rush hour traffic. The King Street pilot demonstrated a way of unlocking this jam by restricting auto access.[55]

Toronto's streetcar/light rail system is poised for more growth. Construction on the 19-kilometre Eglinton Crosstown LRT (Line 5) is nearly finished and the line expected to open in 2022. Proposals for east and westward extensions are being studied. Construction has begun on the Finch Avenue West LRT (Line 6) and will soon begin on the Ontario Line, between Ontario Place and the Science Centre.[56] Plans are also underway for a new Waterfront East LRT line along Queens Quay from the existing Harbourfront line to Cherry Street and into the Port Lands. Moreover, across Ontario, two new LRT systems opened in 2019 in Ottawa and Kitchener-Waterloo, the Hurontario LRT is under construction in Mississauga and Brampton, and Hamilton is moving forward with its east-west LRT.

Figure 1.10 | Three views from a helicopter above Toronto, 1972. Railyards at Spadina and Front (*top*); King Street near Bathurst (*bottom*); and looking north along Spadina (*opposite page*). Photographer: Michael Doucet.

Figure 1.11 | Views from the CN Tower, 1985/6 and 2020. Photographers: Michael Doucet (1985/6) and Brian Doucet (2020).

2

TORONTO IN A GLOBAL CONTEXT

There is a tendency to think of the changes happening in one's own city as unique and different from those in other cities around the world. But the forces of change which have shaped Toronto over the past fifty years are part of wider processes which have affected cities around the world. The impact of these trends is seen in many aspects of city development. The photo sets in this book present the visual manifestations of change; in the next two chapters, we will situate these photos within wider conceptual, theoretical, and policy perspectives.[1]

These larger trends can play out differently in different cities, sometimes in contradictory ways. For instance, while Toronto shares many characteristics with other advanced, post-industrial cities, it is the product of unique historical and political circumstances. In a 1995 book, Toronto journalist Robert Fulford categorized his hometown as an accidental city.[2] In other words, decisions made elsewhere have contributed to Toronto's transitions over the past half century. During this period, Toronto, unlike other cities in the Great Lakes region of North America, has experienced population growth rather than decline. Part of Toronto's growth was spurred by Montreal's decline in the 1970s and 1980s, caused partly by the rise of French separatism and accelerated with the election of the Parti Québécois to office in 1976. Because of this, several prominent companies, such as Sun Life, relocated from Montreal to Toronto, as did many anglophones. The balance of economic power in Canada was shifting from Montreal to Toronto during the 1970s and 1980s; corporations that moved to Toronto between 1970 and 1981 were valued at $31.4 billion in assets. During this same period, Montreal lost $20.5 billion in assets.[3] Concomitantly, approximately 300,000 Montreal anglophones left the city,[4] with most moving to the Toronto region.

Toronto also benefited from changes to the federal government's immigration policies, such as the 1976 Immigration Act, which opened up more possibilities for immigrants from non-white, non-European backgrounds. In the mid-1960s, 80 to 90 per cent of all immigrants to Canada came from Europe or the United States; by 1996 this percentage had fallen to less than 20 per cent.[5] During the same time period, immigration from Asia increased from 20 per cent to more than half of all arrivals to Canada. Immigrants, regardless of their country of origin, have disproportionately settled in the big cities (Toronto, Vancouver, Montreal); despite having one-third of Canada's population, these three metropolitan regions received 71 per cent of Canada's immigrants in 1998 (42 per cent going to the Greater Toronto Area alone).[6] While this decreased to 56 per cent of new immigrants settling in the three big urban regions in the 2016 census (with many now going to cities in western Canada), the vast majority of new immigrants still settle in metropolitan areas in and around big cities and are the major contributor of population growth therein.

Economic Restructuring

Toronto, like many major cities in the Global North, has developed from a major centre of manufacturing and production into a post-industrial city; it is now a city with a diverse economy dominated by financial services, insurance, the creative sector, education, medical science, tourism, government, and retail activities, particularly in the urban core.

In the 1980s, Toronto was undergoing a profound economic shift, with steep declines in manufacturing. Some of the city's biggest manufacturers, including Massey Ferguson, CCM, Swift, and Canada Packers, began struggling in the 1960s. By the 1980s, these companies, which had come to dominate and define many Toronto neighbourhoods in the twentieth century, were virtually broke, and were forced to shut down their manufacturing operations.[7] Other major factory closures during this period included General Electric on Lansdowne Avenue; breweries near King and Berkeley, Dundas and Victoria, and Fleet and Bathurst; Inglis Appliances, and Irwin Toys in what is now the Liberty Village area; the General Motors van plant in Scarborough; and the Goodyear tire plant on Lake Shore Boulevard West in New Toronto.[8] While many of these industries had been in decline for decades, the Free Trade Agreement with the United States (1989) and the North American Free Trade Agreement (1994) dealt a major blow to manufacturing in Canadian cities.

Garment manufacturing, centred around the intersection of King Street West and Spadina Avenue, held on a little while longer, but also eventually succumbed to the combined pressures of globalization, rising downtown real estate values, and changing urban policy. Later in the book, we present photos of the Massey Ferguson (on King Street West at Shaw Street) and the Swift (St. Clair Avenue West at Keele Street) factories in operation, as well as several locations in the garment district (now rebranded as the Entertainment District), and what has become of these sites since industrial production ceased.

These once vast industrial spaces in the city's core, which contributed to making Toronto a wealthy city (even if all that wealth did not trickle down to the employees of those factories), have either disappeared entirely, or their buildings have been repurposed with non-industrial activities such as offices, shops, and lofts. One of the first major conversions of an industrial building to post-industrial use was Queens Quay Terminal. Built as a cold storage facility in 1926, and known then as the Toronto Terminal Warehouse, the building was converted into offices, shops, and condos in the early 1980s. The former Massey Ferguson site is now home to many condo developments. The former meat-packing plants around St. Clair and Keele are now a combination of houses and shops, with the Stock Yards Village shopping centre replacing the last major slaughterhouses and processing plants, some of which remained in business as late as the early twenty-first century. One of Brian's early childhood memories is riding in the family car from our home in North Toronto, to visit his grandmother, who lived near Islington and Bloor. The trip would involve driving along St. Clair Avenue West, which was an excitement in and of itself because it had streetcars (and usually involved a short detour to Wychwood Avenue to see the PCCs awaiting removal to the scrapyards). But a distinct memory is of the smell of the meat-processing plants around St. Clair and Keele as well as the numerous industrial rail sidings that crisscrossed St. Clair, connecting the factories to the main Canadian National and Canadian Pacific freight lines.[9]

While these are prominent sites of deindustrialization within the city, it is easy to forget that industry has not entirely disappeared from Toronto, or its wider region. The geography of deindustrialization is uneven; manufacturing is still important in (and to some extent defines) the suburbs, but virtually none of it is found within the urban core anymore.[10] Automotive manufacturing is the largest economic sector in the Greater Toronto Area behind financial services. Large auto assembly plants can be found in Oakville, Brampton, and Oshawa, as well as further afield in Cambridge, Woodstock, Alliston, and an engine plant in St. Catharines. Additionally, there are many automotive parts plants throughout the Toronto region.

The construction industry is also still very important, especially in the area of home building. Over the past several decades the city has undergone a condominium boom, a topic we discuss in detail in the next chapter.[11] It should also be noted that in the early twenty-first century, the GTA had the third-largest concentration of industrial floor space in North America (behind Chicago and Los Angeles) and more than Vancouver, Calgary, Edmonton, Montreal, Ottawa, and Kitchener-Waterloo combined.[12] Within the city of Toronto, industrial clusters can still be found in the northwest, around Pearson Airport, as well as in parts of North York and Scarborough. These large industrial, warehousing, and logistics activities continue deep into the outer suburbs of Mississauga, Brampton, Milton, and

Vaughan, and include not only automotive production and other manufacturing but massive sorting and warehousing facilities for companies such as Amazon. These suburbs around Toronto are also home to many who work in these sectors. But most manufacturing (and its associated working-class neighbourhoods) has disappeared from the core of Toronto, where our streetcar photographs are concentrated.[13] This can lead to the false impression that all of Toronto has undergone a complete process of deindustrialization, as the most prominent parts of the city and region no longer house major manufacturing, warehousing, and distribution centres. However, the reality is far more complex, particularly when looking at both the entire Greater Toronto Area and the Greater Golden Horseshoe (which includes Hamilton, Guelph, St. Catharines, Niagara, Peterborough, and Waterloo Region).

While there are clusters of manufacturing and industry in the Toronto region, these activities represent a much smaller share of employment than in the 1960s, as discussed in the previous chapter. Many scholars would argue that the 1970s and 1980s, in particular, marked a profound shift away from an international economy towards a more global one.[14] The international economy, characterized by closely regulated trade of goods and services across national boundaries, was conducted by individuals and firms. This was the era of the Fordist mode of production, an era characterized by mass production, mass consumption, and a Keynesian system of welfare and regulation, which emphasized the role of the state to even out the social and spatial inequities of capitalism.[15]

In contrast, the global economy is based on the production and trade of goods and services by a smaller number of global firms that operate worldwide, with far less regulation by the state.[16] The transition to a global economy led to a profound process of deindustrialization in the Global North, particularly in North America and Western Europe, and deep economic and social crises in cities that were predominantly industrial. Even New York and London, today's pre-eminent global cities, were on the verge of bankruptcy and witnessed unprecedented economic and demographic decline during these decades. But their economies were more robust than cities whose primary function was manufacturing (particularly if that was based on one sector, such as the automotive industry in Detroit), or those that lacked existing global connections in trade, transport, and commerce. The so-called Rust Belt cities of the American Midwest, as well as industrial communities in northern England, Belgium, and Germany's Ruhr Area, were particularly hard hit by the loss of industry. So too were many smaller cities in Ontario and Quebec, especially those based around one or two large factories.

As economies in the Global North have shifted away from manufacturing, much of it moving to the Global South, there has been a reordering of the hierarchy of cities, as seen in the rise of a handful of cities that are increasingly connected through intricate webs of global networks. Toronto is one of these cities.

As a rapidly growing city and one of North America's leading financial services hubs, Toronto stands out among many other cities situated on the Great Lakes, which have experienced extreme processes of economic, social and demographic decline. Cities such as Detroit, Cleveland, Buffalo, Hamilton, and Milwaukee have suffered severely because of deindustrialization and have had little to help their economies rebound. There are a number of reasons why Toronto did not turn out like Detroit or similar American cities. One explanation is offered by University of Toronto geographer Jason Hackworth:

> Most important, it is Canada's financial hub. Its emergence in this role was somewhat an accident of history, as it occurred in the 1970s when banks grew squeamish about the potential Quebec secession, and relocated their headquarters from Montreal. The transfer of financial firms created tens of thousands of jobs in Toronto, and the city emerged as a global financial hub just as the Rust Belt region (including in Toronto) was being hit with crippling industrial sector job losses ... Many American Rust Belt cities found themselves isolated from a globalizing economy or larger cities, like Toronto, which might serve as a conduit for them to access it.[17]

While their industrial prowess and technical innovations made them prosperous during the late nineteenth and early twentieth centuries, many Great Lakes cities have been less able to compete in a new economy driven by finance, knowledge, creativity, and globally connected networks. However, today, it can be easy to forget that while Toronto was of little global significance in the early 1960s, Detroit was one of the key pieces of the industrial heartland of America. A modern, cosmopolitan, and prosperous city, Detroit could boast America's best public school system and highest rates of homeownership; it was a major cultural centre and had the fifth-highest population of any city in the United States.[18] How can we explain divergent trajectories and Toronto's meteoric rise from provincial city to Canada's largest and foremost world city?

One of the most important theories developed to explain the rise of cities such as Toronto and how they have diverged or decoupled from other cities in terms of population composition, economic growth, and connectivity to the global economy is the "world city hypothesis" put forward by John Friedmann and Goetz Wolff in 1982, and articulated in much greater detail by Friedmann in 1986.[19] Rather than seeing the city as a result of accidental processes, or urban change as natural or organic, Friedmann conceptualizes urbanization as being linked to global economic forces. The economic structure and hierarchy of a city is therefore related to its position within the spatial organization of the "new international division of labour" (NIDL): the shift in manufacturing from advanced capitalist countries of the Global North (like Canada), to developing countries, today more frequently referred to as the Global South.

Until the 1970s, the main function of large parts of Asia, Latin America, and Africa in the global economy had been to supply raw materials and natural resources to wealthier countries, which were involved in the manufacture of everything from textiles to coffee tables. Better transport and communication infrastructure, combined with changing global trade laws and policies, enabled companies to shift production away from high-wage countries to lower-wage ones. This new global division of labour, sometimes referred to as the "first global shift," meant that the production of goods was no longer primarily confined to places such as Western Europe, North America, and Australia. For cities that relied primarily on manufacturing, such as Glasgow or Detroit, the effects were particularly devastating. For cities with more diverse economies, such as Toronto, New York, and London, the transition was at times painful, but a more robust economy meant that new economic activities in sectors such as finance, insurance, real estate, culture, education, and technology led to new economic growth, although the benefits of such changes were not distributed evenly.

With his world city hypothesis, Friedmann argues that this context shapes the rise of major global cities and accounts for the urban hierarchy of cities both within and between countries. He articulates seven aspects of this hypothesis, each of which is relevant and applicable to Toronto.

His first point is that a city's integration within the world economy and its function within the NIDL can be decisive for any internal structural changes. This means, according to Friedmann, that urban change is largely a response to changes that are *externally* induced as part of a city's connection to the global economy. Changes in both the labour market and urban form can be attributed to global processes related to capital flows, finance, production, and the employment structure of economic activities.[20] Local factors, such as histories of capital accumulation and local and national policies, also shape the ways in which these global economic factors play out in specific situations.

His second thesis is that key cities in the world are used as "basing points" by global capital and that this produces a complex spatial hierarchy of cities. While the geography of this hierarchy has changed significantly since the publication of Friedmann's article, the concept of primary, secondary, and tertiary cities in the global financial network still holds true today. London, New York, and Tokyo are undisputed leaders in global finance; many others, such as Toronto and Frankfurt hold key, yet secondary, positions in the global urban hierarchy. All but two of his original primary world cities were situated in "core countries" (Western Europe, North America, Australia, and Japan – what we today refer to as the Global North). The designation of core places in the global economy has shifted since the 1980s; a more recent classification of world cities reflects this changing geography; it includes among its elite list a number of cities that did not appear in the original classification – such as Beijing, Dubai, and Shanghai.[21]

Friedmann's third thesis is that the global command and control functions of world cities are directly reflected in the structure and dynamics of their production sectors and employment. This means that the driving force behind growth in world cities is a small number of rapidly expanding economic sectors such as corporate headquarters, international financial services firms, global transport and communications, and high-end, advanced business and producer services (accounting, advertising, legal services, and so on). This is often referred to as FIRE (finance, insurance, and real estate).[22] Friedmann also argues that the role of culture is important. These changes can be seen in the visual landscapes of Toronto. The dramatic additions to the skyline in the 1970s and 1980s represented the consolidation of key financial services activities in downtown Toronto, as the city eclipsed Montreal, not only in population, but also as Canada's most important global corporate and commercial hub. Additional changes to the built environment initiated during this time reflect the creation of new housing for those employed in these sectors (such as condos), the upgrading and gentrification of many inner-city neighbourhoods, as well as cultural and recreational facilities, all of which have been stimulated and encouraged by local planning and policy decisions.

The fourth thesis is that world cities are sites for the concentration and accumulation of international capital. Graham Todd argues that in Canada, much of the capital accumulation in this new era of globalization has taken place very locally within the country's largest cities, specifically in Toronto.[23]

Fifth, world cities are destinations for large numbers of domestic and international migrants. As we noted in chapter 1, Toronto, Montreal, and Vancouver are the main centres of immigration in Canada. Toronto is one of the world's most diverse and multicultural cities, with immigration from abroad, rather than domestic migration or an increase in birth rate, being the main source of population growth.[24]

Friedmann's sixth thesis states that world cities bring into focus the major contradictions of industrial capitalism, which include spatial and class polarization. He argues that this takes place on three geographic scales. At the global level, there is a widening gulf in wealth, income, and power between a small group of rich countries and the rest of the world.[25] The second scale is regional or national; world city regions tend to be far wealthier than the rest of their national economies. Although, as he argues, this tendency is more extreme in semi-peripheral countries such as Brazil or Thailand, it can be seen in the ways in which major business investments coalesce around global cities. The online retailer Amazon's 2018 decision to open its second headquarters in both New York and the Washington metropolitan area, two regions already at the forefront of the global urban hierarchy, is a prime example. The final scale is metropolitan; here spatial polarization is a result of class polarization, a situation where affluent enclaves and gentrified neighbourhoods exist alongside areas of extreme poverty and decline in increasingly fragmented urban regions. Friedmann attributed three factors to this class polarization: large income gaps between urban elites and low-skilled workers, large-scale migration to the city from poorer rural areas or from abroad, and the structural trends in the labour market mentioned earlier. While Friedmann was writing several decades ago, these inequities can now be seen in the contrasting outcomes and experiences of different populations in response to the COVID-19 crisis; disparities related to class, ethnicity, and geography are exacerbated by these pre-existing divisions.[26]

This polarization is by no means unique to Toronto; simultaneously growing levels affluence and poverty are found in major cities around the world.[27] This shift in income distribution, from predominantly middle-class cities to ones dominated by extremes of wealth and poverty, is depicted by Peter Marcuse:

> The best image, then, is perhaps that of the egg and the hourglass: the population of the city is normally distributed like an egg, widest in the middle and tapering off at both ends; when it becomes polarized the middle is squeezed and the ends expand till it looks like an hourglass. The middle of the egg may be defined as "intermediate social strata" … or if the polarization is between rich and poor, the middle of the egg refers to the "middle-income" group.[28]

Friedmann's seventh, and final, point is that world city growth generates social costs at rates which exceed the fiscal capacity of the state. This will become obvious to anyone relying on publicly funded services such as affordable housing, adequate health care, and effective transportation. The rapid growth of world cities puts strains on these services. But equally important, there are also competing interests concerning what types of services and amenities to fund; the interests of the poor (or even middle-income residents) sit in contrast to the needs of transnational capital, which often relies on public subsidies or public investment to further attract more capital – such as the subsidies offered to attract Amazon or the redesigning of urban spaces to attract and retain more affluent residents. Public money is commonly spent on tax incentives, hockey arenas, high-end developments, flagship projects, express airport rail links, and other projects and programs which help a city's international competitiveness; meanwhile basic services deteriorate.

Such tensions can be seen in the ways in which the goals of mega-projects have shifted over time. Toronto's waterfront serves as a useful case study for this; earlier mega-projects in the first half of the twentieth century focused on ideas of democratizing public goods and services, fostering progress towards the better provision of infrastructure. The Toronto Harbour Commission invested in infilling and port construction to help make Toronto one of the Great Lakes' major industrial and shipping cities. More recently, however, mega-projects along the waterfront have focused on "further[ing] the aims of global interurban competitiveness rather than those of more local and pressing needs and requirements."[29] In other words, what we see along the waterfront is no accident. It represents the choice to favour consumption and high-end uses, something that is rooted in the competition to become a world city and that comes at the expense of potentially using these spaces to address local needs such as affordable housing. While not directly on the waterfront, the nearby St. Lawrence neighbourhood, developed in the 1970s and 1980s, represented a decidedly different approach to urban redevelopment. It was a concerted effort by all three levels of government to build an affordable, mixed-income, and mixed-tenure neighbourhood, with good urban design in the core of the city. Policy shifts and the end of provincial and federal funding for affordable housing mean that the waterfront redevelopment since the 1990s features virtually no affordable housing within it.[30]

Friedmann's major contribution to urban debates has been his ability to see cities within global, rather than national or regional, networks. Until the latter part of the twentieth century, cities were often compared and ranked within national hierarchies, but as he states, both in his original thesis, and in subsequent writing, these truly global networks are a new phenomenon.[31] Despite being over thirty years old, Friedmann's hypothesis remains important to our understanding of uneven urban development, both within and between cities. Recent scholarship emphasizes that the "overall consensus is [that] capitalist command and control is exercised from a limited set of cities which function as nodes for transnational flows of capital, goods, people and information, from which actors operating from these places draw their power."[32]

What is striking about Friedmann's work is that it remains relevant in describing the major changes which have taken place in Toronto since the 1970s and 1980s. The shift away from manufacturing and the rise of services (both high- and low-end), the consolidation of Canadian financial and corporate power in downtown Toronto, continued immigration, a polarizing and increasingly unequal population and neighbourhood structure, and a hollowing out of middle-class jobs owing to shifts in the global economy are all defining trends which have shaped Toronto. Later in the book, we will explore the ways in which these major forces of change can be visualized. Friedmann's thesis useful conceptual starting point to explain why, for example, the factories that lined King Street in the 1960s have been replaced by bars and restaurants, condos, lofts, and creative hubs.

Another major theorist on the subject of world cities is Saskia Sassen. She expands on Friedmann's work by focusing on what she terms "global cities." Her ground-breaking 1991 book, *The Global City: New York, London, Tokyo*,[33] argues that telecommunications and globalization are the major forces shaping cities and urban space. She points to the emergence of a transnational network of cities centred on global financial markets, advanced and specialized services, and high-end production. Sassen argues that globalization both centralizes and disperses economic activities. Many production activities are able to be relocated to areas with cheaper production costs (this can be either to other parts of the same country where land is cheaper or to other countries with cheaper labour). At the same time, command functions increasingly cluster in a handful of global cities. (While her book is focused on London, New York, and Tokyo, many of its principles apply to other leading cities in the global financial network, including Toronto, Sydney, Frankfurt, Amsterdam, and Hong Kong.)

As the economy becomes increasingly globalized, more and more specialized services are needed to manage these international networks and companies. This "corporate services complex" includes activities such as finance, insurance, advertising, and accountancy; Sassen argues that these services are increasingly clustered in global cities such as Toronto.

One of the ways in which this happens is through a process of national consolidation.[34] In the era before the globalization of the financial services industry, each country would have its own hierarchy of stock markets and financial services centres. While the Paris stock market has always been France's largest, the stock markets in smaller French cities, such Lyon and Marseilles, used to be much more important than they are today. Globalization means an increasing consolidation of the key command functions in these sectors into a handful of cities: Paris wins and Marseilles loses in this scenario. The Philadelphia Stock Exchange, once one of the most important in America and responsible for both financing the westward expansion of the nation and the incredible industrial growth in its home city, ebbed and flowed with the city's economic fortunes, eventually becoming merged with the NASDAQ in 2008.[35] In some cases, this has meant one city eclipsing another; both Sydney and Toronto have seen national consolidations of financial services activities at the expense of once larger and more important Melbourne and Montreal. Toronto has become, over the past forty years, Canada's major centre of finance. This includes not just a consolidation of Canadian firms whose headquarters or major decision-making centres are found in Toronto (such as the Bank of Montreal and Scotia Bank),[36] but also the Canadian headquarters for global financial companies such as HSBC.

Saskia Sassen also contends that the emergence of a financial and services complex brings with it a new economic regime which devalues other economic activities. Because of the financial services sector's ability to generate "superprofits," it has outstripped other functions, such as manufacturing, in their ability to generate profits: Sassen explains:

> This has had devastating effects on large sectors of the urban economy. High prices and profit levels in the internationalized sector and its ancillary activities, such as top-of-the-line restaurants and hotels, have made it increasingly difficult for other sectors to compete for space and investment. Many of these other sectors have experienced considerable downgrading and/or displacement, as, for example, neighbourhood shops tailored to local needs are replaced by upscale boutiques and restaurants catering to new high-income urban elites.[37]

In other words, while many economic activities may generate profit, the tremendous wealth accumulated in the financial services sector has contributed to the displacement of other activities and services. Such trends can be witnessed in many Toronto neighbourhoods. Many of the photos in and around downtown in the 1960s reveal a small, but highly concentrated financial services cluster, around King and Bay, as well as many different activities to be found on adjacent streets such as York and Richmond (including industrial warehousing, and commercial spaces), most of which have entirely disappeared as the financial district has expanded its footprint dramatically. To meet this growing demand for high-end office space for the financial services sector, the South Core (the area south of Union Station) was developed as a southern extension of the Financial District onto what were formerly railyards and industrial spaces.

The concept of world cities has been empirically analysed by the Globalisation and World Cities (GaWC) Institute at the University of Loughborough in the United Kingdom. Their focus is on the external relationships that cities have and the networks and linkages between cities, rather than the internal patterns and processes *within* world cities. Their analysis is based on the presence of advanced producer services (APS) companies in areas such as financial services, accountancy, law, advertising, and management consultancy.

For each of these major firms, they give a value from 0 to 5 for every city in which they have a presence. A score of 5 means that the firm is headquartered there and a score of 0 means that the firm has no presence in that city. They have calculated this for close to two hundred firms in more than five hundred cities around the world to produce a "service values matrix." This analysis is then used to produce a classification of "the world according to GaWC," which measures a city's integration into the world city network. This produces different classifications: Alpha++, for London and New York, which stand out in all analyses as more integrated into the world economy than any other cities; Alpha +, for highly integrated cities which complement London and New York; Alpha and Alpha-, for very important cities which link regional and national economies into the world economy; Beta, for important cities which link regional and national economies into the world economy; Gamma, for important cities whose global orientation is not in APS; and, finally, cities with sufficiency of services, for cities that are not world cities as defined by GaWC, but which are not overly dependent on world cities (such as smaller capital cities and manufacturing centres). In 2018, Toronto was classified as an Alpha world city along with 22 other cities including Chicago, Mumbai, Los Angeles, Frankfurt, Bangkok, Brussels, Mexico City, Kuala Lumpur, and Moscow. Table 2.1 gives an overview of Canadian cities mentioned in the GaWC classification in 2000 and 2018.

What sets Toronto apart from other cities in this category is its dominant role in a national, G7 economy.[38] This is evident in the location of Canada's major bank headquarters (or dominant decision-making centres), which are concentrated in downtown Toronto. One of the first studies of Toronto as a global city was done by Graham Todd. He noted that "Toronto's unrivalled position in the national economy has been as important a factor in propelling the city to the status of a 'second tier' world city as have been global linkages to the world economy."[39]

Much of this transition towards a global city started in the 1980s, which was not only a period of steep industrial decline, but also of tremendous growth in the services sector, particularly in banking and financial services. Much of this growth, particularly in corporate headquarters, was concentrated in downtown Toronto, and is part of a deliberate strategy by a variety of actors to remake the core of the city into an ever more globally oriented space. As Todd outlines, this growth not only included the development of office space for national and global FIRE-based companies, but also the cultural, amenity, and recreational transformations that have taken place along the waterfront since the 1980s, as well as the boom in residential developments such as condo towers on former railway lands. There was also

Table 2.1. Canadian cities in GaWC classification, 2000 and 2018.

City	2000 Classification	2018 Classification
Toronto	Alpha	Alpha
Montreal	Beta	Alpha-
Vancouver	Beta-	Beta+
Calgary	Gamma-	Beta
Ottawa	Sufficiency	Gamma-
Edmonton	Sufficiency	High Sufficiency
Quebec City		Sufficiency
Winnipeg		Sufficiency
Halifax		Sufficiency

Source: https://www.lboro.ac.uk/gawc/world2018.html.

a national consolidation in the media and publishing sectors during this time, at least in English-speaking Canada. Further policy developments, particularly after the amalgamation of the city of Toronto, focused on developing an international strategy to become more global.[40] These market forces and policy ambitions have together shaped the socio-economic geography of the new post-industrial Toronto, as well as influenced changes to its urban form and built environment. But, as an inherent part of world city formation, these forces have also produced conflicts with the declining industrial and manufacturing spaces (and their related workforces), which have been unable to support the higher property prices that have resulted from this transition towards a global city.

Urban Competitiveness, Neoliberalism, and Climbing the Urban Hierarchy

In the context of deindustrialization and the rise of world cities, the economic and social geography of cities has become more uneven. Throughout much of the twentieth century, one of the main roles of central governments was to iron out the spatial differences of capitalism. National governments would operate transfer schemes from rich parts of the country to poorer ones, invest heavily in infrastructure in less developed parts of a country in order to stimulate investment there, or relocate government offices to more peripheral locations. In 1977, the Ontario Progressive Conservative government under Premier William Davis moved the headquarters of the Ontario Health Insurance Plan (OHIP) from Toronto to Kingston and the Ministry of Revenue to Oshawa, resulting in a transfer of 900 and 750 jobs respectively.[41] Liberal premier David Peterson built on this base in the mid-1980s, initially using the strategy of decentralization to stimulate the economies of cities in Northern Ontario. In 1986 some positions at the head office of the Ontario Lottery Corporation were moved to Sault Ste. Marie (145 jobs), the Ministry of Northern Development and Mines was relocated to Sudbury (290 jobs), and positions in the Forest Resources Branch of the Ministry of Natural Resources were transferred to North Bay (200 jobs). A year later 230 jobs in the Registrar General Branch and the Student Awards Branch were moved to Thunder Bay. This relatively small shift of 865 jobs would have almost no impact on the Toronto economy.[42]

Within the city, the role of local government was to provide for basic services, distributed evenly across the city. These services included roads, schools, libraries, parks, public transit, water, sewers, and garbage collection. Large municipal infrastructure projects concentrated on wealth distribution, collective consumption, and social reproduction, with an emphasis on spreading these across the city.[43] In many countries, local governments directly intervened in the housing market in the form of social, or subsidized, housing. By the 1970s, most of the housing stock in cities such as Vienna, Glasgow, and Amsterdam was owned by their municipalities or held in other forms of social, non-marketized forms of tenure; this period was also the high-point for subsidized housing in Canadian cities. Large social housing and urban renewal projects, funded by all three levels of government, were among the major public investments in cities between the 1940s and 1980s.

However, such a socially oriented approach is no longer common, and contemporary civic investments have far different goals and involve more complex sets of actors than in previous eras. The emphasis has shifted from a "politics of income redistribution [towards] a politics of growth."[44] Today's major municipal investments tend to focus on speculative, high-profile developments, the hosting of major events, and the building of places which are attractive for tourists or affluent residents. Broadly speaking, the goal has shifted from wealth distribution towards wealth creation (with the hope that this wealth will then "trickle down").

This transition was outlined by the geographer David Harvey. He called this a change in urban governance from "managerialism" to what he referred to as "urban entrepreneurialism."[45] This meant that, rather than simply managing the resources within a city in order to ensure appropriate distribution of services and adequate provision of housing, education, and other services provided by the local level of government, the role of municipalities changed towards actively stimulating inward investment, taking risks in economic development, and working closely with private sector actors to realize these goals. Harvey conceptualized urbanization as a spatially grounded social process where different actors with different goals interact to produce and reproduce urban space.

As cities such as Toronto transitioned from industrial to global cities, starting in the 1970s, much of the manufacturing and warehousing activity situated along waterfronts, rail corridors, and near downtowns disappeared. Some of these activities relocated to the suburbs, with their ample space and proximity to highways,[46] while other activities moved offshore. Regardless of where the industry went, it was within its former spaces in the inner city that the new post-industrial landscapes were built. In North America, Baltimore provided one of the earliest examples of turning a formerly industrial space into a mixed-use flagship project. In 1978, the city voted in a referendum to allow for the use of city-owned land along the waterfront for a private development. This site, in the old Inner Harbor, became home to Harborplace, which would serve as a model for other cities around the world to emulate in their pursuit of growth-oriented, post-industrial development.[47]

In Toronto, this approach to urban development has been framed under the mantra of building a "competitive city," with an emphasis on increasing the city's economic attractiveness and augmenting its position within the global urban hierarchy. The push towards promoting a competitive city has been accelerating since the forced amalgamation of Metropolitan Toronto in 1998, and numerous policy documents frame Toronto as a global city, one in competition with other globally aspiring cities for jobs, investment, educated workers, and tourists.[48] As Julie-Anne Boudreau, Roger Keil, and Douglas Young explain:

> Governments at the time [1990s] set in motion a set of practices, driven by right-of-centre ideologies, which "liberated" economic growth through marketization and privatization of previously public services through an application of "neoliberal" modes of governance (performance indicators, streamlining, etc.) to bureaucratic processes. These new practices are based on and accepted by new subjects, and collectives have emerged in a new frame of societal reference where individuals and communities expect to "do their share" in protecting themselves against the ups and downs of markets.[49]

The neoliberal turn is broadly associated with the elections of Ronald Reagan in the United States in 1980 and Margaret Thatcher in 1979 in the United Kingdom. In Toronto, this turn would come later, as the 1970s were a particularly progressive period in local politics. While building slowly in the 1980s, with development-oriented local governments, the emergence of full-fledged neoliberal governance would arrive with the election of the Mike Harris Progressive Conservative government in Ontario in 1995, and their platform known as the "Common Sense Revolution." A combination of dramatic budget cuts, downloading of social services onto municipalities, and the forced amalgamation of the municipalities within Metropolitan Toronto ushered in a new era of policy and development.[50]

Stefan Kipfer and Roger Keil argue that the competitive city has three elements: the entrepreneurial city, as conceptualized by David Harvey, which emphasizes strategies of capital accumulation; the city of difference, which focuses on patterns of class formation and the integration of culture and diversity into urban development strategies; and Neil Smith's concept of the "revanchist city," which emphasizes social control to effectively take back the city from the poor. All three elements have one key aspect in common: the pursuit of investments or upwardly mobile households in the name of succeeding in inter-city competition.[51] In Toronto, this was done through specific measures such as criminalizing the practices of homeless people and panhandlers, the violent eviction and removal of homeless people from city parks by the police, as well as privatizing the control of public spaces, which consequently serves to limit and regulate the types of activities that can be done among other practices and policies.[52]

In Toronto, neoliberalism takes form in different political and economic programs and policies. The city has long been operating within a context of fiscal austerity. Place-making and historic preservation are used to support and promote gentrification. Multiculturalism is increasingly viewed as an experience rather than a policy or an ideal intended for the pursuit for social justice. Finally, neoliberalism can be seen in the implementation of a law and order agenda, with its emphasis on everything from zero tolerance of "squeegee kids" in the early 2000s, to more recent policies of carding, which disproportionally affect BIPOC (Black, Indigenous, and People of Colour) populations.[53]

Two sites in particular, both of which feature in this book, represent places where the entrepreneurial, or competitive, city can be found. The first is along the waterfront. The aim of turning the waterfront into a mixed-use site central to the city's global economic competitiveness has been a major goal since the 1990s.[54] As manufacturing, shipping, and warehousing businesses left the waterfront beginning in the 1960s, the area was transformed into what Ken Greenberg called a "terrain of availability": a "screen on which cities project and explore emerging trends and prescriptions for urban development."[55]

The first major development was Harbourfront Centre, which was initiated by the federal government in the 1970s. The reconstruction of the landmark Queens Quay Terminal, one of the first major developments in the area, is discussed later in this book. In 1989, the federal government appointed a Royal Commission on the Future of the Toronto Waterfront, led by former Toronto mayor David Crombie. The commission would eventually make 80 recommendations; these are focused on an "ecosystem approach" to bridge together competing interests under the umbrella of environmentally oriented developments.[56]

The creation of the Waterfront Regeneration Trust followed in 1992, with support from all three levels of government (municipal, provincial, and federal). The trust was one of the main actors in Toronto's failed 2008 Summer Olympics bid. Had the city been successful in hosting these Games, the waterfront would have been a central location for venues and facilities. In 2001, Waterfront Toronto was created as another venture involving all three levels of government, and it continues to have the task of administering the redevelopment along the city's waterfront. Its many projects include Sugar Beach, Underpass Park, Corus Quay, Corktown Commons, the Wavedecks, the redesign of Queen's Quay, Canary District, and the Bentway.[57]

Much of the recent development along the waterfront is framed within cultural and socioeconomic benefits. However, as York University's Ute Lehrer and Jennefer Laidley argue, these benefits are premised by two major rhetorical streams that relate to growth-oriented, competitive city practices. The first is the idea of "world class culture," which aims to make Toronto resemble other major world cities. The second is "world class creativeness," an approach that is rooted in Richard Florida's idea of the creative class.[58] Florida, now based in Toronto, argues that creative individuals – a highly mobile, diverse, and sought-after group – are the drivers of economic growth in cities and that cities should refashion their urban spaces and their policies to attract and retain this talent.[59] While not as explicitly stated in Toronto as in some other cities, the aim is to attract the "right" kind of creative workers to the city: those who are young and well-educated.[60]

The second location where neoliberal urbanism and "competitive city" ideas can be seen is Yonge-Dundas Square.* Yonge Street, the central spine of Toronto, had become rather seedy downtown, particularly after the opening of the Eaton Centre in the 1970s, still Canada's largest downtown shopping centre. Initially, the Eaton Centre turned its back on Yonge Street with stores facing into the shopping mall, rather than onto the street. The construction of a new public square on the southeast corner of Yonge and Dundas was part of a strategy to secure Toronto's position as a world class city.[61] The discount retailers, strip clubs, and cheap fast-food restaurants that were located there were not in keeping with the aim of attracting high-end retail to this central location in an aspiring global city. When constructing the square, several buildings were demolished and the space was modelled after Times Square in New York City, complete with large advertising billboards and LED lighting (figure 2.1). While Yonge-Dundas Square is public space, there are strict rules that govern the square and the management of the space is in private hands. At Yonge-Dundas Square and elsewhere, there have been complaints about harassment of homeless, youths, and others by the private security guards who patrol the space.[62] Activities that do not fit within the intended uses of the space, or detract from its goals or positionality within the aspiring global city (such as begging, rough sleeping, or playing unsanctioned music), are not permitted. These spaces are good examples of Neil Smith's revanchist urbanism mentioned above. Megaprojects and policies like those described above are often spared scrutiny or criticism. Scholars have argued that this is because these current projects do not focus on a single issue (which could attract many critics), but rather on a variety of uses, specializations, and functions, and thus can appear to offer something for everyone.[63] Even progressive and left-leaning politicians tend to support them, particularly when they fall under the umbrella (and politically benign) term of "revitalization." However, as Lehrer and Laidley argue, such spaces are often highly undemocratic, with public consultations rarely extending to marginalized groups and other parts of the city, and are predicated on the inevitability of the development taking place rather than a robust discussion of whether or not such megaprojects are desired.[64] In analysing the broader shifts around governance and urban redevelopment over the past decades, the authors find that, despite the inclusive rhetoric, these new megaprojects do little to actually address the pressing socio-economic divisions, and because of their "quest for urban status rather than the pursuit of urban inclusion,"[65] they may actually further the social and spatial fault lines within the city.

Events, Spectacles, and Other Urban Competitions

Just as cities are trying to build new spaces which attract investment and affluent or creative residents, the competition between cities can also be seen in both the hosting of events and the more recent overt inter-city

---

* In July 2021, Toronto City Council voted to rename Dundas Street, Yonge-Dundas Square, the Dundas and Dundas West subway stations, three parks, and one Toronto Public Library branch because of the role that Henry Dundas played in delaying the abolition of the slave trade. At the time of writing, no new names for these streets and places have been determined.

Figure 2.1 | Yonge-Dundas Square, 1998, 1999, 2005, and 2011. Photographer: Michael Doucet.

competitions for new corporate headquarters. Today, most cities, large and small, actively bid for the hosting of events, as part of economic development, branding, and urban regeneration strategies. London's successful 2012 Olympic bid, for example, was based to some degree on using the Games to regenerate the poorer East End of the city, although how much regeneration has actually happened is questionable and much of the nearby investment has fuelled gentrification.[66] Toronto has bid for (and lost) two Olympic Games, in 1996 and 2008. It managed to win the rights to host the 2015 Pan American Games. For many of the city's boosters, hosting the Olympics would have demonstrated that Toronto was in the upper echelon of elite world cities:

> Toronto will make an ideal host to the world [if it wins the 2008 Olympics], because it is a world-class centre. We are all ready to host the world. (Maureen Kempston Darkes, GM Canada President, 2000)[67]

These were heady times for the businessmen at the helm of the [1996 Olympic] bid. The southern Ontario economy was booming, corporate profits soared through expansion, mergers, and leveraged buyouts, and the universe seemed to be unfolding according to neo-conservative plan. With their dazzling office towers and the new stadium with the retractable roof, the corporate elite [Paul] Henderson had recruited to the bid effort seemed on the edge of turning Toronto into a "world-class city." Surely the Olympics would follow. (Bruce Kidd, former Olympian, 1992).[68]

Many scholars have criticized such approaches to urban development. One of the most prominent critiques, put forward by David Harvey, is that they are based on the ancient Roman formula of "bread and circuses."[69] The idea is that bidding for, and hosting, events and festivals, as well as devising fancy marketing slogans, serve to

> distract those who have taken their place at the bottom of a restructured social stratification ladder, particularly immigrants and substantial sections of the organized-capitalist working class. The refashioning of the city's collective emotion at an aesthetic level is evoked, excited and sustained by the centralized structures of economic power controlling and providing the public with appropriate sites, signs and symbols.[70]

Despite such critiques, we see no let-up in the intense competition to host events, build spectacular iconic museums or attractions (such as the Ripley's Aquarium of Canada or the new additions to the Royal Ontario Museum and the Art Gallery of Ontario), and promote gentrification and high-end spaces and marketplaces for tourists, investors, and potential high-income residents. Meanwhile poverty, inequality, and polarization continue to grow. The competition launched by Amazon for its second corporate headquarters (HQ2) illustrated that this interurban competitiveness shows no signs of abating. When Amazon announced that it would be looking at bids from cities in September 2017, more than two hundred jurisdictions from across North America submitted proposals. Cities offered development sites and tax incentives, and touted their amenities, creative and dynamic workforces, and infrastructure. This list was narrowed down to twenty, including Toronto, in January 2018. Ultimately, Amazon chose to split its HQ2 between Crystal City, in the Virginia suburbs of Washington DC, and in Queens in New York City. When seen from a perspective of world cities and urban entrepreneurial literature, such a move made sense. According to Richard Florida, this was always going to be the case: "Where else are they going to go? It just showed us it was going to go to the biggest and most centralized, most global cities from day one. That's what this was about. Seattle wasn't a big enough place to host this headquarters. It has to be out on the East Coast in the power corridor, the Boston, New York, Washington area and that's where it went."[71] In New York, in particular, Amazon's decision was met with considerable opposition from local residents and community groups over fears of housing affordability, gentrification, and displacement. There were concerns that tax breaks were going to a large corporation, while the city's subway was chronically underfunded. Ultimately, Amazon pulled out of New York and focused on its northern Virginia site.

Moves such as this reinforce and exacerbate inequalities between core and peripheral parts of a country: Amazon chooses the largest and most important cities for its headquarters, partly because of the talented workforce, amenities, and good national and international transport links already present. But this competition also reinforces the idea that if you are talented and want to be ambitious and well connected in the post-industrial world, you need to locate (either as an individual or a business) in a handful of select cities.[72] It is here that the entrepreneurial aspirations of a large number of cities run into the realities of a global urban hierarchy that is increasingly concentrating more wealth, power and talent into a small number of cities.

However, it is interesting to note the growing criticism of the entire model of bidding (with public subsidies) for the corporate headquarters of one of the world's largest companies. Many of the bids, initially private information for the benefit of Amazon, have been made public. Some of the more outrageous offers from local governments included $2 billion in taxpayer-funded incentives from Atlanta, which included an exclusive lounge (complete with free parking) at the city's Hartsfield-Jackson International Airport; $6.5 billion in tax incentives from Montgomery County, Maryland (outside Baltimore), with a further $2 billion in promised infrastructure improvements, committed by the state government; and $4.6 billion in financial assistance (the majority in the form of performance-based grants) from the state of Pennsylvania, to help bids in Pittsburgh and Philadelphia.[73] Toronto, on the other hand, offered no financial incentives, with the head of Toronto Global stating, "Others may provide large subsidies and tax breaks, but like the Province of Ontario, we in the Toronto Region don't want to play that game and frankly we feel we don't need to play that game."[74]

In addition, several cities have rejected the hosting of Olympic Games in recent years through referendums and plebiscites. In November 2018, Calgarians voted 56 per cent against the city's bid to host the 2026 Winter Olympics, with concerns about costs and a general lack of enthusiasm for hosting another major sporting event that would create little in the way of new infrastructure that could be used after the Games were finished.[75] Residents in Innsbruck, Austria, also rejected hosting the 2026 Games in a referendum, and Rome, Hamburg, and Oslo are other cities with high-profile and potentially successful Olympics bids, which pulled the plug on their Olympic aspirations in recent years, either through public votes or changes in political mood.[76]

Changing Governance Forms: Public-Private Partnerships

One of the main characteristics of entrepreneurial urban governance has been the emergence of so-called public-private partnerships, or P3s. The building and operating of new TTC infrastructure are good examples of the shift to these types of partnerships. Going back in history, it has been argued that the creation of the TTC in 1921 came about because of the failure of a far earlier public-private arrangement between the Toronto Railway Company and the City of Toronto for the provision of public transit, something C. Ian Kyer has dubbed a "Thirty Years' War."[77] Much of the subway system, once under municipal ownership, was built under the managerial model outlined by David Harvey: TTC engineers planned and designed the various routes. The TTC was in charge of their construction and, once opened, their operation. Increasingly, however, private partners are being brought in to build new lines and there is much discussion about private companies operating the routes as well. For example, the construction of the Eglinton Crosstown LRT is being overseen by Metrolinx, an agency created by the provincial government in 2006 "to improve the coordination and integration of all modes of transportation in the Greater Toronto and Hamilton Area."[78] The $5.3 billion public-private partnership – the largest transit project in Canada – has, like most large, complex infrastructure projects, been both delayed and over budget. However, as *Toronto Star* columnist Royson James notes, as always, it is the public sector that is bearing the brunt of these cost overruns, not the private contractors hired to build the line.[79] Tensions between public and private partners have also been evident throughout the pandemic; in October 2020, the private-sector consortium building the Crosstown Line sued Metrolinx over rising construction costs and delays due to COVID-19. This is not the first time that the private partner sued the public one; in 2018, the consortium sued Metrolinx for breach of contract. Metrolinx settled out of court for $237 million in order to keep the project on track.[80]

In an era of restricted public sector budgets, public-private partnerships have been promoted as a way to raise increased capital for large infrastructure projects, and to help spread the risk between public- and private-sector stakeholders. Matti Siemiatycki, of the University of Toronto, conducted an extensive study of these partnerships.[81] Among arguments in favour, he noted that they are an important way of providing much-needed funding for key infrastructure in cities that are strapped for cash. However, they tend to reinforce existing patterns of uneven development at various spatial scales; bigger and more prosperous cities tend to see more investment, and within these cities partnerships are more likely to be found in areas that are already doing well and in the construction of premium pieces of infrastructure to connect globally important places (for example, airports). This is partly because they are influenced by market forces and profit motives. In a review of over 900 public-private partnerships, he found that most are concentrated in large cities in wealthy countries, and tend to support road projects rather than transit infrastructure.

This evidence resonates with UK-based urban researchers Stephen Graham and Simon Marvin's concept of "splintering urbanism."[82] Infrastructure and the provision of municipal services, they argue, are becoming increasingly uneven and more fragmented across urban space, and that access to these services – some regular, some premium – aligns with wider socio-economic and spatial trends of inequality. In the past, infrastructure networks (water, waste, transport, communication, information, energy) were the integrators of urban space. Graham and Marvin's research includes a number of poignant examples of how the notion of providing infrastructure evenly across a territory has been replaced by "myriads of specialised, privatised and customised networks and spaces."[83] They cite fibre optic networks initially confined to the centre of London; unequal distribution of drinking water and sewage facilities across cities in the Global South; the privatization of streets and public spaces, and the creation of gated communities, seen in cities in both the Global North and Global South; and pedestrian walkways in downtown cores (such as Toronto's underground PATH network) that often exclude certain kinds of people such as youth, buskers, and the homeless. Graham and Marvin's work, like Harvey's, argues that the idea that "infrastructure operators are assumed … to cover the territories of cities, regions and nations contiguously, like so many jigsaw puzzles" no longer exists.[84] In Toronto, for example, whether your garbage is collected by city employees providing a municipal service, or a private firm seeking to make a profit from this activity, depends on which side of Yonge Street you happen to live on. While P3s are touted for their cost savings, privatized garbage collection costs Toronto taxpayers more money than city employees doing the same job.

The Union Pearson (UP) Express, which opened in June 2015, is another example of the fragmented nature of infrastructure that Graham and Marvin write about. The UP Express is a fast, frequent, limited-stop service connecting the Union Station rail hub in downtown Toronto with Pearson Airport. While for most of its route it uses an existing rail corridor that has been upgraded to allow for more trains, it also involved the construction of a new spur to connect directly with the airport. The total cost was $456 million. It was originally envisioned as a premium service, primarily catering to business people flying in and out of Pearson Airport. Bruce McCuaig, the former CEO of Metrolinx, which oversees the operation of UP Express, stated in 2014, "We're trying to make this an extension of the airline experience, as opposed to a daily commuter service.[85] However, ridership numbers were initially very low, at around 2,000 per day, well short of the 7,000 that had initially been predicted. This was largely due to the expensive fares, especially when compared with the same, albeit much longer, journey by TTC. Less than a year after opening, in an attempt to boost ridership, Kathleen Wynne's Liberal provincial government mandated that the fare be reduced from $27.50 ($19 with a Presto farecard) to $12.35 ($9 with a Presto farecard).[86] As a result, ridership increased dramatically, with many people using it from the intermediate stops, Bloor and Weston, as a faster way of getting downtown. Fares were also lowered from these stops as well, meaning that for just double the TTC fare, one can travel to Union Station in a matter of minutes rather than the better part of an hour. However, as we discuss later in the book, this increased accessibility is one of the factors that contributes to gentrification.

The initial vision behind the UP Express was a classic example of the fragmented, premium infrastructure outlined by Graham and Marvin. It was aimed at the business travel market rather than local residents, and was meant to compete with private taxis and limousines rather than provide an enhancement to the city's public transit network. The fact that it passes close to many parts of the city yet denies access to people living there shows how premium infrastructure has the ability to connect valued users while excluding many others as it cuts through urban space and across existing infrastructure networks. People can be physically close to this premium infrastructure while not having access to it because it either does not stop near them or they cannot afford to ride. However, this premium model failed to attract large numbers of riders and, in the end, something more akin to a regular suburban or commuter train servicing both neighbourhoods and the airport, now operates on a 15-minute headway (30-minutes during the pandemic). While the provincial government still subsidized each ride by around $11 in 2018 to make up for the difference between operating costs and revenue, this is down from a subsidy of over $50 per ride when the line opened.[87]

In this chapter, we have outlined some of the major forces of change that have shaped Toronto and other cities since the 1960s and have also explored some of the policy responses seen in Toronto and elsewhere. These forces of change have resulted in structural, social, and spatial inequalities in cities and urban spaces. In the next chapter, we will explore how these, and other forces, influence the different trajectories of neighbourhood change.

3

NEIGHBOURHOOD CHANGE

As is apparent to anyone who travels around Canadian and other North American cities, urban areas are divided into two distinct parts: one part has straight streets and short blocks, with dense development and little green space, and with housing and commercial uses mixed together, whether as corner stores or as shopping strips with apartments over stores; the other part of the city is more open, streets more often than not are curvy, development is much less intense, and there is a clear distinction among shopping areas (shopping centres), work areas, and residential areas. –John Sewell[1]

The "Streetcar City" and the Automobile City

Before we delve into the different ways in which major structural forces of change influence neighbourhoods, it is worth briefly noting the distinct types of urban form, morphology, built environment, and land-use patterns that exist within a city like Toronto. In North American cities more broadly, we can see rather dramatic changes in these features between parts of the city that were constructed before 1945 and those that were developed after this date.

Why 1945? It was only after World War II that the automobile became the dominant transportation technology and dictated how and where cities grew. Prior to that it was the streetcar, first pulled by horses, and from the 1890s onward, powered by electricity, that influenced how cities grew and expanded. We are by no means the first researchers to examine the city built by the streetcar and the city built by the automobile, and the role that these differences in urban form and morphology play in shaping social, economic, and political divisions. Alan Walks, from the University of Toronto, has cited this boundary of pre-1945 and post-1945 neighbourhoods as a key factor in shaping voting behaviour (see chapter 7). He notes that the urban morphology of post-1945 neighbourhoods produces more car dependency, and contributes to a greater sense of individualism. Areas built before 1945 have higher densities and were originally built around transit. Rather than seeing urban space as a container for processes occurring at other spatial scales, he argues that it is exactly these substantial differences in urban form, morphology, and land-use patterns themselves that have major bearing on political ideologies, the polarization between right and left, as well as the wider divisions within cities and metropolitan regions.[2]

Beginning in the 1860s, the horsecar enabled cities to grow beyond the confines of the walking city. The walking city was small and compact: about as large as one could reasonably travel on foot. It was characterized by narrow streets, a multitude of functions side by side, and social classes existing in close proximity to each other. Moving to suburbs beyond the distance one could easily walk – either for businesses or households – was impractical for all but the wealthiest individuals. Vestiges of this urban form can still be found in old Quebec City.

The streetcar changed all that. Sam Bass Warner's classic book *Streetcar Suburbs: The Process of Growth in Boston, 1870–1900* discusses how the development of the horsecar led to the emergence of new social and spatial patterns of urbanization. He argues that it was the streetcar of the nineteenth century, and not the automobile of the twentieth century, that created the first true suburbs of North American cities, enabling residents and businesses to escape the overcrowding of the city, while still being close enough to be a part of it.[3]

Streetcar suburbs allowed the city to really grow outward for the first time. The areas of the city that developed with the urban form and land use influenced by the streetcar is what we refer to as the "Streetcar City." Edward Relph described this urban form as "high density but low-rise, with single-family detached or semi-detached houses and some walk-up apartments, all within walking distance of a rectangular network of main streets where the streetcars ran as they zigzagged into and out of downtown" (figure 3.1).[4] The Streetcar City is therefore characterized by regular city blocks with neighbourhoods anchored by central streets, lined with retail and (originally) streetcar tracks running down them (figure 3.2 and 3.3). Main streets, with their streetcar lines, were constructed relatively close together, often less than a kilometre apart, ensuring that everyone was within a short walk to a transit stop.[5]

Side streets featured rows of detached and semi-detached houses, spaced close enough together that residents could easily walk to the nearby streetcar stop. Especially before the 1920s, it was assumed that people would walk to the main street. Therefore, sidewalks were also essential pieces of urban infrastructure. In Toronto, this did not mean residential streets lined with duplexes or small apartment buildings, as in Montreal or Chicago, but rather tightly-spaced detached and semi-detached houses (although there are many examples of small, walk-up apartment buildings on residential side streets, the Annex being a neighbourhood where they are relatively plentiful).

It was also not uncommon to see small shops, businesses, and even factories interspersed along these residential streets, as there were few zoning rules dictating what type of land use could take place when these streetcar suburbs were constructed.[6] Today, while many structures remain, very few buildings originally constructed as factories still have manufacturing activities within them. One exception is the Cadbury factory on Gladstone Avenue, between Dundas and College (figure 3.4). In many other cases, these factories have been turned into lofts or offices (figure 3.5).

Over time, the idea of zoning was implemented throughout older parts of the city. Many former shops within neighbourhoods had been converted to residential uses; businesses that exist within areas zoned "residential" have been grandfathered in. However, current planning regulations mean it is virtually impossible to turn a property originally constructed as a commercial space back into this usage if its current function is residential. But there are growing calls for zoning rules to be eased to make it easier for neighbourhood businesses to open and to recreate mixed-use communities within the existing urban fabric. While many planners have advocated for these changes for some time, the pandemic has accelerated calls for living, working, shopping, recreational, and education activities to all be part of neighbourhoods. Paris mayor Anne Hidalgo's vision for a "'15-minute city" –by which all these activities are reached in a quarter of an hour by bike or on foot has inspired cities around the world to focus on neighbourhood-based solutions as part of a COVID-19 recovery plan.[7]

In the Streetcar City, the hub of the transit network (and of the city) was the downtown. Land-use models developed at the time reflected this radial pattern, with land values peaking in the central business districts (CBDs), which were themselves a by-product of the streetcar. What was once a jumble of side-by-side activities in the old walking city transformed into highly specialized clusters of activity. In Toronto, this included a banking and commercial cluster at King and Bay, retail stores centred at Yonge and Queen, and industry along the waterfront and railway lines.

Canadian communications theorist Marshall McLuhan argued that the dominant means of communication of an era (including its transport) influenced urban structure and could explain the patterns of city streets.[8] Between 1870 and the 1920s, the streetcar had a virtual monopoly on urban mobility, and neighbourhoods constructed during this time in Toronto have, to varying degrees, characteristics that fit the description outlined above. Early examples of new suburbs constructed in the streetcar era include the Annex, Riverdale, and Parkdale, which started as affluent suburbs – as evidenced by their large homes – where well-to-do residents could easily travel downtown along the new streetcar lines. Toronto's streetcar suburbs would include housing for a variety of social classes, including many working-class

Figure 3.1 | Broadview Avenue near Riverdale Avenue, characteristic of many residential streets within the "Streetcar City." A small retail building breaks up the row of detached and semi-detached houses. Current planning and zoning rules mean adding more of these commercial spaces is extremely difficult. 22 March 1971 and 20 November 2020. Photographers: unknown, Brian and Michael Doucet collection, and Michael Doucet.

Figure 3.2 | Queen Street East at Lee Avenue, 4 July 1968 and 30 July 2020. Commercial streets, with retail on the ground floor and apartments above, form the heart of neighbourhoods within the Streetcar City. Photographers: Robert D. McMann (John Knight collection) and Brian Doucet.

Figure 3.3 | Queen Street East at Greenwood Avenue, 27 May 1969 and 30 July 2020. Photographers: unknown, Brian and Michael Doucet collection, and Brian Doucet.

Figure 3.4 | Gladstone Avenue between Dundas and College Streets. While most factories closed many years ago, this Cadbury facility is still open. Both the factory and the houses across the street were constructed in an era before the institution of zoning rules and the strict separation of functions and land uses. Gladstone Avenue between Dundas and College Streets. 22 July 2019. Photographer: Brian Doucet.

neighbourhoods, primarily found in the east end, along railway corridors, or in the northwestern parts of the city and suburban municipalities such as Weston. Neighbourhoods were constructed either by speculators, small developers, or individual homeowners. As Richard Harris has observed, the latter was common in much of the former City of York, where it became popular for individuals to buy plots of land and construct their own kit houses, which could be purchased from Eaton's and other retailers. As University of Toronto geography professor Edward Relph has noted, much of the Streetcar City was a product of intense land and property speculation, either by developers or individuals, and was criticized for this well into the 1960s because of a lack of government planning; this resulted in a failure to control land uses (hence the many photos we include that show industry and houses on the same block) and a disregard for the natural environment.[9]

Streetcar suburbs were still being constructed after the creation of the Toronto Transportation Commission in 1921. While the automobile was in its ascendancy during the 1920s, new neighbourhoods such as North Toronto were developed after the Yonge streetcar was extended northwards from just south of St. Clair Avenue all the way up to the new city limits at Glen Echo (just over halfway between Lawrence and York Mills) in November 1922. In the subsequent decade and a half, North Toronto developed as a new neighbourhood all the way up to the city limits, and around half a kilometre on either side of Yonge Street, that being a reasonable distance one could walk to reach a streetcar stop.

In early streetcar suburbs, there was very little provision for the automobile. Off-street car parking, if it could be found at all, was generally via rear laneways, or alleys, rather than parking spots in front of houses. As the automobile became more common, streetcar suburbs of the 1920s and 1930s saw more provisions

Figure 3.5 | Dundas Street West, near Roncesvalles. The Toronto Feather and Down Company has become the Feather Factory Lofts. Another view of this building can be seen in set 58 in Portfolio 3. 19 July 1967 and 28 June 2016. Photographers: unknown and Michael Doucet.

for the car, such as wider roads and private driveways, usually situated between houses so that cars could be driven directly from the street to a garage or parking place behind the house. However, neighbourhoods constructed at this time still retained most of the basic principles of the Streetcar City, regardless of whether or not they ever had a streetcar line running through them. By the time the last streetcar suburbs were being built, such as Lawrence Park and Kingsway Park, some variations of design and planning that would be common in the post–World War II years began to emerge. These included the introduction of some curved streets, which were becoming increasingly popular among modernist planners, and a more rigorous spatial separation of functions between residential, commercial, and industrial uses.[10]

Very little housing was constructed between 1930 and 1945; neighbourhoods that developed during the Depression and World War II, such as Leaside, have design, urban form, and density elements more in common with their earlier counterparts than with what would come after the war. However, a clear progression towards modernist planning as well as more provisions for the automobile are noticeable.[11] After the war, however, Toronto experienced a housing boom that would last for decades. By then, transit played less of a role in shaping urban form and land use than the automobile. Along with keeping pace with growth, one of Toronto's biggest postwar challenges was to address decaying housing stock in the city. As John Sewell articulates, postwar planning in Toronto was heavily influenced by the modernist visions of Ebenezer Howard's Garden Cities, the *Ville radieuse* of Swiss architect Le Corbusier, and the Broadacre City of American Frank Lloyd Wright, all of which proposed solutions to the decay, overcrowding, pollution, and perceived chaos of older cities.[12]. Within the Streetcar City, these ideas would influence slum clearance policies and the creation of Regent Park, Moss Park, Don Mount Court, and Alexandra Park, among other urban renewal projects.[13]

These modernist ideas would also influence the design of the city's booming suburban areas constructed after 1945. The suburbs around the old Streetcar City grew tremendously in the decades after World War II; North York's population grew from 85,000 in 1951 to over half a million twenty years later.[14] While modernist planning principles were evident throughout the region, it was in these suburban developments, more than in the slum clearances in the Streetcar City, where the influence of the automobile would lead to a distinctly different set of spatial patterns, morphologies, densities, and land uses.[15] Whereas main streets, with their streetcar lines and ground floor retail, served to unify neighbourhoods in the Streetcar City, wide arterial roads had the opposite effect in automobile-oriented suburbs in that they delineated different neighbourhoods. The heart of each neighbourhood was a school, and generally a park beside it. Residential streets took on a loop design (and in some cases cul-de-sac, or lollipop) rather than a grid. This was an attempt to create safe and healthy neighbourhoods by discouraging cars from driving through residential streets. This road pattern also helped to separate different land uses and road hierarchies. Within neighbourhoods, lot sizes became larger and houses constructed in the 1950s and 1960s in particular took on a horizontal design, with bungalows, ranch-style, and back- and side-splits being common types of dwellings. Garages and carports were situated prominently and were easily accessible at the front or side of houses, replacing rear lanes and on-street parking. Off-street parking for two or more vehicles was standard in even modest houses built in this period, the assumption being that journeys to work, or major shopping trips, would be done by car. Although in Toronto, a far higher percentage of journeys in these neighbourhoods are made by transit than in either the 905 suburbs around the city or similar communities in the United States, the foundations of the built environment were based around the automobile.[16]

With the automobile being the dominant transport technology, proximity on foot became less important in both the design of neighbourhoods (with their winding streets) and of individual houses; densities decreased as walking distance to transit was less of a mobility concern. Because of the auto-focused nature of land use and mobility, more provisions were given, both in public and private spaces, for the parking of cars.[17]

The first neighbourhood to adopt most of these characteristics was Don Mills, northeast of Toronto, which John Sewell has described as the most influential development in Canada in the twentieth century.

Beginning in 1947, E.P. Taylor began purchasing land far away from the existing urban footprint, separated by ravines and railway lines, in preparation for the creation of an entirely new community. The plan for Don Mills, developed largely by Macklin Hancock, was based on a model that was fundamentally different from that of existing prewar areas. It was focused on five principal elements: the neighbourhood as the primary unit of the development, a discontinuous road system that discouraged using residential streets as through routes, ample green space, the inclusion of new housing designs, such as ranch-styles, side-splits, and bungalows on larger lots, and a strict separation of functions.[18]

Don Mills was divided into four quadrants, each forming a separate neighbourhood with its own elementary schools. At the centre of the development, at the intersection of Lawrence and Don Mills, was the main retail area, Don Mills Plaza, as well as a library, indoor skating rink, curling rink, and post office, themselves separated from the neighbourhoods by a ring road. Apartments, placed largely on main roads, were separated from single-family houses, and residential areas were buffered from other uses by large expanses of green space. Part of this green feel was achieved by replacing sidewalks with grassy ditches, and a network of pedestrian walkways created cut-through paths that provided connections within neighbourhoods, as well as linked them to the central area. Lot widths were considerably wider than in older neighbourhoods, often 60 feet, compared to 25 feet elsewhere in North York.[19] Don Mills would provide the template for subsequent suburban development across Canada, though no other community replicated all the modernist planning principles that can be found there.[20]

Don Mills spelled the end of the grid pattern of streets that was characteristic of the Streetcar City and many of the early postwar suburbs built between 1945 and 1955, such as the central parts of North York between Yonge and Bathurst Streets. Henceforth, wide and angular arterial roads delineated neighbourhoods, with curved streets and cul-de-sacs inside them. This basic road design continues to this day. So too does the strict separation of functions, with plazas and shopping centres (with their ample parking catering primarily to people driving) constructed without any residential units; this contrasts with the retail streets of the Streetcar City, where one would typically see two or three storeys of apartments above the ground floor retail. Toronto's first purpose-built shopping plazas – York Mills Plaza at Bayview and York Mills and Sunnybrook Plaza at Bayview and Eglinton – opened in 1951 and their basic principles have been replicated across the region in the subsequent decades.

Some planning and design elements have changed since the 1950s; lot sizes are smaller, houses are larger, and backyards have diminished in size to reflect the rising land and real estate prices in and around Toronto, as well as a desire for more living space. The Toronto region stands out in North America for having many apartment buildings in its postwar suburbs; 2,000 residential towers were built in the Greater Golden Horseshoe between 1945 and 1984, most of which were constructed between 1960 and 1980 and are situated in suburban areas.[21] As was the case in Don Mills, these tend to be along the main arterial roads, or in large clusters such as Thorncliffe Park, in East York, and Mabelle Avenue, in Etobicoke, rather than interspersed throughout single-family residential streets. However, even here, the basic principles of automobile-based development and land-use patterns are present, with a separation of functions and ample parking facilities. There are some exceptions to this in the post-1945 auto-oriented landscapes, such as pockets of extremely high densities that have developed around some subway stations, or suburban downtowns such as North York Centre, Scarborough Town Centre, and Mississauga City Centre. Some subdivisions, such as Cornell, in Markham, have adopted New Urbanism design principles that focus on increased density, with an emphasis on walkable neighbourhoods, and a return to a grid pattern of streets and garages accessed via rear laneways. However, these represent islands of New Urbanism rather than a paradigm shift in the development of the suburbs, and leaving them tends to require journeys by car.[22]

Defining Toronto's Streetcar City

Streetcars never ran in suburban areas like Don Mills, or Mississauga, so we are unable to visually document their changes in this book.[23] We are, therefore, confined to the boundaries of the Streetcar City. To determine the spatial extent of this part of the city, we examined a map of the streetcar network at its peak, in 1928. While today's network is largely concentrated in areas south of Bloor Street (with the 512 St Clair route being the main exception), in 1928, streetcars ran throughout most of the pre-amalgamation City of Toronto, as well as the southwest corner of Scarborough, much of the old City of York and the southern flank of Etobicoke (the former municipalities of Mimico, New Toronto, and Long Branch, where today's 501 Queen streetcar runs). Examining this network in conjunction with a map of census tracts (CTs) from the 2016 census showed that 203 CTs either had a streetcar line running through them, or were flanked by a line on at least one of their boundaries.

The few neighbourhoods constructed between 1930 and 1945 also retain similar land use and urban form patterns, even if they never had a streetcar line. There are also a handful of CTs in the old city of Toronto, such as in Rosedale, where no streetcar line ever ran, yet they developed contemporaneously to adjacent neighbourhoods with streetcar lines.[24] Therefore, we also included all contiguous CTs where the majority of the housing stock was constructed before 1945.[25] Fifteen additional CTs join the total, bringing it to 218 census tracts.[26] Figure 3.6 shows the extent of what we consider to be Toronto's Streetcar City, overlaid with a map of the network in 1928, when the streetcar system was at its largest, and figure 3.7 shows these boundaries with today's network.

In chapter 1 we examined different parts of Toronto at various moments in time; however, if we aggregate data on the Streetcar City and the automobile city – the city that developed after 1945 – we can see some rather striking trends that largely parallel developments in the former City of Toronto and the remainder of what used to comprise Metro Toronto. Table 3.1 examines these characteristics using data from the 2016 census.

Moving back in time, we aggregated CT-level data from the 1971 census to paint a picture within the same boundaries of the Streetcar City.[27] Table 3.2 outlines the situation in 1971; it shows an urban core that was generally poorer and more immigrant-based than the expanding suburbs. What is quite remarkable about the two cities today is that, despite their vastly different urban forms and morphologies, the percentage of people who take transit to work is remarkably similar: 39.2 per cent in the Streetcar City and 35.5 per cent in the automobile city. While data on journey-to-work mode share was not gathered in the 1971 census, the 1960s saw the first subway expansion beyond the Streetcar City and the development of a frequent grid of bus routes into the suburbs. Jonathan English's research articulates how this bus service expansion in the early 1960s set Toronto apart from other North American cities and produced the high ridership and frequent buses that are virtually non-existent in large American suburbs. It is worth noting that in 1971, only 58.1 per cent of dwellings in the Streetcar City possessed an automobile, while in the automobile city, this figure was 85.5 per cent.

The Jane Jacobs Effect and Planning Visions for Toronto

In Toronto, when people talk of neighbourhoods, the writings of Jane Jacobs often provide the backdrop to the conversation. Her book *The Death and Life of Great American Cities*, published in 1961, shocked the planning, architecture, and urban design worlds. Initially dismissed as an uneducated housewife by these professions, her work has come to dominate these fields today. Jacobs moved from New York to Toronto in 1968 and spent almost forty years living on Albany Avenue in the Annex. She was awarded the Order of Canada in 1996 and passed away in 2006. Her legacy survives in many ways, including in the Jane's Walks movement, which started in Toronto, and involve free, locally organized, citizen-led walking tours intended to bring people together to celebrate neighbourhoods.[28]

Jacobs had no formal training as a planner or architect, yet she is one of the most influential writers in these fields. Her work critiqued modernist planning; she denounced its design, separation of functions, inhuman scale, and the arrogance of its designers, who, she argued, did not respect urban residents. *The Death and Life of Great American Cities* was a rejection of the urban renewal projects being built and designed by famous architects such as Le Corbusier and master builders such as Robert Moses in New York. Moses was an advocate of freeway construction to ease congestion, and of clearing urban slums and

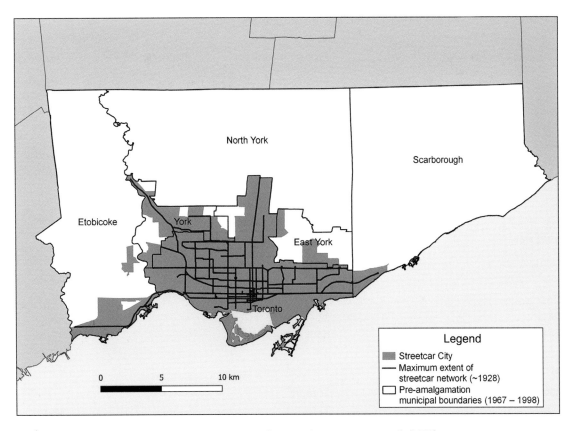

Figure 3.6 | The Streetcar City and the maximum extent of Toronto's streetcar network (1928).

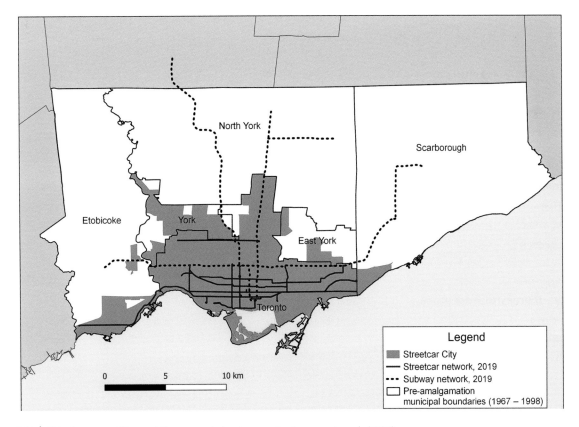

Figure 3.7 | The Streetcar City and the current streetcar and subway network (2021).

Table 3.1. Characteristics of the Streetcar City and the automobile city, 2016.

	City of Toronto	Streetcar City	Automobile city
Population			
Total population	2,731,571	980,155	1,751,416
% change 2011–16	4.50%	8.50%	2.30%
Total visible minority	51.5%	34.2%	61.1%
Not a visible minority	48.5%	65.8%	38.9%
Non-immigrant	49.5%	61.4%	42.9%
Immigrant	47.0%	34.6%	53.9%
Non-permanent resident	3.5%	4.0%	3.2%
Household size			
1 person	32.3%	42.1%	25.2%
2 persons	30.0%	31.9%	28.5%
3 persons	15.8%	12.4%	18.2%
4 persons	13.2%	9.2%	16.1%
5 or more persons	8.7%	4.3%	12.0%
Average household size	2.4	2.0	2.7
Housing			
Average dwelling value ($)	$753,866	$831,143	$709,256
Average monthly shelter cost ($)	$1,682	$1,869	$1,574
% of owner households with mortgage	57.5%	60.8%	55.6%
% of households spending > 30% on shelter	27.4%	25.9%	28.2%
Owner	52.70%	45.60%	58%
Renter	47.20%	54.30%	42%
Income			
Individual income >$100,000	9.6%	14.3%	6.8%
% low-income, total	20.2%	18.4%	21.2%
% low-income, 0–17 years	26.3%	19.0%	29.3%
% low-income, 0–5 years	26.2%	17.3%	30.6%
% low-income, 18–64 years	19.2%	18.1%	19.9%
% low-income, 65 +	17.4%	19.0%	16.7%
Education			
No certificate, diploma, or degree	16.4%	12.7%	17.0%
Secondary school or trade diploma	28.5%	23.6%	31.7%
College diploma	15.8%	14.6%	17.9%
University (bachelor's or below)	26.1%	31.4%	23.4%
University (master's or above)	13.1%	17.6%	9.9%
Mobility – journey to work mode			
Private vehicle	50.5%	37.0%	59.6%
Public transit	37.0%	39.2%	35.5%
Walk	8.6%	16.5%	3.4%
Bicycle	2.7%	6.0%	0.6%
Other	1.1%	1.3%	0.9%

Source: Statistics Canada, 2016 Census, own tabulations

Table 3.2. Characteristics of the Streetcar City and the automobile city, 1971.

	Metro Toronto	Streetcar City	Automobile city
Population	2,089,675	893,315	1,196,360
Mother tongue English	70.6%	61.5%	77.3%
Birthplace Canada	63.4%	56.8%	68.2%
Birthplace outside Canada	36.6%	43.2%	31.7%
Dwelling – ownership	51.0%	44.4%	56.4%
Dwelling – rental	49.0%	55.6%	43.6%
Average income, household	$11,748	$10,399	$12,847
Dwellings with an automobile	73.2%	58.1%	85.5%
Dwellings with 1 automobile	53.8%	47.6%	58.8%
Dwellings with 2+ automobiles	19.4%	10.5%	26.7%
Labour force, by industry			
Primary	0.5%	0.4%	0.5%
Manufacturing	24.4%	23.1%	25.4%
Construction	6.3%	6.3%	6.2%
Transportation/utilities	7.5%	7.4%	7.5%
Trade	16.7%	13.8%	19.0%
Finance, insurance, etc.	7.2%	7.2%	7.3%
Community, business, etc.	24.8%	27.1%	22.9%
Public admin. & defence	5.5%	5.3%	5.6%
Unspecified	7.3%	9.2%	5.7%
Education			
Less than grade 9	40.6%	44.3%	37.8%
Grade 9–10	21.6%	20.0%	22.9%
Grade 9–10 & other training	3.0%	2.8%	3.1%
Grade 11–13	20.1%	17.9%	21.9%
Grade 11–13 & other training	9.5%	8.4%	10.3%
Some university	4.1%	4.1%	4.2%
Some university & other training	1.7%	1.7%	1.7%
University degree	4.6%	5.1%	4.2%
University degree & other	1.0%	1.1%	1.0%

replacing them with modern tower blocks. These new apartments were erected as towers situated among large areas of green and open spaces, an approach that some have called a "tower in a park" or a "slab in a park." In Toronto, this urban form was common for both public housing (Regent Park being the most notable example) as well as privately constructed developments (e.g., St. James Town) in urban renewal schemes that were responsible for the demolition of virtually all the existing buildings within an area comprising many city blocks. It also influenced the design of the apartment complexes that can be found in the inner suburbs of North York, Etobicoke, and Scarborough. Superblocks were created with few routes in, out, or through these complexes and their entire area was developed along a single-use zoning framework that moved businesses, offices, factories, and other non-residential uses elsewhere.

In the decades after World War II, mixed-use areas (i.e., the Streetcar City) were seen as detrimental to modern healthy living, owing to their overcrowding, noise, nuisance, and mix of functions, including industry, and the general appearance of disorder within cities. Therefore, one of the goals of modernist planning was to create order by demolishing these old, mixed areas of the city, replacing them with modern designs and architecture and introducing a discrete separation of functions, people, and activities. A congested urban street, with shops, bars, workshops, and apartments all jumbled together did not fit within the modernist framework for civic improvement.

Urban renewal was not solely focused on building subsidized housing for the urban poor. Slums and other areas were demolished and gleaming new towers were constructed for the urban middle class; Lafayette Park in Detroit, designed by Mies van der Rohe, is one of the best examples of a postwar urban renewal development that remains a desirable address today. Apartments and townhouses were situated within vast open spaces, which were initially praised for their peace and tranquillity (especially compared with the noise and pollution of a busy city street). In the United States in particular, such urban renewal programs have been criticized for their racist overtones, with African American neighbourhoods levelled to make way for highways or new housing developments.

The design of these modernist developments was also criticized. Because there were few reasons to go to these areas (there were no shops to visit, for example), they became empty and devoid of the busyness associated with the dense mixed-use streets and short blocks that they had replaced. There was no one informally keeping watch over these spaces, such as a local shopkeeper, and Jacobs argued that this design reduced the number of eyes keeping watch on public spaces, which actually made them more vulnerable to crime. It was no wonder, Jacobs argued, that modern urban renewal projects (even those for middle- and upper-income households) quickly became unsafe or were perceived as being unsafe. This perspective, however, went against the dominant viewpoint at this time, which associated the chaos and complexity of the inner city with crime and disorder, something that only a redevelopment of the built environment could solve.

Safety, Jacobs argued, came from diversity and density, which could be readily found in the inner-city neighbourhoods that Moses (and similar counterparts in other cities, including Toronto) were rapidly demolishing. Sidewalks and "eyes on the street" were key to ensuring safe neighbourhoods and public spaces. "A well-used street is a safe street," she wrote. "A deserted city street is apt to be unsafe."[29]

In her book, Jacobs contrasted the modern urban renewal projects with her own neighbourhood of Greenwich Village. She noted the constant flow of people on the sidewalks, each of whom played a role in keeping order and safety. A major factor in this was not only the dense housing which faced onto the street, but also the many businesses located on the ground floors of buildings. Shopkeepers not only kept an eye on the street, but had a vested interest in maintaining safety around their premises. Businesses also attracted people to visit them; the more varied the businesses were, the more people were drawn to an area and hence, the safer it was. Rather than seeing this phenomenon as haphazard and random, she likened the movements of people through these dense and diverse neighbourhoods to an "intricate ballet in which the individual dancers and ensembles all have distinctive parts which miraculously reinforce each other and compose an orderly whole. The ballet of the good city sidewalk never repeats itself from place to place, and in any one place is always replete with new improvisations."[30]

Jacobs' writings sent shockwaves through the planning, design, and architecture worlds. She critiqued what had been steadfast beliefs in modernity, progress, and the "planner knows best" mentality. Jacobs battled with Robert Moses to save her Greenwich Village neighbourhood from demolition to make way for a crosstown expressway.[31] When she moved to Toronto, she immediately became involved in a similar, ongoing battle to stop the Spadina Expressway from ripping apart the Annex and other neighbourhoods through which it would pass (figure 3.8). Under such a plan, the vitality of Bloor Street and Spadina Avenue would have been destroyed.

By the early 1970s, some, but by no means all, civic leaders in Toronto were starting to take notice. In the spirit of other contemporaneous movements, like the one to save the city's streetcars,[32] a grassroots organization formed to stop the construction of the Spadina Expressway. Compared with most large North American cities, Toronto does not have many highways running through urban neighbourhoods. In the 1960s, as the city and region boomed and automobile use surged, there were plans for multiple expressways connecting downtown with the rapidly growing suburbs, which, as we discussed above, were built around the idea of automobile-based mobility. These plans included a crosstown expressway to be built parallel to the east-west CPR tracks (which cross Yonge Street at Summerhill Avenue and passes through the north end of the Rosedale neighbourhood), an eastern extension of the Gardiner Expressway through the Beaches and southern Scarborough to connect with Highway 401 near Rouge Hill, and an extension of Highway 400 south towards the crosstown expressway near downtown. While these were officially in the city's plans, a combination of slower growth and changing politics meant that by the early 1970s they were on the back-burner indefinitely. However, constructing the Spadina Expressway, which would run south from Highway 401 towards downtown Toronto, passing through a combination of ravines and existing neighbourhoods, including Jacobs', before ending at Spadina and Bloor, remained a priority of the Metro Toronto government.[33]

Construction had progressed south from Highway 401 to Lawrence Avenue but then funds ran out. In 1971, the Metro government went to the Ontario Municipal Board (OMB) to ask for permission to borrow funds to complete the project to Bloor. Despite the protests of civic groups and community leaders, including Jacobs, who presented testimony against the expressway by focusing on its detrimental effects on neighbourhoods, pollution, and noise, in February 1971, the OMB ruled in favour of Metro and the completion of the Spadina Expressway. The last remaining option was to appeal directly to the Ontario premier, William Davis. On 3 June, the province withdrew its support, and most important the financing, for the expressway, agreeing to fund only the Spadina subway. In the legislature, Davis stated, "If we are building a transportation system to serve the automobile, the Spadina Expressway would be a good place to start. But if we are building a transportation system to serve people, the Spadina Expressway is a good place to stop."[34]

The current Allen Expressway (which is the truncated version of the Spadina Expressway) stops abruptly at Eglinton Avenue; the Spadina subway line, built in the median of the highway, opened in 1978. Where the expressway ends, the subway continues underground,

Figure 3.8 | Jane Jacobs speaking at a rally to stop the Spadina Expressway. March 1971. Photographer: Michael Doucet.

along much of the same proposed routing of the Spadina Expressway, all the way to downtown, linking up with the University Avenue subway and creating the U-shape of the TTC's Line 1 that remains to this day. Partly because of the community efforts to stop highways running into the heart of the city, Toronto's inner-city neighbourhoods did not get ripped apart and, as a result, did not see the same disinvestment, depopulation, and devastation that can be found in many American cities, such as Detroit and Buffalo. The creation of Metro Toronto in 1953, which pooled tax resources of the city and its suburbs, avoided the jurisdictional disparities in income, services - such as schools - and amenities between cities and suburbs that are common in the United States. Finally, Toronto did not see the same extreme racial segregation and "white flight" to the suburbs that defined the postwar relationship between American cities and their suburbs.[35] As we pointed out in chapter 1, another notable difference with many American cities was that Toronto remained a predominantly white city until well into the 1980s.

The 1970s was, in many ways, the pinnacle of progressive politics in Toronto. Premier William Davis, nicknamed "Mayor Bill," not only involved the province in stopping the Spadina Expressway. He also spearheaded the construction of Ontario Place, the Ontario Science Centre, and other recreational and cultural amenities, as well as capital improvements to the city's infrastructure. James Lemon argued that this was part of the province's strategy to promote Toronto as a world-class city, partly spurred on by competition with Montreal, after it hosted its successful Expo in 1967 (although he also noted that this spending on amusement and spectacles came about at the same time as diminishing spending on education).[36] The era was characterized by significant involvement by higher levels of government in Toronto's (and other cities') affairs – this was a period in which the federal government supported the construction of affordable housing, funding around 20,000 units per year across the country.[37] Toronto had two progressive reform mayors in the 1970s, David Crombie and John Sewell, and growing levels of citizen participation. Toronto in this period earned the reputation as a "city that worked."[38]

In addition to stopping the Spadina Expressway, and saving Toronto's streetcars from abandonment (see chapter 5), resistance and activism by citizens of Toronto led to the preservation of Old City Hall (which in initial plans would have been demolished to build the Eaton Centre)[39] and the saving of Union Station. In 1968, the property arms of the CN and CP railways proposed Metro Centre, a massive redevelopment of the railway lands south of downtown. The removal of freight yards along the waterfront (partly arising from the deindustrialization of the area) meant that the remaining tracks could be relocated to the south of Union Station, with a new transportation hub adjacent to the Gardiner Expressway. That would free up the land occupied by Union Station (situated closest to downtown) for new commercial developments. After much negotiation, the city approved the plans, although the idea of extending the subway loop south to the new train station never took hold. It was then that citizens' groups and community organizations formed to oppose the demolition of Union Station. While there were a variety of reasons for opposing Metro Centre, as Mark Osbaldeson notes in his book *Unbuilt Toronto*, there was a zeitgeist that resonated with Jane Jacobs' views on urban planning: "people who saw mega-projects like the Spadina Expressway, the Eaton Centre and Metro Centre as the antithesis of a liveable, walkable, human-scaled city were organizing and organizing successfully ... there was a feeling that there had been little opportunity for input from the public."[40]

An umbrella group lobbied to have the Metro Centre plans referred to the Ontario Municipal Board (OMB), who ruled in the summer of 1972 that the project should have much more green space, larger apartments for families, and a limit to development (these features, which the city's planners had recommended for approval, were rejected by the planning board). This ruling, combined with a more progressive council that had been elected after the city approved the initial plans, effectively killed the development.[41] As a result, Union Station remains a Toronto landmark and much of the intense development along the railway lands has taken place south of the remaining tracks. The only visible outcome of the Metro Centre development proposal is the CN Tower.

The 1970s was also an era that marked a change in the design and approach to public housing as well as a turn away from slum clearances, superblocks, and modernist planning ideas in general. After World War II, public housing projects were typically monolithic superblocks of subsidized housing in the style of Regent Park (before its current redevelopment). Following ideas of Robert Moses and Le Corbusier, internal streets were eliminated and large tower blocks were surrounded by green space or parking. However, while these design principles were popular with planners and architects, they were rarely popular with residents. The displacement, eviction, and expropriation associated with urban renewal were unpopular, and many downtown residents liked the community feel of their neighbourhoods as well as their proximity to jobs, amenities, and recreational facilities. After it constructed Regent Park, however, the city continued to demolish areas it deemed to be slums; Moss Park, Alexandra Park, and Don Mount were all cleared of their old housing, industries, and commercial spaces, and were replaced with new public housing built along modernist principles.

This approach was facing increased criticism that would come to a head around plans for the renewal of Trefann Court. In the mid-1960s, this small area, bounded by Parliament, Queen, River, and Shuter Streets, was slated for demolition and renewal. However, local residents vehemently opposed the city's plan. John Sewell, who would later become mayor of Toronto, was elected to represent the ward in 1969. He was part of a growing reform movement, which moved away from the top-down, modernist ideas about planning. A key policy shift of this movement was that any plans for a neighbourhood had to be developed in consultation with local residents and businesses. This led to a considerably revised proposal for Trefann Court, which retained the existing street network; existing buildings were also retained and refurbished when possible and new ones were constructed at a similar density so as not to overwhelm the neighbourhood. This marked the end of Toronto's modernist urban renewal and the beginning of a shift to a very different form of planning and governance.[42]

The 1970s was the high point of progressive, reform politics in the city. While they all represented different groups of people and interests, as John Sewell noted, there were four interrelated movements that were opposed to modernist planning visions for the city, and all subscribed to one or more of Jane Jacobs' ideas: heritage preservationists working to stop the demolition of the city's historic landmarks, those opposed to highway construction and the Spadina Expressway, residents working to stop urban renewal in their neighbourhoods, and residents fighting against rezoning their predominantly single-family residential neighbourhoods into high-rise zones. For the latter, the most prominent fight was to stop the site at Quebec and Gothic Avenues (near High Park subway station) from being entirely rezoned for high-rises, a battle that was partially won by local residents.[43] The result was amended planning guidelines that favoured new construction that fit within the character of existing areas, rather than starting anew. Downtown, new policies aimed to encourage a mix of uses, particularly at ground level, while prohibiting large plazas in front of buildings. Early postwar office towers, such as the Toronto-Dominion Centre and Commerce Court, placed retail in the basement, had very little street frontage, and were set back from the road to such an extent that vast, empty, and often windy spaces dominated intersections such as King and Bay.[44]

All of these reformist ideas constituted a profound rejection of modernist planning that had dominated for the previous three decades; they reflected the more intimate and human-scale forms of urban development which Jane Jacobs had called for. In housing, emphasis shifted towards a mixture of tenures, styles, and uses (with a focus not just on affordable housing, but on retail, commercial, and community spaces within a development), as well as smaller blocks and buildings that fronted directly onto the street. After Trefann Court, the next successful new urban regeneration project was the redevelopment of the St. Lawrence neighbourhood, located south of Front Street and east of Jarvis Street. As with much of the land around downtown Toronto, by the 1960s deindustrialization had left vacant spaces just east of the St. Lawrence Market. Mayor David Crombie wanted to redress the planning failures of earlier public housing projects, and the city, with the help of federal money, led the redevelopment of this area. There were five key elements of St. Lawrence that marked a fundamental break with modernism. First, the streets replicated and extended the city's existing grid pattern. Second, houses and buildings faced directly onto the streets. Third, while dense, the neighbourhood would be built on a human scale, with a combination primarily consisting of mid-rises and townhouses. Fourth, there were different uses, not just residential, as was common in older neighbourhoods throughout the city. And fifth, the housing stock contained a mix of tenures: condos, co-ops, social housing, and private rentals.[45] St. Lawrence was the largest, but by no means the only development that followed these principles, and has remained a popular and attractive place to live for forty years. From a political-economy perspective, it fits squarely with David Harvey's ideas of urban managerialism discussed in the previous chapter. It used public investment to redistribute resources and iron out the unevenness of capitalism by focusing on the provision of social and affordable housing for local residents. However, this model to build mixed-income communities with ample affordable and subsidized housing required money from higher levels of government. Budget cuts and an end to federal and provincial support for affordable housing in the early 1990s meant that, despite its many successes, St. Lawrence was not replicated on a large scale. By the 1990s, the goals of projects such as Trefann Court and the St. Lawrence neighbourhood fell increasingly out of favour with neoliberal urban policies aimed at attracting affluent residents and enhancing Toronto's position within the global urban hierarchy.

Jane Jacobs' ideas about neighbourhoods and urban design can be seen in the St. Lawrence neighbourhood. But they are best evident in the original fabric of the Streetcar City. As anyone who has ever been stuck behind a streetcar can attest, they work towards regulating and slowing down traffic, preventing roads from becoming fast thoroughfares to drive through, thereby making the streets comparatively more pleasant to walk and cycle along than busy suburban arterial roads, or one-way streets downtown such as Richmond or Adelaide. Within the Streetcar City, the "eyes on the street," mixed uses, density, walkability, good transit, and a variety of old and new buildings are all defining characteristics. The value of older buildings, Jacobs argued, was that they provide cheaper rents and allow for different activities to take place within an area. Her own neighbourhood north of Bloor Street in the Annex is a classic example, although it lost its streetcar in 1966 when the Bloor-Danforth subway opened. Other streets such as Queen, College, Dundas, Broadview, Gerrard, and Roncesvalles all feature a diverse mix of residential and retail activities at a density that encourages not only high transit ridership but also plenty of foot traffic.

While Jane Jacobs' ideas are now widely praised and accepted in planning, architecture, and urban design circles, she has been criticized because her ideas have contributed to gentrification of inner-city neighbourhoods in Toronto and many other cities, with displacement of many of the lower-income residents and neighbourhood businesses which were so vital to her "street ballet." Jacobs was an underdog in the 1960s and 1970s, fighting to save existing neighbourhoods from massive redevelopment. Today, the ideas of Jacobs are a dominant planning paradigm and are frequently invoked to protect and preserve what are now extremely expensive residential areas from even modest increases in density that could contribute to making them more affordable. It takes a lot of money to live in neighbourhoods that follow these design and planning principles today. Herein lies one of the biggest paradoxes in contemporary urban planning: the neighbourhoods considered to have good design and planning, with a range of amenities, walkability, and a variety of mobility options that help to produce healthy, happy, and vibrant communities, are some of the most expensive, and consequently, exclusive communities in Toronto.

A recent study found that mixed-use areas in Toronto were becoming increasingly unaffordable for everyone apart from those working in high-income service occupations (such as those described by Saskia Sassen – see previous chapter), meaning that the benefits of these neighbourhoods – more walkability, better transit and mobility, and a range of amenities – are enjoyed by increasingly affluent households.[46] The authors of the study suggest policy interventions such as inclusionary zoning (a market-based planning tool that requires affordable housing to be included in new residential developments), density bonuses which allow developers to construct more units than otherwise permitted in exchange for building affordable housing, and housing trust funds, where not-for-profit builders can draw on investment funds set up by both public and private donations in order to build affordable housing. We will discuss these ideas in more detail later in the book.

In Toronto, much of the legacy of Jane Jacobs has been enshrined into the city's Official Plan. While the reform movement opposed the large-scale demolition of neighbourhoods and the construction of tall towers beside single-family homes, an extension of this logic means that today it has become virtually impossible to add even modest increases in density, such as townhouses, walk-up apartments, or even semi-detached houses to the neighbourhood streets that Jacobs and many others fought to protect from mass demolition, high-rise development, and expressway expansion. As a result, Toronto's Streetcar City, which contains some of the most desirable neighbourhoods in the city, continues to see the cost of housing rise beyond the means of many people who want, or need, to live there, while very little new housing is built. It is also important to note that in a city and region whose ethnic diversity is growing, the Streetcar City is actually becoming whiter and more Canadian-born, with racialized populations increasingly clustered in the city's inner suburbs, far from rapid transit, jobs, and many of the services and amenities that downtown residents take for granted.

The increasingly unaffordable nature of the urban core presents difficult choices for households: pay very high rents to live in small dwelling units close to work, amenities, and the vitality of the urban core, or move to the suburbs where prices are somewhat (though not always) cheaper, space is more plentiful, but the commute is longer and there are fewer transportation options (in other words, more car dependency). While the trade-offs between city and suburb in terms of lifestyle have been evident for some time, the increased costs of suburban housing, combined with both more expensive and longer commutes mean the cost savings of leaving the city are rapidly diminishing.[47] With the high cost of housing in the urban core of Toronto, and suburbs which are rapidly becoming more expensive, an increasing number of households, particularly those who want the kind of neighbourhood celebrated by Jane Jacobs, are looking farther afield, to cities such as Hamilton.[48] As prices in Toronto have skyrocketed, urban neighbourhoods in Hamilton, 75 kilometres away, have become some of the frontiers of gentrification in the Greater Golden Horseshoe. It remains uncertain whether the pandemic will accelerate these trends in the long run; two of Canada's fastest-growing urban areas are Waterloo and Peterborough. Like Hamilton, they are within 90 minutes' travel time from Toronto. Even before the pandemic, these communities saw the bulk of their population growth coming from intraprovincial migration (i.e., people moving from other parts of Ontario, most notably the GTA), rather than other provinces in Canada, or international immigration. With more people working from home over extended periods of time, and greater demand for more living space (indoor and outdoor), these migration patterns could accelerate in the coming years, though that does not necessarily mean that large cities will decline, or that housing in Toronto will become any more affordable.

Traditional Explanations of Gentrification: A Story of Capital and Demand

As we have stressed throughout this book, urban change does not just happen. Gentrification is one of the main factors to explain both why housing costs in Toronto are so high and why there are so many condominiums that have been developed in recent decades. It is therefore worth unpacking this often used, yet complex, term before we progress to a visual analysis of urban change.

Gentrification has been described as "the most politically loaded word in human geography."[49] In major cities such as Toronto, New York, San Francisco, London, and Amsterdam, it is one of the major forces shaping urban and neighbourhood change. And even in cities further down the urban hierarchy, such as Detroit, Glasgow, Hamilton, or Philadelphia, gentrification still takes place, though neighbourhood downgrading is often a larger overall trend. As we will explain later, neighbourhood upgrading and downgrading can occur simultaneously within the same city and they are often two sides of the same coin.

The term gentrification was first coined by Ruth Glass in 1964. A Marxist sociologist, Glass described a process she was observing in inner-city London:

> One by one many of the working class quarters of London have been invaded by the middle class ... [they] have been taken over when their leases expired, and have become elegant, expensive residences ... [O]nce this process of "gentrification" starts in a district it goes on rapidly until all or most of the working class occupiers are displaced and the whole social character of the district is changed.[50]

Central to gentrification is both an upward class transformation and displacement. The form, style, and geographies of gentrification have evolved and changed since Glass' initial observations, but it is, first and foremost, a social process of class struggle and change, rather than a commentary on a particular architectural style or physical attribute of the city.

Historically, there have been two main theories about the causes of gentrification. While earlier scholarly debates often pitted one against the other, they are both important to understanding why gentrification happens and how it spreads. The first theory was put forward by Neil Smith in 1979. He reacted against what had, until that point, largely been viewed as a back to the city movement of people."[51] Gentrification was about a movement back to the city, but it was a return of *capital*, rather than households, that was driving the process, he argued. Smith looked at the disinvestment which had taken place in the inner city for decades and noted that while the value of its buildings was often very low, the underlying land on which those buildings sat (particularly those close to the downtown core) remained high. This led to the situation in which poor people ended up living on valuable land. His concept of the *rent gap* was based on the difference between the "capitalized ground rent" – the actual economic returns of a property given its present use, and the "potential ground rent" – the maximum economic returns if a property was put to its highest and best use. Smith argued that gentrification was a process which closed these rent gaps by upgrading properties and putting them much closer to their maximum economic potential. This was done by displacing the low-income occupants of these buildings, renovating housing stock, and selling (or sometimes renting) it to more affluent households. The same capital gains could also be made by demolishing smaller and poorly maintained buildings and replacing them with newer, more luxurious ones. According to Smith, the capital invested by different actors such as households, speculators, banks, investment firms, or governments constitutes the key driver of gentrification.

His theory was both praised and criticized. One of the criticisms of the rent gap theory was that it did not factor individual household choices into its analysis. Smith's focus on capital-based explanations was countered by Canadian geographer David Ley. Ley's work was heavily influenced by the post-industrial thesis of Daniel Bell, which focused on the decline of manufacturing and the rise of service-based employment discussed in the previous chapter. Ley's account of gentrification focused on demand-side factors such as the role of preferences, lifestyles, and demographics. His book *The New Urban Middle Class and the Remaking of the Central City*[52] examined this theory with empirical examples from across Canada. He focused on the role of artists and hippies in the 1960s and how they ended up stimulating gentrification; he noted that several hippie neighbourhoods at that time, such as Yorkville in Toronto or Haight-Ashbury in San Francisco, would later go on to become some of the most exclusive areas of their respective cities. The cultural and social amenities of the city, the historic qualities of urban neighbourhoods, and the proximity to jobs and entertainment are some of the demand-based factors that have been used to explain why middle-class, professional households would choose to live in the city rather than in suburbs. Many of the important jobs in financial and professional positions that we discussed in chapter 2 are also found in downtown cores rather than in distant suburbs.

Demographic shifts are also important. The rise of women in the workforce and two-income households changed the daily patterns of families, who now needed to manage commuting, childcare activities, and social lives in different ways; the advantage of the inner city is that these tasks can be managed more easily if long commutes to the suburbs are avoided. The practice of delaying marriage and having children also opened up new possibilities for urban living. In the 1960s, many couples would marry in their twenties, move straight to the suburbs, and start a family. This left little time in one's lifespan for urban living and all its amenities and excitement.

In recent decades, however, many people started having children in their thirties, rather than their twenties, opening up new possibilities for childless urban living. These trends fit into what Markus Moos from the University of Waterloo has dubbed "youthification." His research on Toronto, Vancouver, and Montreal has shown a tendency for young adults to live in high-density urban environments during the stage of their lives between leaving the family home, or finishing university, and starting a family.[53]

Even when raising children, many people prefer the city. In his book *Through the Children's Gate*, Adam Gopnik described raising a young family in post 9-11 New York. He recounted a conversation with a real estate agent who was asked whether New York is doomed. The realtor responded that people will always need New York:

> You can ask a thirty-year-old with children to move to New Jersey, but you can't ask a single thirty-year-old to move out there before he or she has found a mate. He or she would basically rather die … That's the great secret, the key demographic of New York. Kids at twenty-five, cities die; kids at thirty-five, cities thrive. It's just that simple.[54]

While the process may indeed be more complex than Gopnik's realtor friend explained, this example helps to illustrate the demand-side forces shaping gentrification and cities.

Gentrification's spatial locations, and the forms, practices, and policies surrounding it, have not remained static. Jason Hackworth and Neil Smith outlined this progression from one where gentrification was primarily focused on the "sweat equity" of early urban "pioneers," who invested their own time, capital, and energy into their inner-city properties in a handful of post-industrial cities in the Global North, to a process which has become central to government strategies in cities around the world.[55] Gentrification is no longer confined to old Victorian neighbourhoods in the Global North; it can also be found in a variety of different contexts, including in rural communities, suburbs, student enclaves, postwar neighbourhoods, and many more types of spaces.[56]

The global financial crisis of 2008 saw a temporary halt to gentrification in some cases. However, recent years have seen the continuation of gentrification and there are new backlashes emerging. In earlier waves of gentrification, the process opened up new parts of the city to middle-class living and consumption. Well-educated university graduates could find a nice job and be able to afford a house in a gentrified part of the inner city. But many of the first neighbourhoods to gentrify have undergone "super-gentrification" – places where only the super-rich can purchase houses, often without mortgages and at skyrocketing prices. These include neighbourhoods such as Brooklyn Heights in New York or Chelsea in London.[57] In some cases, early gentrifiers have themselves been displaced. This is particularly true of artists, who often make an area hip and trendy, yet ultimately fall victim to the ongoing success (and increased costs) of the neighbourhoods they helped to initially gentrify. In cities such as London, New York, Amsterdam, Toronto, and San Francisco, access to gentrified neighbourhoods is now out of reach even for many middle-class, professional households, particularly those who do not get help from their parents.[58] As a result, new critiques of gentrification have emerged, particularly in the media. These differ from many earlier critiques of gentrification that interpreted the process as one of working-class displacement. In newspapers such as the *Globe and Mail* contemporary discourses and narratives of gentrification critically examine how the process has put the urban dreams of middle-class professionals increasingly out of reach while concerns about working-class displacement have themselves largely been evicted from mainstream public and media debates.[59]

When reading gentrification in the visual landscapes of the city, it is important to remember that it is both the product of capital investments and of residential demand. Renovations or extensions of older homes can be interpreted as both: increasing the value of a house by investing capital into the built form and changing how a building looks to reflect dominant tastes and values. In Toronto's Cabbagetown neighbourhood, early gentrification brought with it the practice of painting houses white; this became so associated with the process that it became known as "whitepainting."[60] Some of the first gentrified houses in Cabbagetown were purchased and renovated by the architect Irving Grossman, who built his career designing suburban neighbourhoods. Painting over the grime and soot was an easy way to beautify these Victorian houses. As John Sewell noted, "To make the houses appear fresh after one hundred years of city life, Grossman painted them white, both inside and out. The houses were rented or sold to young professionals."[61] However, as more and more professionals moved into older working-class neighbourhoods, the need to paint houses white diminished; later rounds of gentrification have seen middle-class aesthetics shift towards restoring houses back to an "original" state by sandblasting and/or chemically removing the paint and dirt off the bricks.

Neil Smith's rent gap can be visualized in another sign of gentrification: the process of de-converting houses from apartments or rooming houses back to single-family dwellings. Many large old houses in neighbourhoods such as Parkdale, Riverdale, and the Annex (some of Toronto's first streetcar suburbs) were initially constructed for affluent families escaping the overcrowded downtown. Over time, as wealthier populations moved out to newer suburbs, these houses were subdivided into apartments and rooming houses. A rent gap emerged between what the houses were currently worth in their present state (as low-rent rooming houses) and what these large, centrally located historic homes *could* be worth if they were renovated and their financial potential maximized. Closing this rent gap involves de-converting rooming houses back into expensive single-family homes. This is one of the reasons why the gentrification of the Streetcar City has resulted in many neighbourhoods seeing a decrease in population since the 1970s.[62] Another visual marker of gentrification involves the conversion of old factories into lofts and condos; the Candy Factory Lofts at Queen Street West and Shaw was one of Toronto's first major examples of this form

Figure 3.9 | Retail gentrification in Little India (Gerrard Street East). Asma Boutique, a ladies and children's clothing store, was replaced with Furballs Pet Stuff. The unit to the right used to be Taj Mahal Kitchenware and became The Pantry, a specialty cheese shop. 19 February 2011 and 14 April 2014. Photographer: Brian Doucet.

of gentrification. This process of loft creation can be seen along Dundas Street West and King Street, in particular.

While gentrification has been primarily associated with residential class transformation, it also affects the retail spaces in neighbourhoods.[63] The promotion of vibrant shopping streets is part of the development of "creative city planning" that is central to making cities appealing to the middle classes that politicians and planners are keen to attract and retain.[64] Commercial gentrification represents both an increase in capital, through the renovation of business properties as well as the introduction of the values, habits, and lifestyles of the gentrifiers. Retail activities that are part of the gentrification landscape include specialty food stores, pet food or pet grooming businesses gourmet burger restaurants, yoga centres, cafés, bars and restaurants, galleries, organic grocery stores such as Whole Foods, and fancy ethnic restaurants that cater to a middle-class clientele rather than to immigrants from those particular communities. The presence of a Starbucks franchise has long been associated with urban areas that are gentrified or gentrifying; more recently, however, it has been the arrival of independent cafés that signifies the spread of gentrification.[65] In the case of neighbourhoods that are still in transition, one often sees bizarre juxtapositions, with new amenities such as trendy cafés existing side by side with businesses such as laundromats, pawnshops, and restaurants catering to working-class or immigrant populations (figures 3.9 and 3.10).

For some, retail and amenity changes such as these represent a major gain and an enhancement to their quality of life and enjoyment of the city. However, for others, particularly low-income residents, these retail changes represent conflict, displacement, and loss. In many cases, non-gentrifiers have little need for many of the new amenities, nor can they afford to use them.[66] As a result, such changes can lead to a sense of alienation and a "not-for-us" mentality.[67]

The idea of displacement has gradually shifted away from a narrow interpretation of direct and forced relocation from one place to another. Quantitative researchers still use these criteria in order to measure this one-time process of outmigration from a spatially delineated area, such as a census tract. However, many critical urban scholars have expanded upon this in order to capture the totality of change and disruption that gentrification and displacement bring to low-income populations. Peter Marcuse outlined different forms including direct or last-resident displacement, chains of displacement over many years, displacement pressure, and indirect or exclusionary forms (such as new housing being too expensive for existing residents).[68]

New research is also expanding the meanings and interpretations of displacement by moving away from economic terms, that is, low-income residents' inability to pay for housing or amenities in gentrifying space, to include experiential, psychological, and phenomenological forms of displacement. Seen from these perspectives, displacement is a process rather

Figure 3.10 | A bistro opposite a coin laundry, Gerrard Street East at Greenwood Avenue. 31 May 2019. Photographer: Brian Doucet.

than an event, and can be conceived in broader terms that encompass not only being forced out of one's home or community, but also experiences of loss (of friends, businesses, amenities, etc.) while one remains in situ. Terms such as "loss of place," "slow violence," "un-homing," and "symbolic displacement" have been used to attempt to capture residents' diminished sense of connection to their changing community as familiar spaces of belonging are replaced by spaces of exclusion.[69] These types of displacement can be very difficult, if not impossible, to measure through statistical analysis and therefore require other approaches and methods such as interviews, ethnographies, oral histories, and visual methodologies.

Conceiving of displacement as more than being physically forced out of a neighbourhood is important when studying how commercial spaces are remade and the struggles, conflicts, and tensions that exist over who decides how neighbourhoods develop and who is included and excluded from those conversations. In Toronto, one of the most important studies of commercial gentrification, conducted by Katharine Rankin and Heather McLean, takes this more holistic view. The study did not focus on retail changes within the super-gentrified downtown core, with its high-end shops, cafés, and amenities, but rather in what is today considered a frontier of gentrification: Mount Dennis, an inner-suburban neighbourhood centred on the intersection of Eglinton Avenue West and Weston Road. It houses a major cluster of Somali-Canadians and many have opened businesses along the retail strips. This neighbourhood is included in our definition of the Streetcar City; Weston Road had a streetcar line until 1948.

Rankin and McLean argue that while these retail spaces in Mount Dennis are typically envisioned as areas of poverty and decline, the combined context of new transit infrastructure (the UP Express, Kitchener GO line, and the Eglinton Crosstown LRT (Line 5) will all connect in Mount Dennis), a growing population, a thriving real estate market, and proximity to vacant industrial lands (a former Kodak plant was situated within the neighbourhood) mean the area faces significant redevelopment and gentrification pressures. They noted two visions for redevelopment/gentrification from local boosters, politicians, and business leaders: one focused on real estate, which was centred on a narrative that the existing population was unable to support a thriving commercial street and therefore the arrival of a more affluent population would bring about new businesses through market forces. The other was centred on a green and cultural economy, with an emphasis on cycling infrastructure, a farmers' market, and the development of a new creative hub.[70] The new Artscape Weston Common is one outcome of this vision.

They argue that both visions "find common ground in explicit demarcations of Weston Road as empty and deficient as well as in explicit aspirations for gentrification and a form of "social mix" that would bring consumers with disposable incomes and storefronts catering to more "downtown' sensibilities."[71] However, both these visions sat in contrast to the perspectives of local business owners and community leaders, who saw this redevelopment as a racialized class project with little room for their own practices and values in the visions ascribed to the neighbourhood by planners, real estate developers, white homeowners, and artists. In Toronto and elsewhere, the role of race and ethnicity in driving gentrification is an emerging area of research that is rapidly becoming more central to our understanding of what shapes urban change.[72]

The Policy Side of Gentrification: Toronto's Condo Boom

Gentrification is not just about old houses and trendy retail. It is also not only about capital investment or middle-class demand. The products of gentrification (capital investment, middle-class consumption, improvements to urban aesthetics) have been central to planning, policy, and development goals, and aspirations of cities around the world for several decades. There are many visual indicators of the policy and planning side of gentrification. In Toronto and other cities, the development of large, new-build projects that dramatically reshape both the social geography of the city and its skyline, is one of the most important ways in which contemporary gentrification can be visualized. In Toronto, this new-build gentrification has been primarily associated with condominium construction.

While condominiums are most identified with the tall towers that cluster around downtown and other key nodes of development in the city, they are, in fact, a form of ownership and tenure, not an architectural style. A condominium is essentially a double ownership; individual units are owned by each homeowner (which can be bought and sold) and collectively all unit owners own the common spaces in the building and are responsible for its maintenance and upkeep.[73] This form of tenure has been legal in Ontario since 1967 and can, in theory, exist in almost all types of urban (suburban or rural) form and building design.

While not everyone considers condos to be part of the gentrification process, British geographers Mark Davidson and Loretta Lees have made a strong case to include such spaces as part of the "mutation of gentrification," based on their empirical work along London's riverside.[74] They gave four main reasons for including new-build developments as part of the conceptualization of gentrification: they constitute a reinvestment of capital, bring about social and class upgrading, and involve physical and visual landscape changes, as well as displacement. Residential displacement is likely to be indirect or exclusionary (prices in new-build developments are too high for low-income residents) because very few people lived on sites where new condo and high-rise developments take place. However, as DePaul University geography professor Winifred Curran's research has shown, direct displacement does happen, although in the form of the loss of industrial, rather than residential, spaces. Gentrification pressures can lead to the displacement of dynamic, varied, and vibrant urban manufacturing districts. As surrounding neighbourhoods gentrify there is pressure to redevelop these spaces, and there are substantial profits to be made for developers who convert former factories into lofts or redevelop industrial spaces (such as waterfronts) into high-end housing. Curran argues that small manufacturers have been actively and directly displaced through mechanisms such as buyouts, lease refusals, zoning changes, and increasing rents.[75] Policies to promote a post-industrial, creative, entrepreneurial city can also contribute to planning visions that are based on a gentrified future for these industrial spaces. While Curran's work has largely focused on industrial districts in inner-city Chicago and New York, similar processes can be seen in Toronto; her research serves as an important reminder that we should be careful not to assume that this inner-city industrial decline has been natural or inevitable.

In Toronto, urban policies that focus either explicitly or implicitly on fostering and promoting gentrification are intertwined with wider ambitions towards global city status, climbing the urban hierarchy, attracting the "creative class," or engaging in other forms of interurban competition. They are one major spatial manifestation of the drive to become a world city. Therefore, the almost unbroken wave of condominium development that has taken place in Toronto since the late 1990s cannot be fully understood without first understanding the city's policy shifts away from the reform politics of the 1960s and 1970s, towards an emphasis on the need to make Toronto competitive on a global scale. The condo is, therefore, not just a form of housing but rather part of the wider remake of the city that Leslie Kern aptly describes: "In the context of Toronto's global city aspirations, condominiums are expected to fulfill expanding housing needs, curb suburban sprawl, lift the spaces of deindustrialization to their highest and best use, respond to a cultural shift in favor of urban living, and stimulate the economy by providing sites for capital investment."[76]

Toronto's condominium boom has stimulated one of the most dramatic transformations of any skyline in North America. Toronto is North America's largest condominium market,[77] and much of this is concentrated in the downtown core. Between 2000 and 2010, of the 554 condominium projects (totalling 97,631 units), 335 of these projects (55,149 units) were in the old City of

Toronto.[78] Ute Lehrer and Thorben Wieditz call this process "condofication,"[79] Dutch architect Hans Ibelings has dubbed it the "condominiumization of Toronto,"[80] and Gillad Rosen and Alan Walks have referred to the process as "condo-ism," which they describe as "an interlocking nexus of political and economic agendas regarding urban economic development, a planning philosophy that favours intensification, downtown living and densification and the cultural promotion of high-rise living as both sophisticated and environmentally friendly" (figure 3.11).[81]

The intensification and redevelopment of former industrial spaces in and around downtown was seen as the solution to many of the problems associated with the deindustrialization and decline of the urban core, including the lack of urban vitality, the proliferation of underused spaces, and the lack of capital investment and accumulation.[82] But, as we will argue, policies promoting intensification also spurred the further gentrification of the urban core and are part of what Neil Smith has called a "global urban strategy" of gentrification that is "no longer about a narrow and quixotic oddity in the housing market but has become the leading residential edge of a much larger endeavour; the class remake of the central urban landscape."[83]

All three levels of government have enacted legislation that directly or indirectly relates to how dense new developments can now be found on formerly industrial spaces. Federally, it has been argued that Canada's policy of immigration, which sets levels of 300,000 new immigrants per year, is essentially an urban population policy because the vast majority of new immigrants settle in Canada's biggest cities and their surrounding suburbs, and are responsible for the lion's share of growth in these places.[84] Between 2011 and 2016, 216,830 people moved to Toronto from outside of Canada, compared with only 184,120 internal migrants from within Canada (141,135 from elsewhere in Ontario and 42,985 from other provinces).[85] The city of Toronto actually has a net loss of intraprovincial migration of –9.9 per thousand, meaning more people move out of Toronto to other parts of Ontario than move into the city from the rest of the province.[86]

While the federal government does not play a role in where this population settles within a metropolitan area, provincial legislation has directly contributed to more intensification within the urban core. Two provincial pieces of legislation, enacted by the previous Liberal government, contributed towards steering where this growth takes place within the GTA and the Greater Golden Horseshoe. The 2005 Places to Grow Act called for intensification in existing areas in order to curb sprawl. In the same year, the passage of the Greenbelt Act protected farmland and countryside around the GTA from development.[87] This type of measure is related to what Noah Quastel, Markus Moos, and Nicholas Lynch call the "sustainability-as-density model," which provides the philosophical justification to limit sprawl on the fringes of a region, while encouraging increasing density within its core.[88]

Policies enacted by the city also played a major role in facilitating the condominium boom. The amalgamated City of Toronto's first Official Plan removed density and height restrictions in many places (though this has not led to an intensification of single-family residential neighbourhoods) and simplified land-use categories. There were two key arguments for this: first, the number of units or storeys does not matter as much as the way in which a building "fits" into its surroundings; and second, intensification of the core was necessary in order to stem continued sprawl at the region's fringes. However, this radically changed the ways in which planning debates were framed; debates around numbers switched to those around vague notions of aesthetics, and as a result the space for ordinary residents to participate in the planning process has been greatly diminished.[89]

In the mid-1990s, the first two areas of the city to see specific policy deregulation to allow for new mixed-use developments (with condos playing a large part) were the formerly industrial lands to the east and west of downtown, along King Street. Nicknamed the "Two Kings," the areas around the intersections of King and Parliament and King and Spadina had become quiet and underutilized, as industrial decline meant factories were closing and zoning restrictions prohibited other uses. Many old buildings were demolished to make way for surface parking lots. To reverse these downward trends, zoning was changed from singular uses, such as warehousing, office, or industry, to mixed use, which enabled multiple functions (residential, commercial, retail, etc.) on the same site and in the same building. The only major requirement was that the relationship

Figure 3.11 | The "condofication" or "condominiumization" of Toronto. Bathurst Street at Fort York Boulevard, 2004 and 2019. Photographer: Brian Doucet.

between the existing buildings and the streetscape be maintained in order to preserve character and feel of the area.[90] This change in zoning triggered a new era of downtown investment, as old factories (particularly the remaining garment manufacturers that had survived the 1980s period of deindustrialization) were turned into lofts, new condominium towers were erected, and property values rose dramatically.[91] This displacement of industry has parallels with what Winifred Curran described along the Brooklyn waterfront.

Later in this book, we will present several photo sets of the Two Kings. The amount of development and intensification has indeed been impressive; as of December 2015, 46,000 residential units were built or were in the pipeline and 20,000 new jobs had been added since the policy was introduced. Those behind the success of the Two Kings pointed to two key lessons: first, simply removing zoning guidelines does not guarantee the kind of transformations seen along King Street. These areas are situated close to downtown, have good transit, and are highly walkable; not every formerly industrial area is endowed with these key features. Second, the success of the Two Kings has led to issues of affordability and the need for policy measures to manage this growth.[92]

Two subsequent reports by the newly amalgamated City of Toronto emphasized the need for sites within the city that could attract development and international capital, as well as reflect the city's desire to transition away from a manufacturing city towards one focused on creativity.[93] Both of these reports worked towards providing the rationale for supporting the rezoning of former industrial and underutilized space (i.e., surface parking lots) for denser, mixed-use projects. It is important to reflect on the fact that, unlike in Vancouver, or in many European cities, no provisions were made for affordable housing during this period and, as a result, virtually all of the high-rise development seen since 2000 has been market-rate and market-driven. By 2008, the city had fully embraced Richard Florida's idea of the "creative class" and was celebrating the success of the Two Kings and other parts of the downtown in turning formerly industrial districts into "vibrant, authentic places ... critical to attracting the best talent in the world."[94]

Toronto's condo boom can be seen as a combination of many factors which we describe in this book. The restructured economy of the city created many new, well-paying, professional post-industrial jobs in the downtown core. Changes in housing preferences meant that many young people in particular have rejected the suburbs in favour of an urban, rather than a suburban, lifestyle, at least for part of their adult lives. Deindustrialization created the spaces in which new and dense development could take place, without demolishing entire neighbourhoods. And finally, the policies mentioned above created the justification and the regulations by which very dense, tall, new developments could be built. As we will demonstrate later in this book, this can be observed within changing visual landscapes.

Until the provincial government led by Doug Ford changed planning regulations with Bill 108, one of the few avenues for community benefits from this development boom were agreements made under Section 37 of the Planning Act. This enabled developers to increase height or density in exchange for providing public space enhancements and other community benefits. These investments needed to be spent locally, in the municipal ward where the development was taking place, rather than distributed across the city. As a result, the downtown core and other major nodes or corridors saw the lion's share of Section 37 investments, further exacerbating differences in amenities between the gentrified urban core and the inner suburbs, where comparatively little investment in new condos took place. This reality resonates with Stephen Graham and Simon Marvin's concept of "splintering urbanism" mentioned earlier in the book; services and amenities that used to be paid for through general taxation and distributed across the city on the basis of population or need today increasingly rely on the investment decisions of developers in order to be financed. Because where developers choose to invest is highly uneven across the city, this leads to a greater fragmentation of services and amenities.

The result, according to Gillad Rosen and Alan Walks, was "a 'two-track' system in which the majority of new public infrastructure and amenities were dependent upon, and paid for by new high-density development, while local government austerity meant those neighbourhoods not receiving new development had to fight for increasingly scarce tax dollars."[95] In terms of development, this also led to a "let's make a deal" planning model, where existing zoning regulations are merely a starting point for negotiations, rather than a set of defined guidelines and a rulebook by which all parties must play.[96] Hence, many of the incredibly tall condominium towers have increased their density from what was initially permitted, in exchange for extras such as parks, community centres, spaces for amenities such as daycare centres, or even cash.

However, the Progressive Conservative government of Ontario, elected in 2018 and led by Premier Doug Ford, passed sweeping changes to Ontario's planning system through Bill 108, the More Homes More Choice Act. The bill, which received royal assent in December 2019, affects provincial regulations on affordable housing, density, planning appeals, heritage preservation, and development charges. The cost of new parks and community infrastructure has been removed from development charges, and municipalities will need to pay for more of the hard infrastructure that services new developments. Section 37 benefits have been replaced by an optional community benefit charge, which can be used to pay for facilities, services, and other costs to redevelop or develop the area. Before such a fund can be established, local governments will need to develop a strategy and identify the types of facilities and services that will be funded, and the amount they can charge a developer will be capped at a percentage of the land value.

Local governments in Toronto and across the province are still coming to terms with the full ramifications of Bill 108, though there is a worry that while this legislation may make development easier, this will not necessarily lead to lower housing costs, and there are genuine concerns about how planning appeals have been tipped to favour developers, rather than the public, local governments, or community groups.

A new planning tool in Ontario is inclusionary zoning, which requires that a certain percentage of units in a new development be affordable. Toronto Council approved its inclusionary zoning plan in late 2021. The previous Liberal provincial government, under Premier Kathleen Wynne, began developing inclusionary zoning policies in 2016. Doug Ford's Progressive Conservative government restricted it to areas around transit stations. Toronto has until July 2022 to designate at least 180 Major Transit Station Areas (MTSAs), most of which will be eligible for inclusionary zoning. MTSAs extend 500–800 metres from a transit station and are intended for high-density and mixed use. Within MTSAs, there are fewer opportunities for cities, or citizens, to challenge development proposals. Inclusionary zoning has been popular with municipalities faced with budgetary constraints and little funding from higher levels of government for new affordable housing. Had Toronto enacted inclusionary zoning a decade ago, one estimate suggested that almost 75,000 people could have obtained affordable housing as part of the city's condo boom. However, there are drawbacks to this strategy, including vague rules about what constitutes affordable housing, and for how long a unit must remain affordable. It also relies on private developers to build a lot of for-profit units. As Martine August and Giuseppe Tolfo concluded, inclusionary zoning should be part of broader strategies to address housing affordability rather than the silver bullet that it is often seen as being by planners and policymakers.[97] It has been shown that other changes enacted by the Ford government, such as cancelling rent control on new buildings, have not increased the supply of affordable housing.[98]

While many scholars agree that condominium developments on former industrial lands comprise part of the changing nature of gentrification, there remains considerable debate. Markus Moos' work brings some nuance and complexity to this matter by stressing that not all condo dwellers are affluent; many rent or share units with roommates in order to access employment downtown and avoid lengthy commutes.[99] Many condo units are small and are designed for young professionals. The advertising and marketing of these condos is also aimed at young people, by selling an exciting urban lifestyle that sits in contrast to the monotony of the suburbs and also as a first step onto the property ladder.[100] We see these condominium developments as part of Toronto's gentrification landscape; they represent tremendous capital investments in the city, they are bringing many professional and well-educated residents into new areas, they are the product of municipal policies aimed at climbing the urban hierarchy, and they provide no social or affordable housing, thus limiting who has the opportunity to live in the city's core.[101]

It is important to note that while condo units are individually owned, not all those living in condos are property owners themselves. As new purpose-built rental buildings have become scarce, especially in the downtown core (although there are some indications this trend is starting to reverse, with several high-profile purpose-built rental developments, including at the former Honest Ed's site at Bloor and Bathurst), condos have become the de facto rental spaces, particularly for young professionals who are unable to pay the increasing cost to buy property. While it is difficult to ascertain how many condos are bought by investors as rental properties, the Canadian Mortgage and Housing Corporation estimated that 23.6 per cent of condos across the city were rented in 2011; a city of Toronto survey a year later estimated that 45 per cent of downtown condos were rented, and another study claimed that approximately 80 per cent of new downtown condos were bought by investors.[102] In the years before the pandemic, these were also highly popular with national and international investors who purchased units for the purpose of renting them out on short-term rental sites such as Airbnb. Condominiums therefore also represent a source of capital gains, speculation, and investment into the built environment, which, as Neil Smith articulated forty years ago, are just as much a part of gentrification as trendy boutiques and restored Victorian homes.

Contemporary Frontiers of Gentrification

While much of the older, single-family housing stock in Toronto's core is gentrified or gentrifying and new-build condos represent another major upward class transformation of the inner-city, two types of housing in the core – older high-rise private rental apartments and subsidized housing projects – had remained impervious to gentrification for many years.[103] Many remaining pockets of poverty or low income residents – what Elvin Wyly and Daniel Hammel called "islands of decay in seas of renewal"[104] – are currently undergoing gentrification, leading to the displacement of some of the remaining low-income residents, immigrants, and working-class households still residing in the urban core. In Toronto, older apartment buildings, such as those that can be found in Parkdale, St. James Town, and in other locations scattered throughout gentrified parts of the city, are "a final frontier for gentrification, remaining as last bastions of affordability amidst landscapes of gentrified retail and low-rise housing."[105]

Gentrification has long been associated with owner-occupiers. Individual households buy and renovate properties to live in themselves. Small developers buy and then "flip" rundown houses by renovating them and reselling them, funding their next projects with the profits. Large developers construct condominium towers which are bought either by residents or investors. However, a new trend has shifted the frontiers of gentrification into the rental market. This wave of gentrification is not just about the physical upgrading of rental properties but about how they have become financialized investment products.

Seen from the perspective of Neil Smith's rent gap, there is considerable profit to be made by evicting low-income tenants, renovating their units, and renting them out at much higher rates and to more affluent tenants. Such "renovictions" have become more common and are partly fuelled by the continued financialization of the housing market. Financialization refers to the structural changes in the economy, businesses, governments, and individual households that result from the increased dominance of financial actors, markets, practices, measurement, and narratives.[106] While less visible in our repeat photography, the financialization of multi-family housing in Toronto and elsewhere has become a central piece of the contemporary gentrification landscape and has become increasingly important in understanding urban trends, particularly the increasingly unaffordable nature of housing in big cities. Since 2009, a number of "financialized" landlords – private equity and asset management firms, as well as real estate investment trusts (REITs) – have entered Toronto's real estate market, including the Swedish-based, Bahamas-registered Akelius Canada Ltd. Akelius and other REITS such as MetCap Living and Timbercreek have bought up low-rent apartment buildings in areas adjacent to gentrifying neighbourhoods with the aim of "repositioning" them through upgrading to appeal to higher-end renters. This involves remodelling units, building-wide renovations, and rent increases, with the intent of forcing low-rent tenants out and replacing them with more affluent tenants who can pay higher rents.[107] This reduces some of the last remaining affordable rental stock in the gentrifying core. Parkdale is one of the last frontiers of gentrification in the Streetcar City, but this process is now taking place in many of its high-rise apartment buildings, which, for decades, were skipped over in previous rounds of gentrification. Tenants have tried to resist evictions both through the courts and by organizing rent strikes. However, apart from the very small Parkdale Community Land Trust, there have been few efforts that have managed to stem the tide of gentrification in Parkdale and other similar pockets of affordable housing that remain within the Streetcar City.[108]

But as Martine August and Alan Walks have found, it is not just in the inner city where these financialized landlords operate. Their research identified a different strategy active in suburban locations. Here, rather than upgrading to attract gentrifiers, the idea is to squeeze profits from these buildings by reducing maintenance and raising rents. In a sad twist of irony, many of the tenants displaced from buildings in gentrifying neighbourhoods have no other choice than to relocate to these "squeezed" buildings in the inner suburbs.[109]

The second type of area that has, until relatively recently, resisted gentrification are the large subsidized housing projects built in the decades after World War II. Here, policies that promote gentrification can also be seen in the ways in which social housing has been restructured over the past decade. The most notable of these neighbourhoods is Regent Park. Constructed in the years immediately after World War II, Regent Park was Canada's first and largest social housing development. It replaced a former part of Cabbagetown, which, under the 1944 City Planning Board's neighbourhood classification, was considered to be the city's largest slum.[110] Bounded by Parliament, Gerrard, River, and Dundas Streets, the entire area was demolished (except for an old church) and replaced with the type of modernist apartment buildings and street design that Jane Jacobs would later criticize (the whole neighbourhood was turned into a superblock, with no through streets, effectively demarcating it and cutting it off from its surrounding areas). Its tenure was 100 per cent subsidized housing, with 2,000 rent-geared-to-income units that would later be managed by the Metro Toronto Housing Authority.[111] Its design was based on ideas from both the Garden City Movement of Ebenezer Howard and the modernist, rational vision of the Swiss architect Le Corbusier.[112] While Regent Park was initially seen as a model community, a combination of underfunding, poor management, and changing populations meant that, by the late 1960s, it had become undesirable and unsafe, earning it a negative reputation, particularly in the media. Much of the blame for this was placed either on the project's design and layout, or the "culture" in what had become an overwhelmingly non-white community.[113]

In the early 2000s, plans were formulated to redevelop Regent Park into a mixed-income community. Plans for the renewal were based on ideas that had become popular with planners and policymakers starting in the 1990s: demolish social housing and replace it with a mix of ownership and tenure styles, employing architecture and urban design in keeping with surrounding areas. This became a common planning approach to address what were considered to be "problematic" neighbourhoods in many countries throughout the world. In the United States, the HOPE VI program initiated in the 1990s spearheaded the demolition of many subsidized housing projects, replacing them with mixed-income and tenure communities. Cabrini Green in Chicago was one of the most notable examples of this program. In other countries such as the Netherlands and the United Kingdom, national governments contributed vast sums of money to restructure problematic housing estates, many of which were originally built quickly on the peripheries of cities in the decades after World War II. In the North American context in particular, much of the logic for these social-mixing policies was centred on theories which advocated poverty deconcentration, new ideas about urban design (largely influenced by Jane Jacobs' "eyes on the street" concept), and a belief in neighbourhood effects. This is the idea that one's neighbourhood has either a direct or indirect effect on an individual's behaviour, outcomes, or employment prospects. The theory argues that by having richer households move close to poorer ones, the latter's life chances will improve.[114] However, despite entrenched policy and planning beliefs in the idea of social mixing, there has been little academic research substantiating that economically and socially mixed communities offer better opportunities for low-income residents than the housing and neighbourhoods it replaced. These policies have been critiqued as constituting "gentrification by stealth," a situation in which "the movement of middle-income people into low-income neighbourhoods causing the displacement of all, or many, of the pre-existing low-income residents" is "rhetorically and discursively disguised as social mixing."[115]

The redevelopment of Regent Park follows this rationale and features a mix of subsidized housing units and market-rate condominiums. In this context,

> the condo owner and the social housing tenant [previously separated by space and class] are brought together in the redevelopment discourse, to forge a normalised, healthy community in which the stigma of poverty is neutralised … According to the official discourse of public housing revitalisation, the condo owner will promote social cohesion and reduce territorial stigmatisation, but their presence also marks a contestation over urban space, in which private property and middle-class interests dominate.[116]

In recent decades, some of the most intense (and contentious) spaces of urban redevelopment have been in these remaining pockets of affordable inner-city housing: the large, modernist subsidized housing estates constructed in the middle part of the twentieth century. Therefore, many scholars see this as part of the continued development and spread of gentrification. University of Waterloo planning professor Martine August has argued that "for cities, restructuring has been manifest in 'entrepreneurial' strategies to attract investment … and to attract consumption dollars by reimagining central areas into safe spaces to play, consume and work. Gentrification is the 'leading edge' of this restructuring process and contributes to the production of polarized urban landscapes as the poor and racially marginalised are shuffled from coveted to undesirable spaces according to shifting geographies of potential real-estate value."[117]

After more than a decade of cuts from higher levels of government, the Toronto Community Housing Corporation announced the redevelopment of Regent Park in 2002. All 2,087 of the original rent-geared-to-income units would be demolished in stages and replaced by 1,877 new on-site subsidized units and a further 5,400 market-rate condominiums. The remaining subsidized units would be built in three other locations away from Regent Park. The rationale for this form of redevelopment was driven by both financial motivations and a corporation-wide belief in the benefits of social mixing.[118] Owing to a lack of funding, the Toronto Community Housing Corporation entered into a public-private partnership with the Daniels Corporation, a private developer, to implement the redevelopment.[119] This type of public-private partnership is increasingly common (see chapter 2), particularly after both the federal and provincial governments stopped funding affordable housing, leaving cities with few financial resources to either maintain existing housing stock, or build new units. Therefore, it fits squarely within the framework of the neoliberalized entrepreneurial city.

The redevelopment of Regent Park is ongoing and has been praised in many circles as being visionary.[120] However, a critique has also developed which focuses on the diluting of subsidized housing and the removal one of the last barriers towards the complete gentrification of the inner core of Toronto. Now that the stigma of Regent Park is largely gone, there is little stopping the continued gentrification of the eastern side of downtown and the subsequent elimination of many of the area's low-income residents (most of whom benefit from their proximity to downtown) and their affordable, non-market housing. Condo sales in Regent Park have been very brisk. Because of this, the percentage of private market condos has gone up from 40 per cent of the housing stock in phase 1 to 74 per cent of the housing stock in phase 2, with projections even higher for phase 3.[121] In any event, social-mixing policies do not adequately or directly address the root causes of poverty, which are embedded in wider processes of economic, social, political, and racial exclusion. In their critique of Regent Park, Ute Lehrer, Roger Keil, and Stefan Kipfer reference Engels' statement that the bourgeoisie's "solution" to the housing crisis was to "move it elsewhere," and that much of the rationale behind social mixing is "to disperse, hide and micro-manage low-income populations with pre-functionalist design, income mixing and ownership housing instead of concentrating them in large, homogenous and visible segregated housing tracts, as in the post-war era."[122]

While Regent Park was the first public housing project to be built in Canada, and one of the first to be redeveloped with social-mix principles, similar approaches can be found in other neighbourhoods comprising social housing. However, as Martine August explains, using the profits from gentrification (i.e., selling new condos in mixed-income communities) to fund subsidized housing only works in areas with potentially valuable real estate that is attractive for middle-class households; thus, this approach mostly succeeds in reinforcing existing socio-spatial divisions within the city.[123] Whether this model would work in the declining neighbourhoods of the inner suburbs remains to be seen. August's work also found that tenants in subsidized housing projects such as Regent Park did not have the same negative experiences that were often projected onto their neighbourhoods by outsiders. Although they did feel the effects of decades of mismanagement and poor maintenance, there was often much more community spirit, appreciation of the old layout and design of the neighbourhood, and sense of connectivity than the literature promoting social-mixing policies suggests. They also appreciated their central location, and having the connectivity of being on two streetcar lines.[124]

Toronto: An Increasingly Divided City

The dramatic shift away from a manufacturing economy towards one based on services in the latter part of the twentieth century has had a profound impact on the social geography of cities like Toronto. Increasingly, those working in the higher-end services tend to live in more central and amenity-rich parts of the city, with lower-end service workers, new immigrants, refugees, and those still working in manufacturing residing in more peripheral neighbourhoods.[125] These shifts, particularly the gentrification of the core of the city, combined with neighbourhood downgrading in inner-suburban areas (typically built between 1945 and 1970), have dramatically transformed the city over the period of time studied in our book.

Socio-economic and spatial divisions within Toronto and other Canadian cities have received significant attention over the past decade, largely through the work of David Hulchanski and his team of researchers. In 2010, Hulchanski published a report entitled *The Three Cities within Toronto: Income Polarization among Toronto's Neighbourhoods, 1970–2005*.[126] Its findings became front-page news. As the title of the report suggests, it tracks long-term change across the city's neighbourhoods. Hulchanski painted a picture of three distinct patterns of development. The first, which he calls City 1, are the neighbourhoods that are upgrading, where average incomes increased by 20 per cent or more, compared with the overall regional average. In 2005, this comprised 19 per cent of the city's census tracts. City 1 is primarily found along the central north-south spine of the city, anchored by the Yonge subway, traditionally affluent areas such as central parts of North York, and much of the area south of Bloor Street and Danforth Avenue, including the waterfront, which, as we described above, are neighbourhoods where gentrification is taking place. City 2 consists of neighbourhoods where very little upgrading or downgrading is taking place. They represent 39 per cent of the city's neighbourhoods and can primarily be found between City 1 and City 3, for example, in much of the former municipalities of York and East York. City 3 comprises neighbourhoods that have been downgrading since 1970, where average incomes have decreased by 20 per cent or more compared with the regional average. City 3 represents an astonishing 39 per cent of the city's neighbourhoods. They can generally be found in the northwest and northeast parts of the city, including the vast majority of neighbourhoods in Scarborough and north Etobicoke.

In addition to mapping and analysing long-term trends, the report also paints pictures of inequality and polarization within the city at different moments in time. Most striking are the maps of 1970 and 2005, which vividly illustrate the concept of the "egg to hourglass" social structure outlined in chapter 2. In 1970, 66 per cent of all Toronto neighbourhoods were middle income (average incomes within 20 per cent of the city average). Most parts of Scarborough and Etobicoke were solidly middle class. Affluence was concentrated in the north end of the city in both Toronto and North York, as well as a large pocket in Etobicoke north of Bloor Street near the Humber River. Low-income neighbourhoods (those with average incomes between 20 and 40 per cent below the city average) were primarily concentrated in the old City of Toronto, and the wider Streetcar City. Most neighbourhoods south of Bloor Street and Danforth Avenue were classified as low income, and there was also a large pocket of low-income neighbourhoods north of Bloor Street on the west side of the city, which stretched all the way up to Eglinton Avenue. While not considered ghettos compared with their American counterparts, and not seeing anywhere near the same levels of property abandonment or racial tensions either, inner city Toronto in the 1970s was not the affluent space that it is today. These are similar to trends we noted when we aggregated data for the Streetcar City from the 1971 census. These areas were a mixture of working-class communities and immigrant enclaves, with different ethnic groups carving out their piece of Toronto. It is also worth noting that in 1971, only 1 per cent of all neighbourhoods throughout the city were classified as very low income, with average incomes of more than 40 per cent below the city's average.

In 1970, the subway was being extended up Yonge Street beyond its original terminus at Eglinton into wealthy areas in North Toronto and North York, and had just been extended to middle-class suburbs in Scarborough and Etobicoke. The streetcar network was concentrated in the parts of the city built before 1945, and served many low-income neighbourhoods. In the 1970s, manufacturing jobs were still plentiful and comparatively well-paid, and were housed within older parts of the city; one could leave high school, get a factory job, and make a decent living. Large parts of what are now considered to be the inner suburbs, areas constructed in the decades after World War II, featured modest single-family homes, occupied by blue-collar workers employed in nearby factories who attained middle-class status. For them, being able to leave the overcrowded neighbourhoods and older, rundown housing of the inner city and move to a new suburb, like Scarborough, represented a real sign of progress and a major step up the economic ladder.

The picture painted in 2005 could not be starker, both in terms of income distribution and of the spatial locations of wealth and poverty. By 2005, only 29 per cent of the city's neighbourhoods were middle income (down from 66 per cent in 1970). This suggested a city that was becoming increasingly polarized. The biggest downgrading occurred in Scarborough, the northern half of Etobicoke and western North York; neighbourhoods that had been solidly middle class in 1970 had become low-income districts by 2005, with pockets of very low-income neighbourhoods around Rexdale, Jane-Finch and throughout Scarborough, Thorncliffe in East York and along Weston Road in York. The affluent areas along Yonge Street expanded and new high- and very high-income neighbourhoods emerged south of the Bloor-Danforth subway line (where the bulk of today's streetcar network is concentrated). The number of very high-income neighbourhoods doubled during this period and the number of very low-income neighbourhoods jumped from 6 tracts in 1970 to 67 in 2005, representing 14 per cent of the city's neighbourhoods. Hulchanski noted that these trends are "both surprising and disturbing. Surprising, because 35 years is not a long time. Disturbing, because of the clear concentration of wealth and poverty that is emerging."[127]

Hulchanski and his team have continually updated their analysis to reflect the most recent data available. His findings since his initial report are ominous. By 2015, the processes of inequality and polarization had only accelerated. In the old city of Toronto, south of the Bloor-Danforth subway, only a handful of low- and very-low-income neighbourhoods remain; many more are middle-income (which often denote areas where gentrification is starting to emerge) and high- and even very-high-income areas.[128] Within 45 years, the neighbourhoods served by Toronto's streetcar network have transformed from some of the city's poorest to its most desirable, and affluent, districts.

Further analysis shows that the biggest decade of change, particularly in the growth of low-income districts, was the 1990s. As we have previously articulated, this was not inevitable. It was in this period that major economic, political, and policy shifts were reorienting Toronto towards becoming a globally focused financial city. In addition to many factory closures, the 1990s also saw the emergence of neoliberal urban policies that focused less on ironing out the social and spatial inequities of capitalism and more on promoting post-industrial economic growth and development. Deindustrialization, which was accelerated at this time through the signing of the Free Trade Agreement with the United States in 1989 and the North American Free

Trade Agreement with the United States and Mexico in 1994, not only meant a loss of jobs but also a loss of relatively well-paying jobs, particularly for those with low levels of education. As factory workers got laid off, many took lower-paid, more insecure, and non-unionized jobs in the service sector.[129] Also influential during this time were provincial and federal government decisions to stop providing funding for affordable housing. These trends all contributed to a context where both inequality and polarization grew dramatically in this period, shaping the geography of Toronto (and the wider region) that we see today. The fall in the construction of new rent-geared-to-income houses post-1995 is particularly dramatic. A total of 23,800 units were added between 1971 and 1981, 11,800 between 1981 and 1991, and 11,310 between 1991 and 1996. But between 1996 and 2001, only 1,300 units were added, as higher levels of government stopped funding its construction.[130] Waiting times for subsidized housing, unsurprisingly, have continued to grow longer; there were over 95,000 households on the active waiting list in 2018 and it can take upwards of a decade to receive subsidized housing.[131]

While Toronto prides itself on being a multicultural and ethnically diverse city, a worrying trend has also emerged that sees new immigrants and visible minorities increasingly clustered in City 3, predominantly in high-rise apartment towers constructed in the decades after World War II. These are far removed from job opportunities in the downtown core and lack good-quality transit. These neighbourhoods were constructed with the idea that people would use cars as their primarily mode of transit, but for many, owning a car is too expensive; bus routes are overcrowded and, while service may be frequent, especially compared with that of many American suburban areas, they are often very slow and subject to "bunching" because they get stuck in traffic. There are also very few good quality bike lanes and walking can be a hostile and dangerous experience.

Segregation outside of the city's core is a growing problem for visible minorities, particularly among Toronto's black community,[132] a trend that is especially distressing in a city whose official motto is "diversity our strength." In a city in which visible minorities make up more than 50 per cent of the population, the geographical split of where whites and visible minorities live is striking, with strong correlations to the trends in income inequalities. Most census tracts that are majority-white are found within the Streetcar City, largely synonymous with Hulchanski's gentrifying City 1.[133] To compound these inequalities, race- and income-based statistics on the impact of COVID-19 in Toronto showed that, by August 2020, 83 per cent of reported cases were among racialized Torontonians, although they constitute only half the city's population.[134]

So far in this book, we have outlined some of the major trends that have shaped Toronto since the 1960s. While it is impossible to discuss every trend, we have attempted to paint a picture of a Toronto that has undergone profound change, from being predominantly an industrial city to Canada's major global metropolis. This is rooted in wider economic shifts that have shaped all cities, and recent policy responses have worked to promote the city as a creative and globally competitive place, rather than to address the structural causes of poverty, inequality, and polarization. At a neighbourhood level, while the core of Toronto is becoming increasingly affluent and gentrified, there are worrying trends about the ways in which the wider city and region are becoming fragmented and unequal, particularly as housing becomes more expensive overall, but even more so in neighbourhoods with high quality amenities as can be found within the Streetcar City. We will reflect on these trends in our final chapters. We now turn to the visual aspects of our analysis and the questions of what these transformations look like, and how we can visualize Toronto's emergence as a gentrified, global city. But before examining these images of Toronto, we must first introduce the approaches and visual methodologies which have guided our analysis.

4

VISUAL METHODOLOGIES AND REPEAT PHOTOGRAPHY

Photography helped to constitute the modern city by picturing previously unseen spaces and people. It shaped ideas about modern urban life and played a role in the identification and management of urban issues. In this respect, it was central to the project of envisioning the modern city and influenced plans for its development.[1]

The Power and Limitations of Photography

Images of the city are inherently political and the question of what is worthy of being photographed reveals ontological choices that confirm values, social relationships, and identities.[2] Aesthetics and politics are difficult, if not impossible, to separate.[3] Therefore, urban photography has a lot to contribute to scholarly, policy, political, and planning debates. Photography and visual analysis have the potential to open up new ways of analysing, interpreting, and imagining cities. This chapter begins with an overview of visual methodologies and urban photography and then examines the specific visual approach – repeat photography – used as the basis for our analysis.

Urban scholars often use photographs to illustrate a key point, process, or place. Most of the time, however, these images are used to support an argument that is derived from other forms of analysis, such as statistics, interviews, or document analysis. However, the use of photography as the central method of analysis is becoming more established within urban studies.[4] As University of British Columbia geographer Elvin Wyly states, photographs have the potential to "begin conversations" about urban space.[5] Nevertheless, much of the theoretical scholarship on visual imagery is inherently critical of photography. Wyly notes that "in some parts of the critical literature, it almost seems that the main result of theoretical innovation is an outright hostility to vision itself."[6] This critical visual scholarship was necessary, however, in order to debunk the idea that photographs offer us "precise records of material reality."[7] Critical visual scholarship is, therefore, as much about contextualizing photographs within the wider social contexts in which they are produced and viewed as it is about analysing the actual images themselves.

Until the 1970s, it was commonly thought that photographs were an accurate depiction, or "record" of a time, place, or person, representing an indisputable truth. Critical approaches to visual imagery, which emerged out of the pioneering work of Roland Barthes and Susan Sontag, began to question this objective and neutral interpretation of photography and examine how photographs themselves are socially constructed. Barthes stated that photography is violence because its perceived accurate representation of a person, moment, or place denies any sort of counter-memory, or chance to refute the past. Sontag's main critique of photography focused on the inherent prejudices and influences that are part of every photograph. As Kelly Lemmons, Christian Brannstrom, and Danielle Hurd summarized, "Sontag and Barthes address two sides of the same coin: photographs make us feel that they are absolutely, impartially and universally true, when they are in actuality only shadowy reflections of one individual's view of the past."[8]

These theories have been built upon by Gillian Rose, a professor in cultural geography at the University of Oxford and one of the leading contemporary thinkers on visual methodologies. Her critical approach is based on three principles: (1) take images seriously, (2) think about the social conditions and effects of visual images and how they are distributed, and (3) critically reflect on your own way of looking at images. To interpret and situate photographs, she argues, four "sites" are important to understand. The first is the site of production, which includes how a photograph was made and why. The second is the site of the image itself: what it actually shows. This will be embedded in social practices. The third is the site of circulation, which is focused on the many changes an image undergoes between being created and being viewed, including quality, composition, editing, and social, cultural, political, and economic considerations. Online, these also include the algorithms used to determine which images appear and the order in which they are presented. Finally, the fourth site is the audience, or "audiencing." This includes the social practices and social identities of those looking at the images.[9]

For Rose, photographs are not a representation of reality, but rather a site of interpretation.[10] She argues that "visual imagery is never innocent; it is always constructed through various practices, technologies and knowledges."[11] Her own empirical work employs a feminist approach to interpret photography; in her analysis of 1930s images of East London she argues that interpreting visual imagery requires an explicit focus on what constitutes sexual difference. Therefore, for Rose, photography's potential as a tool for critical analysis is to demonstrate power relations, particularly gendered ones, through its inscribing of sexual differences onto the bodily.[12]

Many urban researchers, however, are less interested in theoretical debates about what photographs are and are more concerned about the possibilities of generating new knowledge through visual analysis. Rather than focusing on the limitations of photography, Elvin Wyly asserted that we can move "beyond disillusionment," and, instead concentrate on the possibilities of photography to begin conversations about our understanding of urban space. This does not mean that we should uncritically accept photographs for what they are. He suggests that there are three elements of what pictures do not tell us that are important to understand, acknowledge, and situate when using visual imagery as a tool for analysis.[13]

The first is the condition of possibility, which involves understanding the contexts and conditions in which an image was created. These are the "unseen and therefore virtually unknown contexts by the image maker and the material and social conditions that influenced those decisions."[14] Using a critical constructive approach, Wyly suggests that photographs are a promise, or debt, to the subject and to potential viewers, and that repaying that debt involves contextualizing the image and explaining the conditions which made it possible. This involves not only describing what happened but what did not happen as well. Therefore, the task is to "recognise that the photograph was torn from the context of infinite continua of time and space – and that we need to tell a story that describes at least part of that context with care, honesty and integrity."[15] In this book, we do this by contextualizing Toronto's changing geography and situating the photographs used in our analysis.

The second thing that pictures do not tell us, according to Wyly, is what exists outside the frame. Any photograph is subject to displacement from its wider contextual history. This displacement occurs because a photograph is a "neat slice of time," as Susan Sontag states, but it rarely tells us anything about what is taking place outside of the specific time-space seen in the image. Therefore, we need to displace ourselves when we interpret photographs and ask critical questions about what is visible, not only in the image, but also at other times and in other places.

Finally, the third thing that pictures do not tell us, according to Wyly, is the relationship between power and representation. Stereotypical images can be used to reinforce existing inequalities and power relations, and social constructs can easily be replicated in visual imagery, leading to "significant gaps and historic amnesia" in what is represented.[16] Wyly's critical constructive approach encourages us both to be aware of these power relations and to use whatever power we may have for good. He also argues that if those of us who are committed to social justice simply retreat from the world of visual communication because of the idea that all images depict stereotypes, then visual imagery will likely be more conventional and aligned with a conservative agenda.

Wyly concludes that photography is necessary to save cities; it can be used to challenge the logic of capital accumulation, consumption, and privilege of the contemporary city, and to also draw attention to inequality, uneven development, and injustice. Linking to Peter Marcuse's three pillars of critical urban theory, photography is able to powerfully *expose* a particular urban condition. In other words, photography can be a central tool to "analyse the root of a problem and [to make] clear and [communicate] that analysis to those that need it and can use it."[17] Photography can also play a role in Marcuse's other objectives of critical urban theory: to *politicize* an issue and to *propose* alternatives. This sense of optimism about the power of photography is echoed by many in planning, geography, and sociology.[18]

Urban Photography

Photography as we know it dates from the middle of the nineteenth century. By the end of that century, urban photography was well established.[19] Urban, or street, photography can provide visual "data" that can be analysed in systematic and rigorous ways just like other types of data. Urban photography's key advantage is that it has the potential to provide a "ground-up representation of inner-city change to counterbalance more distant and large-scale accounts."[20] However, as was discussed in the previous section, care must be taken when examining and analysing these images.

Some of the most famous examples of urban photography have been centred on documenting various social aims or conveying messages related to specific aspects of the urban condition. Early urban photographers, such as Jacob Riis and Lewis Hine, were pioneers in the genre of social reform photography, or social photography, which focused on documenting the impoverished conditions of the American ghetto. Riis, for example, aimed to alleviate the poor housing and living conditions of many New Yorkers by exposing and politicizing these places through photography. His book, *How the Other Half Lives*, would contribute to housing reforms, such as the Tenement Law, in the early twentieth century. In Toronto, photographs taken in the early twentieth century by the city's first official photographer, Arthur Goss, helped to both identify and address the problems of the modern city.[21]

While much historical urban photography was taken by official, archival, and documentary photographers, ordinary individuals – pursuing their own interests and objectives – also took many important and insightful photographs of urban contexts and conditions. One of the most famous urban photographers, who became widely known only after her death, was Vivian Maier. Born in 1926, she never married and worked primarily as a nanny in Chicago. Between the 1950s and 1990s, she took over 100,000 black and white negatives, most of which were never printed and remained virtually unknown until just before her death in 2009. In 2007, several Chicago-based collectors acquired some of her images; these were auctioned off after her storage locker was closed due to non-payment of rent. The bulk of her collection came into the hands of John Maloof; he first posted them on the internet and later produced a book of some of her work.[22] He described Maier, who never displayed her images, as someone who lacked confidence in her photographic abilities. However, more recent scholarship has revealed a photographer who was much more skilled and conscientious about how her images were produced and who made conscious decisions never to display them. In her biography of Maier, Pamela Bannos juxtaposes Maier's desire for privacy with her posthumous popularity and acclaim for her work. Bannos' meticulous research paints a profile of a photographer who was much more aware of her skill and much more focused than was initially portrayed by Maloof and others who "discovered" her photographs upon her death.[23]

The photography of Vancouver-based Fred Herzog has also gained international attention in recent years. In the 1960s, while most streetcar-subject photography was shot in colour, most art photography (including street and other urban photography) was in black and white. Herzog's work was a major exception. Born in Germany in 1930, Herzog moved to Canada in 1952, eventually settling in Vancouver. A medical photographer by profession, he photographed Vancouver between the 1950s and 1990s, focusing particularly on the city's downtown and working-class districts. Largely because of his use of Kodachrome colour slide film, he received little notice from the art community until relatively recently. Herzog carefully observed the day-to-day experiences and practices in a Vancouver that is almost unrecognizable today. His images range from landscape views of Vancouver's harbour, streets, and skylines, to intimate photos of building details, storefronts, and pedestrians. The latter were made possible by his use of slide film and the inconspicuous nature of his Leica camera.

Herzog's work stands out in this era, particularly for its depiction of an industrious working-class city. The old Vancouver he photographed began to be demolished, redeveloped, and replaced, starting in the 1970s, by concrete and steel skyscrapers, reform politics, and private capital investments. Because of his use of colour, which gives a far more vivid impression of these spaces than black and white photography, his photography is unique. His focus, like that of Maier, was on ordinary spaces and on ordinary people in everyday situations.

Herzog's background was similar to that of many streetcar photographers. He had a day job that provided the means to support his family; his photography was something done in his spare time as a hobby, rather than as a source of income. As he was virtually unknown in the art world and seemingly uninterested in publishing his work, he was able to work on his own time, to his own exacting demands, without the pressure of deadlines or editors. His position as the associate director of the University of British Columbia's Department of Biomedical Communications provided sufficient income so that he could afford to shoot in colour (approximately two rolls per week).[24] Also, because of the nature of Kodachrome (a slow film requiring longer exposure times, and expensive to buy and develop), Herzog needed to be much more careful and formal in his photography than others using faster or cheaper black and white film. And while the wider photographic world was relatively unaware of Herzog's work, he did give slide shows to various groups and organizations around Vancouver, much like the streetcar photographers who showed their slides to fellow enthusiasts at various forms of meetings.

While the work of Maier, Herzog, and others focuses on daily urban life and ordinary spaces in cities, far more common types of photography in this era (and to a large extent today) were the photographs taken around the home and other familiar areas, as well as photographs of iconic images within cities. The former created memories at an individual level, while the latter worked towards defining and reinforcing the iconic images of a city. Despite their different subject matter, both these types of photography are said to be predictable, conservative, and repetitive.[25] But with the familiar being primarily photographed, one of the challenges for urban researchers is how to analyse the many ordinary and unremarkable spaces that largely went unphotographed.

In cities, most tourists photograph the same iconic locations: the Eiffel Tower, the CN Tower, Big Ben. Or, if they are not iconic in and of themselves, they represent wider interpretations or expectations of a place, such as a typical Parisian café. These images contribute towards the expectations that visitors have when searching for an "authentic" experience, part of what John Urry called "the tourist gaze."[26] The repeated use of these images produces what Ayona Datta called the "iconic city." This repetition of the iconic and the familiar shapes what visitors expect to find in a city, and help define its image.[27] Because of this, they appear frequently in official city marketing campaigns. The use of these images contributes to a particular identity of a place. But, in turn, they also reinforce the dominant narratives, images, and expectations about cities and spaces: every image of the Eiffel Tower serves to reinforce it as the central landmark of Paris. Visitors to Paris subsequently seek out the Eiffel Tower and other well-known landmarks they have seen in photographs in order to both experience the city and to photograph their own version of the image.[28]

These iconic images which are produced and reproduced do not need to be of spectacular architecture or cosy cafés. Detroit, for example, is a city which often gets reduced to simplistic images of ruins and abandonment. There are a handful of famous ruins which are found on most urban photographers' lists of places to visit and photograph in the city. These include the former Michigan Central Railway station and the Packard Automotive Plant. It is not just amateur photographers who seek out these images; these abandoned buildings feature prominently in news stories about the city, thereby actively reinforcing and reproducing the narrative of Detroit as a failed city. In successful cities such as New York or Paris, or struggling ones such as Detroit, a similar type of repetition exists. In Detroit, photographers search for familiar ruins and other imagery that they have seen on the internet or in glossy photo books.[29]

A focus on these singular views of cities, as represented through their iconic spaces, produces one-dimensional images which belie their complexities. Detroit's Michigan Central Station is a good case in point. Detroit historian Dan Austin wrote of the station, "no other building exemplifies just how much the automobile gave to the city of Detroit, and how much it took away."[30] However, by fixating on particular visual imagery, such as ruins, many commentators fail to understand the complex power relations and structural forces which shape them. The work of Joshua Akers, from the University of Michigan, has added needed context to the photos of the station. After being abandoned as a railway station in 1988, the site was acquired by the same individual who owns the nearby Ambassador Bridge, the busiest border crossing between the United States and Canada. He had no plans to revive it as a railway station and let the building deteriorate. However, its location was important to his business interests in transportation and international logistics as the tracks in front of the building are still used by international freight trains.[31] Photos of the Michigan Central Station as a ruin are themselves now history; in May 2018, the building was purchased by the Ford Motor Company, which is renovating the site to become the future home of advanced vehicle research.

The repetitiveness of ruin photography does not go unnoticed by its residents. In a study conducted by Martin Zebracki, Brian Doucet, and Toha de Brant, Detroit residents were asked what they felt about the ways in which Detroit is visually represented and what images they would use to represent their city. Referring to the thousands of pictures of the abandoned train station, one respondent stated, "I probably feel the same way as people in Paris taking pictures of the Eiffel Tower. I would be like 'wow now you took a picture of the Eiffel Tower, there are 900 billion pictures of the Eiffel tower.' Add one more picture to the pile!"[32]

As the recent changes to the Michigan Central Station attest, buildings and cities are constantly changing. Photographs, however, are moments frozen in time and space. Therefore, one of the enduring challenges of photography, particularly for those who use it as a methodological tool to analyse cities, is how to document change. As Susan Sontag states, "Photographs may be more memorable than moving images, because they are a neat slice of time, not a flow. Each still photograph is a privileged moment turned into a slim object that one can keep and look at again."[33] This is both an asset – photography is able to provide memorable images of a particular moment – and also a challenge – an image is of one brief moment. Therefore, in order to make full use of photography as a method to interpret urban change it is necessary to move beyond the use of individual, isolated images. The next section will introduce the technique of repeat photography, the methodological approach that we have used in this book.

Repeat Photography as a Method to Interpret (Urban) Change

As described in the previous section, visual data, just like any form of data, need to be properly contextualized and situated, including a robust discussion about what they show and what they do not show. Many scholars, ourselves included, use photographs as part of our teaching and research to illustrate a particular trend, phenomenon, or experience. However, in order to go beyond merely using photographs as a somewhat "redundant visual representation of something already described in the text,"[34] pictures need to be used as the central analytical tool to interpret and evaluate urban change. Repeat photography – the practice of rephotographing the same spaces at different moments in time – is one such technique that elevates visual imagery to the position of key data that are systematically analysed with the same methodological and theoretical rigour as other data sets (e.g., interviews, surveys, statistics) (see figures 4.1 and 4.2).

Repeat photography involves creating a "temporally ordered photographic record of a particular place, social group or other phenomenon."[35] The practice has its origins in the natural sciences. In Canada, the Mountain Legacy Project has been rephotographing the Geological Survey of Canada and the Dominion Land Survey, taken between the 1880s and 1950s, to measure changes in the Rocky Mountains. Since the project started in the 1990s, they have amassed more than 8,000 pairs of images. Their work shows that, rather than being fixed in time, Canada's Rocky Mountains are continually shifting and changing, with treelines moving significantly further up the mountains over the past century (figure 4.3).[36]

One of the earliest research projects to use repeat photography was the Rephotographic Survey Project, led by the geologist Mark Klett.[37] In the late 1970s, he and his team rephotographed images of the American West taken by Timothy O'Sullivan and William Henry Jackson in conjunction with the US Geologic Survey of the 1870s. These images were further rephotographed in the late 1990s to illustrate three distinct moments in time.[38] Images that show the glacial pace of change at which the natural world evolves are juxtaposed with images that show the often striking alterations made to the landscape by humans. It is not always the case that new developments came with subsequent photos; Mark Klett's work shows nineteenth-century mining

Figure 4.1 | Repeat photograph at Bathurst and Fleet Streets in 2004, 2015, and 2019. Photographer: Brian Doucet.

Figure 4.2 | Like the CLRV streetcars, Honest Ed's at Bathurst and Bloor was a Toronto icon for decades. It is pictured here in 1984. By late 2019, the building had been demolished and the construction had started on a new apartment building complex. In late 2021, this construction was well underway and LRVs had replaced the TTC's iconic CLRV streetcars. Photographers: Robert D. McMann (image courtesy of Robert Lubinski) and Brian Doucet.

Figure 4.3 | Two images from the Rocky Mountain Legacy Project, taken in 1927 and 2009. Courtesy of Library and Archives Canada and Mountain Legacy Project.

boom towns that were abandoned and desolate spaces a century later. Klett has also used this approach to analyse urban scenes, rephotographing images taken after the devastating 1906 earthquake and fire in San Francisco one hundred years later.[39] What stands out in Klett's work is the exacting nature of his photographs; he meticulously photographs from the same locations, with precise repetition in angle, depth of field, focal length, and even time of day and year. However, as repeat photography has moved towards urban or social subjects, this exacting replication is generally less necessary than when trying to measure changes in glaciers, coast lines, or vegetation.[40]

Repeat photography as a tool to examine urban or social change emerged later than in the natural sciences. The sociologist Jon Rieger was one of the pioneering social scientists involved in repeat photography. He outlined the best practices for the technique. They include visual measurements taken at successive points in time; the content within these photographs becomes the basis for analysis. The researcher then compares the photographs from different moments in time, and analyses the changes, or lack thereof, that are visible within them. Theory is used to help interpret and situate these changes (hence our attention to urban theory in the previous chapters). In order to make a meaningful comparison, there must be some elements of continuity; at minimum photographs should be taken at the same location.[41]

Rieger's own work spanned several decades and focused on communities in Michigan's Upper Peninsula. What began as a one-off photographic survey project beginning in the 1970s developed into a multi-decade process of continually revisiting and rephotographing the area in order to document different aspects of social change. His images show the decline of small industry and Main Streets, the abandonment of railroads, and the deleterious effects on communities that have experienced long-term and severe economic decline. He noted that his conclusions drawn from repeat photography analysis are "very much the same impression we got from examination of the conventional kinds of statistical data available on the area and from conversations with local residents. The photographs are powerful 'shorthand' for conveying [this] gradual collapse [of the Upper Peninsula]."[42]

While Rieger focused on rural communities and small towns, a few urban scholars use repeat photography in order to analyse cities. One of these is Geoffrey DeVerteuil, a Canadian now based at Cardiff University, who brilliantly documented the changing landscapes of southwest Montreal, once the epicentre of the industrial revolution in Canada. His first sequence of photographs, taken as a graduate student, dates from the early 1990s and shows the low point of deindustrialization. A decade later, he rephotographed several dozen of these images, and his subsequent analysis reveals a level of detail and complexity regarding the spatially uneven nature of decline and revival that are difficult to capture in statistics. In his conclusion, he urges us to move beyond the "hype of revitalisation" that characterized much of the discussions about southwest Montreal, to consider the "full panoply of landscapes that typify an inner-city area in transition, from gentrified hot spots to islands (or oceans?) of decay and stability."[43]

The most extensive urban repeat photography work comes from the Chilean-born photographer and sociologist Camillo José Vergara. His critical photography centres on documenting the changing nature of the American ghetto, and combines his skills and precision as a photographer with rigorous sociological analysis. Since 1970, he has been focused on documenting a new type of "American ghetto" that is no longer characterized by the overcrowding of the late nineteenth and early twentieth centuries, something that early urban photographers such as Jacob Riis and Lewis Hine sought to expose. Instead, Vergara's photographs depict the abandoned and depopulated spaces of many inner-city neighbourhoods in formerly industrial cities. To document this new American ghetto, Vergara "attempted to make visible the process of disinvestment that was creating a landscape of empty lots, deserted storefronts and closed factories in what had only recently been among the most densely populated and productive urban places in the world."[44]

Vergara systematically and meticulously returns to the same cities and the same spaces, year after year, to photograph how they change. He has travelled, and continues to travel, from his home in New York to places such as Detroit; Gary, Indiana; Camden and Newark, New Jersey; and Los Angeles to photograph their changing inner cities.

Vergara's book *Tracking Time* depicts many different locations throughout the past four decades and reveals much about how urban America has changed over that period. His portrayal of Fern Street in Camden, New Jersey, shows a series of nine photos taken between 1979 and 2014. Collectively they show the street changing from a dense and intact block of small terraced houses to one of almost total abandonment, each photo gradually showing less life and vitality, and more abandonment and ruin. Vergara's photos also show renewal; in the South Bronx, for example, a series of photos on Vyse Avenue since 1980 depicts a four-storey early twentieth-century apartment, first occupied, then abandoned, then demolished. Then the site becomes a vacant lot before being turned into new two-storey, single-family homes.[45] Like many of us who use repeat photography, Vergara challenges viewers to question the one-dimensional nature of cities, in this case, as spaces of failure, by documenting the minute details of ordinary spaces. In many of Vergara's images, he is able to document new spaces that are created when mainstream capital abandons inner-city neighbourhoods. His work in Detroit depicts the decline of neighbourhood banks, the subsequent abandonment of these buildings by their former owners, how these structures "outlive" most other buildings on their block (which end up being demolished), and their subsequent reuse as everything from churches to restaurants and strip clubs.[46]

Vergara combines his photo collection with observations, background research, and interviews. On the subject of photography, he quotes MIT professor Anne Whiston Spirn, who says that it is "a disciplined way of seeing ... a way of thinking about landscape, a means to read a landscape, to discover and display processes and interactions and to map out the structure of ideas."[47]

Jon Rieger, Geoffrey DeVerteuil, and Camillo Vergara are fortunate enough to have had either the foresight to have taken photos that could be used as part of repeat photography or the ability to witness the passage of time so that they could rely on their own images as data for visual analysis. DeVerteuil's images spanned a decade, while Rieger and Vergara have spent most of their working lives systematically photographing and rephotographing the same spaces. This type of repeat photography is, by its very nature, slow scholarship.[48] Not intended for quick scholarly publication, this work, and the analytical possibilities of repeat photography, only improve over time.

But what happens when researchers do not have the option of using their own images? In some ways, this both limits and expands the possibilities of repeat photography. It limits opportunities because we are confined to the photographs that others have taken, which can restrict creativity or geography from the repeat photographer. Relying on other people's photographs as a starting point for a repeat photography analysis also requires a careful consideration of the positionalities and power relations we discussed earlier in this chapter. There could potentially be long periods of time between images; Rieger argues that this makes it more challenging to fully assess changes that have taken place (figure 4.4).[49] Many images also lack any detailed information about either the motivations of

Figure 4.4 | Four images looking south along Bathurst Street at Queen Street West: 1965, 2005, 2015, and 2019. The long time-span between the first and fourth images tells us little about when the dramatic change took place (although careful inspection of the new buildings points to their relatively recent construction). Very little change took place between the 1965 and 2006 views. By 2015, the brick building on the corner has been restored, there are new towers in the background, and a large construction site has appeared. In 2019, the product of this construction is evident. Photographers: unknown and Brian Doucet.

the photographer, or about anything that appeared outside of the camera's view. Relying on official or archival photographs often means having to deal with specific geographies of what was (and was not) photographed, as well as the intentions and motivations of other photographers.

However, relying on the photographs taken by others can expand our possibilities; there are countless visual images that can form the basis of a repeat photography analysis. This requires research and investigation to better understand the context of these primary images: when and why they were taken, their geographies and social conditions, and who the photographers were. While we are the first researchers to use streetcar subject photography taken by enthusiasts as the starting point for urban repeat photography, other scholars have also employed a variety of types of historical photographs that, when updated, are brought into dialogue with contemporary issues and debates.

Some well-known repeat photography projects include William Wyckoff's use of images taken by road engineers in 1920s and 1930s Montana. He paired these with more recent photographs he took himself in order to analyse the changing roadside landscapes.[50] In the early 1980s, Thomas and Geraldine Vale used early post–World War II photographs from George Stewart's famous drive across America to recreate images along US Route 40.[51] Daniel Arreola and Nick Burkhart have relied on historic postcards as their starting point. Their work analysing the Mexican border town of Agua Prieta shows the changing economic landscapes associated with the growth of cross-border tourism.[52] Colleagues of theirs concluded that using historic postcards is "more than a then-and-now exercise"; it "enables inspection of a place in motion, shifting its economic focus ... while clinging to its historical roots."[53] In Canada, Amy Metcalfe has used archival images from the University of British Columbia, as well as photographs from old university magazines, budget documents, annual reports, and archived issues of student and campus newspapers, as the basis of her repeat photography. Her ongoing work uses this visual data set to ask important questions about the changing nature, purpose, and role of the university.[54]

While these studies have spanned several decades, Google Street View offers researchers the potential to explore and analyse more recent changes over the past decade. Street View images are regularly updated and many streets have yearly or biannual images dating back as far as 2007. Jackelyn Hwang and Robert Sampson analysed a series of Street View images in Chicago to look for indications of the spread of gentrification that might not necessarily show up in statistical data. They coded block faces to look for signs of neighbourhood investment that would contribute towards gentrification, including upgrading and upkeeping of properties, visible beautification efforts, lack of order and decay, and changes to public spaces.[55] This approach has been used by other researchers studying cities such as Cincinnati and Hamilton. Google Street View analysis confirms what statistics tell us about broader changes while also revealing fine-grained detail that is not always apparent in official data (such as the census), particularly when processes such as gentrification are in their early phases.[56]

Through these, and other empirical examples of repeat photography, it is clear that this methodological approach offers potential to shed new light on contemporary cities and how they have changed. Because streetcars play such a central role in both transportation and the identity of Toronto, it is only fitting that we use photographs taken of them as the starting point for our repeat photography. While all images are socially produced and context-dependent, as we discuss in the next chapter, streetcar-subject photography offers the possibility to go beyond the parts of the city that were regularly photographed in order to provide a wider range of images of ordinary, everyday spaces that, apart from the work of streetcar enthusiasts, largely went unphotographed.

5

PHOTOGRAPHING STREETCARS; PICTURING TORONTO

Railways and photography developed in tandem in the early nineteenth century. Less than fifteen years separated the first public passenger trains in England and the emergence of photography. As Anne Lyden noted in her book *Railroad Vision*, the railway was a vehicle for social and political change while at the same time photography was the key visual medium enabling people to see themselves in that society.[1] Railway companies used this new technology for a variety of purposes. Some of the earliest images of railways were used as promotional tools by train companies to attract financial backing or in ads for the burgeoning tourist industry. One of the earliest significant North American railway images depicted a train travelling over the Niagara Suspension Bridge, designed by the engineer John A. Roebling (designer of the Brooklyn Bridge). Not only was the bridge a major engineering achievement, it enabled Niagara Falls to become a prominent tourist destination.[2]

Another early genre of railway subject photography was the depiction of newly constructed engines or equipment; companies commissioned photographs to both document and celebrate their engineering achievements and to use as promotional material for potential buyers from around the world.[3] In the United States, and to a lesser extent Canada, railway photography also played a major role in documenting the westward expansion of both the railway lines and European settlement. By the early twentieth century, a more aesthetic approach to railway photography began to emerge, as the public was already familiar with railways and did not need to be educated about them through photography.[4]

Early images of horsecars and streetcars also tended to be official photographs taken by street railway companies, or streetcar manufacturers. Place promotion also played a role in early streetcar photography. In the late nineteenth and early twentieth centuries, postcards of a town with a streetcar, trolley, or interurban visible signalled modernity, connectivity, and progress.[5] Streetcars thus became a familiar feature of early urban images.

However, it was only in the 1930s that photographing railways and streetcars as a hobby began to emerge. This coincided with steep declines in the railway and trolley industries, which were largely privately owned and suffering the combined forces of the Depression and new competition from trucks, buses, and automobiles. Many small towns and cities abandoned their street railways during the 1930s in favour of buses, and the majority of electric interurban lines (or radial railways, as they were known in Canada) also went out of business before World War II. Photographing streetcars was one of several ways of preserving and documenting them.

The railway and trolley preservation movement emerged during the 1930s. The Seashore Trolley Museum, the world's first museum dedicated to preserving and operating historic streetcars, was founded in Kennebunkport, Maine, in 1939. The reason for its creation was to save a local trolley car from the recently abandoned Biddeford and Saco Railroad that would have otherwise been scrapped. Trolley museums spread across North America in the subsequent decades. Seashore's collection remains the largest and now includes more than 250 vehicles from around the world. A recent acquisition was one of Toronto's CLRV streetcars. In Canada, a similar story led to the creation of the Ontario Electric Railway Historical Association (OERHA) and its operating museum, the Halton County Radial Railway (HCRR). They were founded in 1954 by a group of men who were eager to save a former Toronto Railway Company streetcar, number 1326, from being scrapped. The TTC donated this vehicle and former Toronto Civic Railways streetcar number 55 to the new organization, and it grew from there. The museum has many examples of historic Toronto streetcars, as well as vehicles from across Canada, and offers rides on a mile-long track near Milton, Ontario.[6]

Photography was another way in which these disappearing vehicles could be documented and preserved by enthusiasts. For those who had the means, technological advances in cameras in the 1930s made them cheaper and more portable than earlier versions. Ironically, the growing popularity of the automobile also enabled enthusiasts to travel to different cities to ride and photograph streetcars; railway photographers could follow, or "chase" trains across the countryside in their cars, photographing the same train at several locations along its journey.

Enthusiasts' groups started to emerge in the 1930s, with large organizations such as the Central Electric Railfans Association (CERA), based out of Chicago, or the Electric Railroaders' Association (ERA), in New York, holding regular meetings where members could gather to swap photos and share stories. Both of these associations established their own publishing businesses (still active today), which produced photo and history books about trolleys. The Canadian Railway Historical Association (CRHA) is the second-oldest enthusiast organization in North America and was founded in 1932. In 1950, it acquired its first rolling stock and it now operates Exporail, the Canadian Railway Museum, outside Montreal. The Upper Canada Railway Society (UCRS) was another major enthusiasts' group in Ontario during the second half of the twentieth century. It organized charters and at one time owned two private railway cars. Today, the Toronto Transportation Society (TTS), founded in 1973, is an active enthusiast organization that holds regular meetings and produces a monthly newsletter. Additionally, the Toronto Railway Historical Association, founded in 2001, operates the Toronto Railway Museum at Roundhouse Park.[7] There are also many railway museums across the province, in cities and towns such as St. Thomas, Palmerston, Capreol, and Smiths Falls.

As Gillian Rose reminds us, a critical approach is needed in order to analyse visual imagery, one that takes into account the agency of the image, and the social practices and effects of viewing them.[8] We argue that streetcar-subject photography provides a different interpretation of cities than many other urban genres, largely because streetcar photographers were not concerned with documenting particular urban conditions in order to advance a particular cause, as was the case with many urban photographers. However, it too comes with its own power relations, habits, and agency, all of which need to be carefully reflected upon. Streetcar and transit enthusiasts were, and are today, predominantly men; photographers had sufficient income to purchase good camera equipment and film, and to travel to different cities to pursue their hobbies. By the 1960s and 1970s, many of the neighbourhoods where trolleys still ran were poor, and, in the United States in particular, home to increasing numbers of African Americans, who were using transit systems that were either on the verge of bankruptcy or relying on dwindling public funds for their survival. Returning to the ideas of Elvin Wyly, we also need to consider the things that these images do not tell us: many streetcar-subject photographs are deliberately devoid of people, as they were not of much interest to the photographers and detracted focus from the vehicles. However, documenting the politics of transit, economic shifts, and racial divisions within cities was not the intention of these photographers; the urban landscapes that they recorded on film were by-products of their primary gaze. In this sense, they are "accidental archivists," individuals who have unintentionally documented changing urban landscapes.

In exploring the practices and geographies of streetcar photography there is a danger in uncritically accepting these photographs as unconditional truths. We therefore return to Rose's four sites and modalities for interpreting visual materials: the site of the image itself, the site of production, the site of circulation and the site of audiencing, applying them to the genre of streetcar photography. For each of these sites there are three different aspects (which she calls "modalities"): technological, compositional, and social.[9]

Streetcar- and Railway-Subject Photography

By the 1930s, advances in photography allowed for a train, travelling at speed, to be photographed in focus.[10] Prior to this, film required too long exposure times to record moving objects and most early railway-subject photography involved static vehicles posed for the photographer. One of the most famous posed shots was Andrew J Russell's "East and West Shaking Hands at Laying of Last Rail," depicting the driving of the "last spike" of America's first Transcontinental Railroad in 1869. Despite the celebratory nature of the engineers and workmen in the photo (it has been dubbed the "Champagne Photo"), everyone needed to pose for some time in order to be recorded on film.

In the 1930s, technological advances in cameras and film allowed for the "site of the image" that we associate today with railway-subject photography to emerge. Most, but by no means all, railway and streetcar-subject photography follows repetitive, predictable, and normative sites of the image that build on photography that emerged in the 1930s. Jeremy de Souza, in his book *Digital Railway Photography: A Practical Guide*, outlines the "standard" way to capture a picture of a railway vehicle: "a three-quarter shot where the front and the side of the vehicle are in direct sunlight and are the main focus of the image."[11]

The standard "three-quarters," or "wedge" shot has its origins in the 1930s through the pioneering railroad photography of Lucius Beebe. Beebe would later be joined by his business and life partner, Charles Clegg, and the two were at the forefront of a new genre of photography that coincided with, and actively contributed towards, a growing interest in railways as a hobby.

In the three-quarters shot, the background forms part of the overall composition of the image. For our purposes, a three-quarters shot can yield a lot of useful information about cities. The streetcar is the centre of attention, but it does not comprise the entirety of the frame. Higher depths of field are generally used, where more of the background is in focus, rather than deliberately blurred (a bokeh effect) by using a low depth of field. This more normative approach to composition provides more clarity in the spaces around the primary subject. Large crowds of people, particularly if they are walking in front of the streetcar or train, would generally detract from the photo, so many images lack people in them (part of this is also due to the fact that most photographers were amateurs who had regular jobs and largely pursued their hobby on weekends) (figure 5.1).

A second type of traditional railway-subject image is the "roster" shot, which is more of a documentary photograph of the vehicle itself. Roster shots follow strict guidelines: a 90-degree side-view angle, no background distractions, and a stationary vehicle.[12] While still popular with many enthusiasts today, especially those who try to collect an image of every vehicle in the fleet, these roster shots are of little use to us as urban researchers.

Beginning after World War II, a new generation of railroad-subject photographers moved away from the repetitive and normative compositions developed by Lucius Beebe. Jim Shaughnessy was one of the leading photographers in this group. He was an active railroad photographer from the 1940s until well into the twenty-first century. While he photographed during the steam era, Shaughnessy, unlike many of his contemporaries, also embraced diesel engines as well. His lifetime's work spans more than 60,000 images. In a book about Shaughnessy's railroad photography, Kevin P Keefe discusses the movement away from normative and repetitive images that had dominated the field until World War II:

> These young photographers began to make their mark, and the postwar years fairly burst with a startling new level of creativity, insight, and technical prowess ... These artists, exemplified by [Shaughnessy] for the most part stopped taking repetitive 'train pictures' and instead began to concentrate on seeing and portraying the entire railroad world. It was about trains, of course, but it was also about the surrounding industrial and cultural landscape, the people who worked and rode on the trains, the economic and social milieu that gave the railroad a reason to exist.[13]

Streetcar-subject photography has yet to expand into these approaches (figure 5.2), and many streetcar photographers still prefer more traditional and formulaic compositions. However, the artistic and holistic approach pioneered by Shaughnessy and others is now well developed, with many books featuring the work of photographers such as J. Parker Lamb, O. Winston Link, Jack Delano, Ted Benson, Joel Jensen, Wallace W Abbey, and David Kahler, all of whom have moved beyond the three-quarters or roster shot. In Canada,

Figure 5.1 | A typical three-quarters streetcar image. Roncesvalles Avenue at Grenadier Road, 23 July 2016. Photographer: Brian Doucet.

one of the most well-known railroad photographers is Greg McDonnell, who has photographed vanishing railway scenes from across the country over a period of many decades.[14]

Susan Sontag wrote that "in teaching us a new visual code, photographs alter and enlarge our notions of what is worth looking at and what we have a right to observe. They are a grammar and, even more importantly, an ethics of seeing."[15] For streetcar photographers, the question of what is worthy of being photographed, stored, and displayed is very much focused on the vehicles themselves.

For streetcar photographers, recording urban economies, city life, social change, land use, or urban form was not the primary goal of their work. They did not intend to capture Pittsburgh on the verge of the collapse of the steel industry, or Toronto in the days before it became one of the world's most multicultural cities. These images are, first and foremost, the product of (predominantly) men involved in their hobby of photographing trains, trolleys, and streetcars. For many, acquiring images of streetcars is one way of "owning" some part of them. For a streetcar or railway enthusiast, it is virtually impossible to actually possess a vehicle, although one enthusiast, Alex Glista, purchased a CLRV from the TTC and moved it to his family's farm northwest of Toronto (figure 5.3)![16] Most enthusiasts, however, need to rely on acquiring secondary objects such as photographs or slides, models, maps, timetables, or pieces of a vehicle, such as a destination sign.[17] For some, trying to possess an image of every vehicle in a fleet is a goal unto itself. This can be done by either acquiring images taken by others, or by attempting to photograph every vehicle in a fleet. It is interesting to note that many enthusiasts will archive their photo collections based on vehicle number, rather than location or date, emphasizing the importance

Figure 5.2 | A less conventional type of streetcar-subject photography. The vehicle is not the central focus of the image and it is not entirely visible within the composition. Bathurst Street and Fort York Boulevard, 6 December 2019. Photographer: Brian Doucet.

Figure 5.3 | CLRV 4187 in its new home on the family farm of Alex Glista, near Priceville, Ontario. 17 August 2020. Photographer: Brian Doucet.

of collecting images of the vehicles themselves. The *Digital Railway Photography: A Practical Guide* outlines several of the reasons why enthusiasts might want to take pictures of railway vehicles:

> You may want to save a memory [of] your journey. You may want a photo of every particular locomotive in a certain class, or belonging to a particular operator. You may prefer steam, diesel, electric, metro, miniature ... You may wish to catalogue styles of signalling, wagon types, station architecture, or even the staff who run the railway.[18]

For many enthusiasts, images that show deviations from normal operations are highly sought after and prized. This includes photographing a class or type of streetcar running on a different route than it would normally operate on, or using track that was not regularly in service. In Toronto, before the opening of the Bloor-Danforth subway in 1966, most classes of streetcars travelled along the same route every day. While to most people, they all appeared identical, there were 14 separate types of "Red Rocket" PCC cars, constructed between 1938 and 1951 (each variant given a class number, A-1 to A-14). Until the mid-1960s, each of these classes was assigned to a regular route. For example, the one hundred A-7 class PCCs, ordered new by the TTC in 1949 and equipped with couplers and multiple-unit controls (which allowed their operation as two-car trains), were used exclusively on the Bloor-Danforth route until the subway opened in 1966. Second-hand A-13 cars purchased from Birmingham, Alabama, ran on the Long Branch and Dundas routes. A deviation from these normal schedules would attract local streetcar photographers to record cars on different lines; consequently, on today's streetcar slide and negative market (on sites such as eBay), images of these deviations, or diversions onto tracks not normally used in regular service, are far more valuable to collectors than those of ordinary routes.[19]

Because of the normative and predictable nature of these images, we are left with archives of very similar types of photographs (with the photographer standing on a street corner with a three-quarters view of a streetcar, surrounded by background that is largely in focus) across a wide range of urban spaces. This predictability and repetitiveness, combined with varied geographies, are assets when using these images to interpret and understand urban change.

Producing Streetcar Images

Referring back to Gillian Rose's ways of interpreting visual imagery, "the site of production" refers to how a photograph was made, its genre, and, in social terms, for whom, when, and why. Most of the historic images we use in this book were taken in the 1960s, another decade that witnessed advances in photographic technology that made colour street photography easier and more affordable.[20] Good-quality cameras became lighter and cheaper, meaning more people had access to cameras and it was easier to walk along streetcar routes photographing on the go. Kodachrome, while shunned by many artistic photographers of the era, was the slide film preferred by many streetcar- and railway-subject photographers; it became more readily available and more affordable during this time. Kodachrome was introduced in the 1930s; in the early 1960s, Kodak released Kodachrome II and Kodachrome 64, which allowed for sharper images and faster speeds.

The geography of streetcar photography – both between cities and within them – imposes constraints but also offers possibilities for new insights into urban life and urban change. Early enthusiast photography was taken in an era long before large forces of change such as deindustrialization, the decline (and in some cases gentrification) of inner-city neighbourhoods, the rise of suburban shopping malls, the financialization of the economy, and the development of expressways. In cities that retained trolley systems into the late 1960s or 1970s, many of these trends become more visible, but, particularly in the 1940s, 1950s, and early 1960s, streetcar-subject photography reveals many ordinary spaces that had not yet been impacted by these major forces of change.

At an inter-city level, there are many interesting and quixotic geographies of streetcar photography that are related to the abandonment and retention of trolley networks across North America. At one time, virtually every city and major town had a streetcar network. In Canada, there were street railways in every province except Prince Edward Island. Small Ontario cities such as Sarnia, St. Thomas, Peterborough, and Cornwall had electric railways, while larger cities such as Toronto and Montreal had networks that criss-crossed almost the entirety of the pre–World War II city (see chapter 3). Most small systems were abandoned and converted to buses before World War II, and before film and cameras became portable and cheap enough for enthusiasts to photograph them. As a result, there are very few non-official photos of these systems; images of these small streetcar networks that do exist tend to be official archival photos, company photos, or pictures in newspapers or postcards.

Many large cities across North America retained streetcars well into the postwar years; several dozen cities even purchased new streamlined PCC streetcars between 1936 and 1952 (see chapter 6). However, suburban expansion, a growing bus industry, and the predominance of the automobile made trolleys an outdated, inefficient, and inconvenient mode of transport for growing metropolitan regions.[21] They were almost never expanded into the postwar suburbs, which, as we discussed earlier, were designed with the automobile as the dominant transportation technology.

Although the streetcar and railway industries were in decline, the late 1940s, 1950s, and early 1960s was the heyday of railroad and streetcar photography, owing to widespread networks and more accessible photography. Railroads of the era were resplendent in colourful paint schemes. Chicago, for instance, had more than a dozen separate roads running through the city, giving plenty of variety for photography. It also had its famous elevated trains and, until the late 1950s, colourful red trolleys and green PCCs. In the 15 years after the end of World War II, most big cities in North America still had trolley networks, as well as suburban and inter-city trains, offering plenty of opportunity for rail-based photography.

But the end came quickly; the second half of the 1950s saw the abandonment of streetcar operation in Chicago, Detroit, Montreal, Minneapolis, New York, and Kansas City. While colour film was still expensive and slow, there are substantial collections of images of streetcars in these cities, and no shortage of books depicting them.[22] However, the time of the trolley was coming to an end and the 1960s saw the demise of trolley service in Washington, Baltimore, St. Louis, Los Angeles, and other cities. By 1968, only Toronto, Boston, Cleveland (Shaker Heights), Pittsburgh, Philadelphia, Newark, New Orleans, El Paso, and San Francisco had trolleys in regular service, all of which operated only a fraction of the networks that had existed two decades earlier. The El Paso network was abandoned in 1973 (though trolleys returned in 2018, utilizing some of the same PCCs that had been kept in storage in the Texas desert for 45 years). The few American cities that retained skeletal streetcar systems did so mostly because they had private rights-of-way or tunnels that provided quick rush-hour journeys to and from downtown, or difficulties in converting them to bus operation.

This geography of streetcar abandonment created interesting patterns of streetcar photography. In the late 1950s, streetcar enthusiasts flocked to Johnstown, Pennsylvania, because it was the last small-city trolley network in North America. It had a population of around 60,000 at that time. Johnstown's trolley system survived until 1960; in 1948, the Johnstown Traction Company even purchased 17 new streamlined PCC streetcars (the smallest city to do so). This mix of older vehicles from the 1920s and newer PCCs attracted trolley enthusiasts and photographers from across North America. As a result, this small city, better known for its industrial decline and history of catastrophic floods than its role in American transportation history, has a wealth of images depicting ordinary street scenes from the late 1950s (figure 5.4).[23]

Pittsburgh retains a skeletal light rail network that serves neighbourhoods and communities south of the downtown that is the remnant of a much larger system that survived largely intact until the mid-1960s. By this point, very few cities had trolley lines left, and Pittsburgh, with its hundreds of PCCs, varying topography, roundabout routes, and small, winding streets, became a haven for rail enthusiasts. Between 1964 and 1967, however, the bulk of this system was abandoned in favour of buses. But because it was the last great American trolley city, there are thousands of slides and negatives from these final years.[24] At this time Pittsburgh was also about to undergo a long and painful process of deindustrialization; the steel mills that made the city prosperous began closing not long after most trolley lines were abandoned. As a result, not only do we have a wealth of images of the unique geography of the city, but also intimate and detailed accounts of what different parts of Pittsburgh looked like before economic restructuring radically transformed the city.

Within cities, there are also distinct patterns related to the site of production. While there are images taken by enthusiasts along every route in Toronto, we need to be aware of the geographies of streetcar photography when analysing what is visible and what is not visible within these images. Three types of locations feature prominently in streetcar-subject photography, in Toronto and elsewhere.

The first are downtowns (figure 5.5); these are popular places to photograph because of the density of routes and frequency of service. They are also convenient places for both out-of-town visitors staying in downtown hotels and local photographers who work there. They pose their own specific challenges; shadows cast from tall buildings can make it difficult to photograph these streets on sunny days. Traffic congestion adds to the challenge. While our collection

Figure 5.4 | Johnstown, Pennsylvania, the last small city trolley system in America. Before its abandonment in 1960, trolley enthusiasts from North America came to this industrial town to ride and photograph its trolleys. Photographer: unknown, Brian and Michael Doucet collection.

comes from streetcar enthusiasts, it is common to see ordinary tourists taking pictures of these vehicles. Downtowns, especially today, are major tourist zones, and in cities such as Toronto, San Francisco, and Amsterdam, the streetcar, cable car, or tram is a prominent icon.

A second popular location for streetcar enthusiasts is an intersection where different lines come together, or where there is a lot of activity, such as around a carhouse, depot, or storage yard. In Toronto, the intersection of King, Queen, and Roncesvalles is such a hub; it sees frequent service on multiple streetcar lines and is home to the Roncesvalles carhouse. It has therefore long been a popular spot for local, and out-of-town, photographers (figure 5.6). Other common locations in Toronto include the intersection of Gerrard Street East and Broadview Avenue, where the 504, 505, and 506 lines cross, and King Street West and Bathurst Street, which features a "grand union" – an intersection where streetcars can turn in all directions – as well as two busy lines, routes 504 and 511.

The third common location type is the end of a line. A streetcar enthusiast would ride to the end of the line, take a few photographs, perhaps also walking back along the line for a few hundred metres to take a photograph of the next streetcar approaching the loop, before getting back on the streetcar and riding to the other end of the line. High Park loop, at the western end of the 506 Carlton streetcar line, and the Neville Park loop, at the eastern end of the 501 Queen route, are two popular locations (figure 5.7).

Excursions, or charters, by groups of enthusiasts also provided opportunities for photography. In Toronto, the Upper Canada Railway Society ran numerous charters in the second half of the twentieth century. Today, the Toronto Transportation Society (TTS) regularly charters historic streetcars for trips around the city. Visiting American enthusiast societies have also done this for decades. Such charters were particularly common if a certain class of vehicle was about to be taken out of service, or if a route was about to be abandoned. On these rides around the city, the streetcar is posed at specific locations, sometimes on track not normally used for regular service. Whenever possible, the route and destination signs are changed to reflect the location at each photo stop. Because everyone on these charters was interested in streetcars, and most had cameras, many images exist of these events, though,

Figure 5.5 | Downtown Toronto, Queen Street West and York Street, 25 June 1967 and 28 July 2020. Photographers: unknown, likely John F. Bromley, and Brian Doucet.

Figure 5.6 | The intersection of Queen, King, and Roncesvalles in Toronto's west end has been a favourite spot for streetcar enthusiasts for generations. The Roncesvalles carhouse is located immediately west of this busy junction. 1964, 2020; 1964, 2020. Photographers: John F. Bromley © All rights reserved, Brian and Michael Doucet collection, used with permission; Brian Doucet; unknown; Brian Doucet.

Figure 5.7 | Neville Park at the end of the 501 Queen line, 1968 and 2014. Photographers: unknown and Brian Doucet.

Figure 5.8 | An example of "walking the streetcar line" in which the photographer stops to take a picture when a streetcar approaches. Westbound streetcars between Roncesvalles and Dufferin, on King Street West. Clockwise from top left: Wilson Park, Dunn, Dufferin, and Gwynn. 23 July 2016. Photographer: Brian Doucet.

like other streetcar photographs, they are generally devoid of people. Today, many of these images can be found online.

A particularly useful approach for the purposes of our urban analysis comes from photographers who walked along a route from end to end, stopping to photograph each car as it passed by (figure 5.8). This approach produces images at more varied locations than the three types mentioned above. In some cases, the specific place a photograph was taken would be random and coincide with the arrival of a vehicle. In other cases, a photographer would stop and wait at an interesting location, a spot with good sun angles or where there was little chance of being obstructed by cars or pedestrians. In either event, the TTC's former slogan, "always a car in sight," a reference to the high frequency of service on streetcar lines, meant that the photographer never had to wait very long on most routes. The approach of walking the lines helps to give us a wide cross-section of locations throughout the city, not just favoured spots. While the genre of enthusiast photography tends to be rather repetitive and normative in terms of its composition, it should be noted that many streetcar enthusiasts are excellent photographers who use high-quality equipment and carefully select the best light, sun angles, and colour when composing an image.

Figure 5.9 | Two streetcar-subject photography images by Roberta Hill. (*left*) Queen and Roncesvalles, 1967, and (*right*) Pape and Danforth, 1965. Photos courtesy of Rob Hutchinson.

One photographer who regularly walked the entire length of streetcar lines was Roberta Hill, one of the few female streetcar-subject photographers. She lived in several cities in the United States, but visited Toronto annually between 1962 and 1974 to ride and photograph streetcars. While very little is known about her life, she was a well-known figure within the enthusiast community and regularly travelled to cities that still had streetcar networks: Boston, Pittsburgh, Philadelphia, San Francisco, and Toronto. Legend has it that she quit her job in order to ride on a last-day charter in Pittsburgh.[25] Her photographic composition was more varied and creative than that of many other photographers and would include flowers, store signs, or even men in uniform within the frame (figure 5.9). While there was a streetcar in every image they were not always the centre of attention. She also wrote detailed notes on the back of her photographs that had little to do with the vehicles themselves. Her camera was not the best quality and she never took slides, only photographs. Upon her death, her daughter was going to throw away her collection, but sold it to Ken Josephson, who would resell many prints online to enthusiasts.

There tended to be a spike in photographic activity whenever a streetcar route was going to be abandoned. Between the summer of 1965 and February 1966, countless photos were taken along Bloor Street and Danforth Avenue, in anticipation of the opening of the new subway and the abandonment of the streetcars, which took place on 26 February 1966. This event also attracted many American photographers, as most US cities had abandoned their streetcars by this point. The Rogers Road streetcar operated in the suburban municipality of York between 1924 and 1974. However, few enthusiasts photographed the line before 1973. It was far from downtown and had less service than other routes (i.e., not always a car in sight), making photography more difficult and time-consuming. It was only after the TTC announced the route would be converted to trolleybuses that it was heavily photographed by enthusiasts, and most images date from 1973 or 1974.

The technique of walking through the city photographing streetcars gives us unique insights into ordinary urban spaces. It is a credit to the photographers who, in pursuit of their hobby, "accidentally" photographed these urban images in order to fulfil their goal of documenting the city's streetcars.

Exploring cities through their streetcar networks gives new perspectives on them. For us, the pursuit of our hobby has taken us to cities we would not normally visit on holiday (figure 5.10). Even in some of the world's biggest tourist cities, spending a few hours exploring via their tram, trolley, or streetcar networks is likely to take you to places that few tourists visit, and offers much more vivid experiences of the day-to-day life in major cities than just staying in tourist hotspots (figure 5.11). They remind you that, even in global cities such as Paris and Rome, there are places where ordinary people go about their daily lives in ordinary ways.[26]

Viewing Streetcar Images

The third site for interpreting visual materials, according to Gillian Rose, is the "site of circulation," which involves how an image physically or conceptually moves from the "site of production" to the "site of the audience" ("audiencing"), which is her fourth, and final, site. Both these sites will be discussed in this section. Rose argued that there are two important aspects of audiencing: the social practices of the spectating and the social identities of the spectators.[27] The social practice of viewing an image in an art gallery (remaining quiet, walking around the gallery in a certain way) is different than viewing that same image at home on the internet. According to Rose, this is increasingly important because of the enhanced mobility of images today; images can be shared instantly on the internet, and old images (paintings, slides, negatives, and so on) can be digitized and copied far more easily than ever before. Because of this, how an image changes from its initial creation to the image viewed at the site of the audience is also very important, particularly in the digital age.

To illustrate the key differences in both "audiencing" and how an image "travels," let's briefly consider three of the ways in which streetcar images can be viewed: at an enthusiasts' meeting, on the internet, and in this or other photography books. Each one of these represents a different compositionality of the way in which images are viewed and the different technologies used to view them. Streetcar, trolley, and railway societies exist throughout the world and many hold regular meetings where images (slides or digital stills, as well as video) are presented on large screens. In Toronto, the Toronto Transportation Society has an active website and newsletter and regularly holds meetings where members and invited guests give pre-organized presentations. While many associations still host slide shows, most of these are now presented digitally, with scanned versions of slides and negatives, or digital photographs, projected from an LCD projector onto a large screen or wall. These images are likely to have been edited by cropping, or straightening, or manipulating light, colour, shadow, exposure, and so on. When presented in such a venue, the projection of these images is also very large and prominent and takes place within a room of people who are all looking at the same thing.

The spectators in the audience of a transportation club tend to be relatively homogeneous in terms of social identity; most are men and all have a general interest in the subject matter they are looking at. Presenters, who either acquired these images for their collections or took them themselves, dictate which images are selected and the speed at which they are viewed. While the audience is generally quiet, it is not uncommon to hear comments about particular types of vehicles, or expressions of nostalgia with regard to old places, signs, or vehicles, or of excitement towards oddities in route assignments or seeing vehicles they might have ridden or even driven. For many of the older spectators, there is a certain bond that exists as they will remember not only the vehicles presented but many of the locations from this era as well.

Viewing the same images on the internet entails different modalities. Several large transit enthusiast websites, such as New York Subway Resources, date from the early days of the World Wide Web, and offer an opportunity to virtually share photos and information about transit systems with enthusiasts from around the world. For Toronto enthusiasts, the website Transit Toronto is the best, and largest, resource. It was established in 1997 by Aaron Adel, and has been edited since 1999 by James Bow, and is, according to their website, a page "maintained by transit enthusiasts, for transit enthusiasts." It contains pages on vehicle history, routes, and hundreds of current and historical images.[28] While these older sites still exist, social media and photo-sharing websites such as Flickr, Instagram, Twitter, and Facebook provide quick and easy ways to share and view both historic and contemporary photos,

Figure 5.10 | Two streetcar images of cities which are not major tourist draws: Darby, Pennsylvania, just outside Philadelphia (Main and 6th Streets, 2004) and Duisburg, Germany (Kaiser-Wilhelm-Straße and Maternastraße, 2013). Photographers: Michael Doucet and Brian Doucet.

Figure 5.11 | Paris (*top*, Porte de Vincennes, 2013) and Rome (*bottom*, Via dei Castani, Frassini, 2010) as seen through their tram networks. Photographer: Brian Doucet.

though these generally lack the detailed commentaries and histories associated with enthusiasts' sites.

The composition of images on websites is often displayed in a gallery format, with several dozen photos in thumbnail size, which can be expanded to cover the full screen. This often involves decreasing the resolution of the images, which is an aspect of its digital circulation. The viewer selects the images they wish to see and the speed at which they move through them. This method of viewing images is a highly individual affair with no face-to-face interactions, though on social media, interests and likes can be shared. These images can also be taken and viewed anywhere one has connection to the internet, therefore bringing people into contact with other enthusiasts from around the world.

Finally, viewing these images in a photobook (as you will do, or already have done if you have gone straight to looking at the images upon opening the book for the first time) also involves very distinct modalities in both circulation and viewing. As mentioned earlier, the railway photography book emerged in the 1930s, with early publications such as *Train* (1934), by Robert Selph Henry, and *Along the Iron Trail* (1938), by Frederick Richardson and F. Nelson Blount.[29] However, the images in these initial books consisted of either stock company photography (Henry was a railway executive) or static engines posed for the camera, and did not show other elements of railways such as stations, bridges, roundhouses, or people.

Lucius Beebe – the pioneer of the three-quarters shot –helped establish the railway photography book genre through his 1938 publication *High Iron: A Book of Trains*. The book was unique because it featured many of Beebe's own dramatic photographs, rather than relying on a collection of stock images taken by railroad companies, as earlier books did. Reflecting on the influence of *High Iron*, Tony Reevy, a leading expert on railroad photography, stated that it "was an important innovation in publishing. The railroad pictorial album ... is a medium that has given pleasure to millions, has gotten many young people interested in railroads, and has been published in many forms since *High Iron*."[30] Beebe's background was high-society Boston, which tended to disapprove of his lifestyle, including his same-sex relationship with Charles Clegg. The publication of *High Iron* was initially met with surprise by audiences more familiar with Beebe's writings on New York café society; however, his enthusiasm for trains and his skill and creativity as a photographer helped to widen the genre's appeal.[31] Beebe would publish nineteen railroad photography books, many together with Clegg. Clegg would influence the 1945 book, *Highball: A Pageant of Trains*, and they spent several years travelling the country (largely in one of their two privately owned railroad coaches), photographing and researching for their masterpiece *Mixed Train Daily* (1947), which depicted steam operations on rural branch lines across the United States. Beebe's wealth afforded them the means to travel across the country to document the disappearance of local shortline railroads.[32]

Railway publishing has a long and diverse history, and encompasses everything from books solely featuring roster images to those with highly artistic and creative illustrations. *Trains* magazine, based in Milwaukee, helped to establish more artistic, environmental, and humanistic images of railroads in the late 1940s and early 1950s. It became a source of inspiration for Jim Shaughnessy and others who were moving beyond Beebe's three-quarters compositions to depict all aspects of railways including people, buildings, and bridges, as well as the railroad within its wider social context.[33]

The Center for Railroad Photography and Art, based in Madison, Wisconsin, regularly publishes books of beautiful and evocative railroad-subject photography, primarily from the United States.[34] The Center also has a large online archive, comprising donations from private collections that have been made accessible to the public. The University of Indiana Press has an extensive list of publications about railways, including several books depicting the work of important photographers, such as Lucius Beebe and Charles Clegg, Jack Delano (who produced one of the few before-and-after railway-subject photography books depicting trains in Puerto Rico), Wallace W. Abbey, Don L. Hofsommer, and William D. Middleton.[35] Middleton was also an active streetcar photographer; his work regularly appears in trolley books and he authored more than twenty books, including the seminal electric traction histories *The Time of Trolley* (1967) and *The Interurban Era* (1961), both issued by Kalmbach Publishing, another press that continues to focus on railway-related books.

Many early railway photographers sought to depict the final days of steam in the 1940s and 1950s, as well as the end of rural branch lines, as the industry was undergoing a profound period of contraction, mergers, abandonment, and restructuring. The speed at which steam was abandoned in favour of diesel engines in the 1940s and 1950s caught many enthusiasts and photographers off guard and they would scour North America in search of these vanishing engines, swapping knowledge with each other, often assisted by sympathetic railway workers who would give them tips on local steam operations.[36] Well-known railway photographers of this era include H. Reid, who documented the last days of steam engines in Virginia, Otto Perry, who took more than 20,000 images across America, Don L Hofsommer, who recorded the end of branch lines in Iowa and nearby states, and O. Winston Link, whose black and white images of the final years of steam on the Norfolk and Western Railway have become some of the most famous railroad images in the world. Link, in particular, was meticulous in the planning of his photographs, which could take several hours to prepare, especially his famous night images, which involved use of multiple flashbulbs.[37]

Railway photography continues to evolve and expand. Books such as *A Passion for Trains*, featuring the work of Richard Steinheimer, *One Track Mind*, by Ted Benson, and *Steam: An Enduring Legacy* seek to capture the beauty, brutality, bleakness, power, and people of historic and contemporary railways.[38] *Railroad Noir*, a collaboration between photographer Joel Jensen and Linda Grant Niemann, who worked as a brakeman and conductor on the Southern Pacific Railroad, powerfully depicts exhaustion, grit, honesty, and isolation of western railroading towards the end of the twentieth century.[39] Niemann received a PhD in English and worked at several California universities before quitting academia to work on the railroad for twenty years. She published three books about these experiences before returning to an academic position teaching creative nonfiction in Georgia.

While creative and artistic elements of streetcar-subject photography are not as developed as their railroad counterparts, there is no shortage of books about streetcars and trolleys. Most North American cities have at least one book written about their streetcar systems; these books provide detailed histories of streetcars and street railway networks, usually with hundreds of photos. Other books are primarily photo-based, with little background information or history; they serve as a visual tour of a particular city, vehicle-type, or collection of images.

Two recent Toronto-focused books have been written by Kenneth Springirth: *Toronto Streetcars Serve the City* and *Toronto Transit Commission Streetcars*. Like many other enthusiast books, these feature an extensive collection of both roster and three-quarters photographs, arranged by route or vehicle type.[40] While there has not been a definitive history of Toronto's streetcars or transit published for many decades, John Bromley and Jack May's *Fifty Years of Progressive Transit*, published in 1973 by the Electric Railroaders' Association, remains an important work; it combines photos and text that provide detailed descriptions of TTC operations and development between 1921 and 1971. Other important Toronto streetcar books written by enthusiasts include *Mind the Doors, Please! The Story of Toronto and Its Streetcars*, *The Witts: An Affectionate Look at Toronto's Original Red Rockets* (both by Larry Partridge), and *The Toronto Civic Railways: An Illustrated History*, by enthusiast and former TTC streetcar operator J. William Hood.[41]

On the whole, the vast majority of streetcar, trolley, and tramway books are focused on the vehicles or transit systems. Very few authors have attempted to weave broader social histories into their text or photographs. A notable exception to this is the work of Maurits van den Toorn, whose books about the history of public transport companies in Rotterdam and The Hague blend a passion for trams in these cities with a wider understanding of the changing geographies, histories, and politics in two of the Netherlands' largest cities.[42] In a similar vein, Oscar Israelowitz and Brian Merlis' *Subways of New York City in Vintage Photographs* presents a series of images from the early twentieth century that depict everyday life revolving around the theme of rapid transit; transit vehicles are in the background – or even absent – from most of the images in their book.[43]

Other types of enthusiast books about streetcars focus on particular vehicle types, such as the PCC car, developed jointly in the 1930s by North America's major transit companies (see chapter 6), or collections of various photos from across the continent.[44] There are very few streetcar photography books that focus on the work of one individual photographer. Colin Garratt's *The Golden Years of British Trams* depicts the photography of Henry Priestley, who extensively photographed trams throughout the United Kingdom and Ireland in the 1930s, 1940s, and 1950s. He was also instrumental in the founding of the National Tramway Museum in Crich, Derbyshire, where his extensive collection of photographs resides. In North America, Robert Halpern produced *Great Lakes Trolleys*, featuring the work of Eugene Van Dusen, who actively photographed streetcars in many cities, including Toronto, dating from the late 1930s.[45]

Van Dusen is one of many well-known streetcar and trolley enthusiasts who photographed cities across North America during the middle decades of the twentieth century. Other prominent photographers include Joe Testagrose, Joe Jessell, George Krambles, Charles Houser, Gerald Landau, and Richard Vible. Prominent Toronto-based photographers active during this time include John Bromley, Robert McMann, Lewis Swanson, Jack Knowles, Harvey Naylor, Ivor Walsh, and Ted Wickson. Some of their work can be found in various streetcar-subject photography books, as well on enthusiast websites. There are many enthusiasts still photographing Toronto's streetcars today. A small, but by no means exhaustive, list of contemporary photographers includes Rob Lubinski, Rob Hutchinson, Jesse Goulah, James Bow, and Damian Baranowski (who primarily takes video, rather than still, images).

As with other sites of audiencing, books involve distinct power relations and conditions that frame how they are viewed and what is shown. Jack Delano's *From San Juan to Ponce on the Train* is one book that uses repeat photography to depict Puerto Rico's railways in the 1940s and 1980s; however, to the best of our knowledge, our book is the only one which uses streetcar-subject repeat photography in an urban setting as part of a systematic analysis that goes beyond the vehicles themselves. To do this, we have selected photographs that depict particular urban trends, or places, rather than vehicle types or routes.

Railway and streetcar books can be viewed not only by individuals but by small groups of friends exploring the history of their city, or by parents or grandparents sharing their recollections of the city with their children or grandchildren. When Brian was a young child the two authors of this book made a regular practice of looking through streetcar books together, a particular favourite being Bromley and May's *Fifty Years of Progressive Transit*, which, at the time, Brian referred to as "the yellow book." This intergenerational practice continues today; pointing to many of the above-mentioned books at home, Brian's young son, Hugo, could say the words "train book" from a very early age.

Another way to engage with such books is to put them "on display," as with photo and coffee table books found in people's homes. This kind of display says something about us, our values, and how we want other people to see us. As Rose says, the use of particular images "performs a social function as well as an aesthetic one."[46] The same social function applies to other streetcar-related memorabilia on display in Toronto homes: paintings, posters, photographs, replica (or for a few lucky individuals, original) route signs, coffee mugs, and so on. As we discuss later on, the streetcar is a Toronto icon that extends well beyond the realm of the transit enthusiast.

The social identities of different readers of this book will influence how they interpret the images presented within it. Those who are first and foremost interested in Toronto's streetcars will most likely look at the images first. Enthusiasts will probably take note of the different types or classes of streetcars and any oddities in route assignments or among particular vehicles (such as different paint, misplaced route names/numbers/destinations). Those with a nostalgia for Toronto's past may look at the streetcars in the wider context of the changing streetscapes around them. Old storefronts, signs, landmarks, street lamps, people, cars, and other objects may grab their attention. For younger readers who live in the city but were not alive when the historic images were taken, the photo sets may be interpreted by looking at the familiar, contemporary images first and then comparing them with the historic images of what was. Finally, urban scholars and students may look at the changes to the urban form and morphology as visual evidence of wider processes of economic and social change, following in the practices of Elvin Wyly, Jon Rieger, Geoff DeVerteuil, and Camillo Vergara.

These social identities of potential readers are not discrete categories (as authors of this book, for example, we are both academics with lifelong interests in streetcars as a hobby). We hope that all readers will explore the visual images we have presented through a variety of different meanings and interpretations. There is also the question of how you read through the book. Do you go straight to the pictures and read the text later (if at all)? Or do you read it cover to cover, starting with the text and then using that information to help interpret and contextualize the visual imagery? We have structured the book so that the chapters preceding the photos introduce and contextualize the major trends and forces which have shaped Toronto, as well as the theoretical and methodological backgrounds for interpreting urban and repeat photography. The photographs that follow then form the basis of the visual data that we analyse. After examining these photo sets, more text is needed to fully interpret the images, as well as to encourage the reader to draw conclusions, raise new questions, and consider alternatives. However, we do not prescribe a set order for reading through this book and encourage readers to navigate the volume in whatever way that befits their own curiosity and interest.

Photographing Toronto

Toronto enjoys a rich photographic history. The earliest known photographs of Toronto date from 1856, and comprise of a set of 25 views produced by the civil engineering firm of Armstrong, Beere and Hime. They were part of the submission from the city to Queen Victoria regarding the selection of the permanent capital of the United Province of Canada. She chose Ottawa, but the photos remained in London in the Foreign and Commonwealth Office Library.[47] These photos later would serve as an integral part of Michael Redhill's wonderful novel *Consolation*.[48] By the early twentieth century, Toronto as a photographic subject became well established. William James is acknowledged as the city's first photojournalist. A British immigrant who arrived in Toronto in 1906, James took some 12,000 images of the city, including 5,800 lantern slides, between then and the time of his retirement in 1939.[49]

In 1911, the City of Toronto appointed Arthur Goss to the newly created position of chief photographer, and he became an important chronicler of life in the growing municipality.[50] The story of Toronto's photographic history has been brilliantly documented by Sarah Bassnett in her book *Picturing Toronto*. She argued that the use of official photographs in the early twentieth century built on the tradition of social photography, common in other cities at the time. Photographs helped to define the modern city by documenting places that had previously been unseen and also played a part in solving the problems of the modern city.[51] Like many of the other scholars we have discussed in this chapter, she criticized much of the urban history literature that uncritically took historic photos as "self-evident records of real places in the world,"[52] and instead focused on the ways in which a variety of groups of photographers in early twentieth-century Toronto used photography to construct the modern city and influence the built environment. Therefore, photographs taken by Goss need to be seen as part of wider cultural and political histories, rather than simple documentations of a long-gone era. Goss produced thousands of images of the city until he passed away in 1940.

Bassnett's book illustrates the ways in which photographs were used to document a variety of different processes and address urban challenges by different groups of people concerned with social reform. They were used to create records of places, people, and events, but also to influence public opinion by showing civic improvements in Toronto and other cities, with the aim of convincing the city's population to support such initiatives. Photographs of the Ward, for example, which was one of the most impoverished neighbourhoods in the city – where New City Hall now stands – were used to raise concerns about overcrowding and poverty, with the ultimate aim of enacting reform measures.[53] These early photographic collections from Goss and others were safely secured for posterity in Toronto, when City Council created the position of City Archivist in 1960.[54]

Spanning much of the time period we cover in this book, the work of Boris Spremo provides one of the most exhaustive collections of Toronto images. Spremo was a photojournalist who spent much of his career with the *Toronto Star*. His collection has been digitized at the Toronto Public Library Archives and includes thousands of images of buildings, landscapes, and people throughout Toronto between the 1960s and 1990s.[55]

There is no shortage of illustrated books about Toronto that contribute to our understanding of many particular elements of the city. While more than 30 years old, James Lemon's *Toronto since 1918: An Illustrated History* combines academic research with powerful images and captions that vividly portray Toronto's changing geographies throughout the twentieth century.[56]

Many books tend to focus on one topic or neighbourhood. For example, books about the city's architecture have explored demolished structures, historically significant buildings, unbuilt projects, prominent architects, and construction materials and styles.[57] Toronto's neighbourhoods and main streets also have been explored in book form.[58] As we mentioned earlier, enthusiasts have written several books about Toronto's streetcars and transit system. Repeat photography, in the form of before-and-after photo collections, is commonly found in popular books on cities. In Toronto, one of the leading authors employing this technique is Mike Filey. He has produced a series of books, including *Toronto Sketches: The Way We Were* and *Toronto: Then and Now*, which provide a popular historical account of what Toronto used to look and feel like. One of the most comprehensive displays of repeat photography in Toronto is *Full Frontal T.O.*, by Shawn Micallef and Patrick Cummins. Cummins spent thirty years photographing houses, storefronts, garages, and other block faces throughout Toronto; through the use of these photographs the book vividly depicts a variety of changes to the city's vernacular architecture. These books illustrate change over a period of time, and visually document the city, while sometimes waxing nostalgically about the way things used to be.[59]

Methodologies Used in This Book

Building on the strong traditions and principles of both critical urban scholarship and repeat photography, our book uses historic and contemporary images of Toronto to critically analyse the changing urban, economic, and social geographies of the city. Rather than getting bogged down in the minute details of particular places or transit vehicles, or reminiscing about particular buildings, events, or icons, we seek to paint a broad, yet also fine-grained, and comprehensive picture of how economic and social forces manifest themselves in changing urban form, public space, and daily life over a span of many decades.

We have acquired a large collection of original colour slides of Toronto streetcars, largely taken between 1960 and 1980. Some have been acquired at enthusiast gatherings, auctions, and shows, while others have been purchased on eBay, where there is a large market for trolley slides and negatives. Because Toronto retained a large streetcar network into the 1960s and beyond, TTC streetcar slides appear frequently on eBay. Several large collections from (deceased) photographers have been sold on eBay for many years, either individually or in small lots of slides. Our collection of historic, original TTC slides numbers in the hundreds; while our collection is not the largest, we have more than sufficient data to interpret different parts of the city at different moments in time.

We have tried, as much as possible, to determine the photographer of the images in our collection in order to give them credit for their work. In many instances, there is no name recorded on the actual slide or negative, making identification difficult. Other enthusiasts have helped us to attribute some previously unidentified work, however some images remain unidentified, and the challenge of putting names to these images has increased as many photographers active in the 1960s have passed away.

Images that we have been able to identify generally come from a small number of Toronto-based photographers. John F. Bromley (figure 5.12) was active as a streetcar and tram photographer beginning in the early 1960s. He took thousands of slides of Toronto streetcars throughout the 1960s, particularly before the Bloor-Danforth subway opened in 1966 and when all 14 different classes of PCCs were in regular service. Bromley was meticulous in the way he recorded information on each slide about the streetcar fleet number, its route and destination, the location of the streetcar, the date the image was taken, and his camera settings (figure 5.13). He was the author of two books: *TTC '28: The Electric Railway Services of the Toronto Transportation Commission in 1928* and *Fifty Years of Progressive Transit – A History of the Toronto Transit Commission* (co-authored with Jack May). Bromley's visual work extended to Pittsburgh in the mid-1960s, before that city's trolley system was largely abandoned. He also travelled frequently to Europe, making over forty trips throughout his life,

Figure 5.12 | The late John F. Bromley. Photograph courtesy of Margaret Bromley.

Figure 5.14 | Robert D. McMann on the Queen Street bridge over the Don River at some point in the 1970s. Photographer: unknown, courtesy of Robert Lubinski.

Figure 5.13 | A scan of a John F. Bromley slide-mount, showing details of what information was recorded.

most with his wife, Margaret, who he met on one of his trips. His photography covers many cities, including Brussels and The Hague, both of which ran their own versions of PCC trams. Bromley was less active as a photographer between the 1970s and late 1990s, however his digital photography since 2003 covered many tram and streetcar systems around the world. He was particularly fond of night photography and his skills photographing streetcars at night were exceptional.

His final international trip was to Australia, to visit the Gold Coast, Melbourne, Bendigo, and Sydney. Bromley amassed a very large collection of slides and negatives from across North America and was an active buyer and seller of streetcar imagery on eBay. Parts of this collection, including many of his own images, can be seen on Flickr, under the username bear4922. John F. Bromley passed away in late 2019 at the age of 80.

Robert D. McMann (figure 5.14) was another streetcar enthusiast, active between 1961 and 1993, who exclusively photographed Toronto. As with Bromley, most of his work was shot using Kodachrome slide film. The only slides he took outside of Toronto came from a trip to San Francisco in 1980. McMann was active in the Upper Canada Railway Society, editing its newsletter in the early 1970s. Like other photographers, he gave slide shows at enthusiast meetings; one notable presentation featured a slide taken at every stop on the Queen route! He would regularly purchase the newest Nikon camera body and usually had multiple cameras on the go. He worked in a lab at Toronto General Hospital, never married, and had no children. Upon his death in the mid-1990s, his slides were purchased by John Bromley, who would later sell many to other enthusiasts. Today, many of his streetcar slides have been scanned, and some can be seen on the Transit Toronto website.

Ted Wickson (1944–) worked for 31 years at the TTC, many of which were spent as its official archivist. He is also a local historian, documentary photographer,

Figure 5.15 | An exacting replication of this image, taken on 29 May 1954, would be unable to capture to full extent of the new building under construction on Queen Street East near Coxwell. Therefore a slightly different frame was used in 2019, zooming out when compared to the original image. Photographers: unknown and Brian Doucet.

and active public speaker, as well as the author of two books on transportation in Canada. His images have appeared in many books about streetcars and transit, as well as on official TTC postcards and for other purposes by the TTC.

Harvey R. Naylor (1929–2018) spent most of his life in Toronto and extensively photographed the city's streetcars. He also travelled to many cities across North America and his work includes images from several US cities as well. When the Peter Witt streetcars were retired from service in the early 1960s, he made audio recordings of these vehicles that were pressed into a record. He also made audio recordings on an "air electric" model of PCC streetcar. He retired to the Annapolis Valley in Nova Scotia in the 1990s. While many of his slides have been bought and sold by various enthusiasts and collectors, a number of his images of Toronto have been scanned and are part of the City of Toronto Archives, under Fonds number 1526.

As we started acquiring these images and putting this book together, we began visiting the locations of our slides to replicate the pictures taken years ago. We have walked almost the entire length of the streetcar network, stopping at more than two hundred sites to provide updated views of these historic images. Rather than taking a second series of images all taken in the same year, our second views span the years between 2014 and 2021, a period of profound change in the city and its streetcar network. In some instances, we have included third, or even fourth, views to provide more detail and information. But since our original images date from a variety of years, and we are looking for long-term trends, we are less concerned with having precise start and end dates. We try to respect the overall composition of the original photography, and stand at the same location, but we are not fixated on duplicating the conditions of the original photographs with the exactitude of Mark Klett and other natural scientists in their use of repeat photography. As Toronto's skyline has grown upward over the past fifty years, in many cases, it has been necessary to zoom out, or change the perspective in order to fully photograph the extent of change that has taken place in the intervening decades (figure 5.15). As with the original photographs, we also use high f-stops, to ensure that the background, and not just the streetcar, is in focus.

We have selected images that represent a cross-section of different parts of Toronto to help demonstrate the trends and aspects of change that we have outlined in this book. While we have included images from popular locations, we have also selected less photographed places, particularly if their backdrops help us to visually interpret urban change. There are, for example, comparatively few images of the Long Branch line along Lake Shore Boulevard in Etobicoke, which, like Rogers Road, was largely ignored by streetcar photographers in the 1960s, owing to its distance from downtown and the comparatively low frequency of service. But we felt it important to show the former Goodyear tire plant, between Islington and Kipling, as an example of the heavy industry that characterized much of southern Etobicoke.[60]

The visual imagery in this book is structured into three portfolios, each focusing on a different theme: downtown, deindustrialization, and neighbourhoods. Some streets, such as King Street, appear in all three portfolios. Accompanying each image is a brief caption that helps to explain what is in the photo (and sometimes what is outside of the frame). Unlike most streetcar books, we do not feel the need to include the vehicle number, or any additional information about the streetcar, in the captions; those interested in that information can find it within the image itself. We have also decided to provide a full attribution for each image embedded within the caption, rather than a list of photographers as an appendix. This choice was made in order to respect and acknowledge the work of these photographers in a way that makes it easy for the reader to know who took each image. When we are uncertain as to the photographer, we have left this attribution as "unknown." It is also important to acknowledge that both analogue and digital images featured in this book have undergone some degree of editing during their site of circulation. All slides have been digitized using a high-quality Nikon film and slide scanner. We have tried to keep the digital editing of the images to a minimum, performing only basic colour and light adjustments, as well as cropping and straightening where necessary. We have not digitally removed anything from the images. Before turning to the photographs themselves, we will explore their larger context by providing a short history of streetcars in Toronto, with particular attention paid to the social and economic geographies within its network.

6

A SHORT HISTORY OF TORONTO'S STREETCARS

Toronto's streetcars … remain a very distinctive feature of the old city, much celebrated, complained about, photographed, and painted.[1]

In 1973, John F. Bromley and Jack May wrote a book entitled *Fifty Years of Progressive Transit: A History of the Toronto Transit Commission*, which described the first half century of the TTC.[2] Their title was certainly apt for that period of Toronto's transit history; the newly created TTC modernized its streetcar network in the 1920s to make it one of North America's most up-to-date and efficient systems. In the post–World War II years, the TTC further invested in the newest streetcar technology and, as numbers of passengers increased, built and expanded the subway network to keep pace with the city's growth.

The Toronto Transportation Commission was created on 1 September 1921 as a municipally run and operated organization serving the entire city and many of its suburbs. It took over from the privately run Toronto Railway Company (TRC), whose 30-year franchise had expired the day before, and several other smaller companies, including the city-run Toronto Civic Railways and some radial railway companies that served suburban and rural areas. Until the TTC took over, transit in Toronto was an uncoordinated affair. In the 30 years after the TRC was created, the size of Toronto grew dramatically as the city annexed adjacent communities, such as North Toronto, and the neighbourhoods along St. Clair Avenue West, Bloor Street West (beyond High Park), and Danforth Avenue. But the TRC refused to extend its lines to these new parts of Toronto and stuck to the terms of its original 1891 charter, providing streetcar service only to parts of the city included at the time its 30-year franchise was granted. This left large, growing areas without streetcar service. The Toronto Civic Railways and the privately owned radial lines filled some gaps, but passengers still had to transfer from one company to another (and pay a separate fare) to travel downtown. This, combined with a lack of maintenance and investment by the TRC, turned public and political support towards the idea of a municipally run commission to operate all public transit within the city. In a plebiscite that was held on 1 January 1918, Torontonians voted 39,979 to 3,769 in favour of the public takeover and operation of the transit system.[3]

It was in this context that the TTC was born. Public transit in Toronto became a municipal asset, rather than a private one, decades before municipal ownership became common in other North American cities. Many American cities only took over their bus, subway, and streetcar companies in the 1960s, when suburbanization and competition from the automobile meant that they were no longer profitable enterprises. In Toronto, by comparison, the biggest boom in public transit occurred while the system was under public ownership. The TTC's first task was to modernize the track and rolling stock. A new generation of Peter Witt streetcars was purchased (figure 6.1). The Peter Witt was named after a Cleveland Railway commissioner who designed a model of streetcar that would become commonplace throughout North America and beyond. The Witts, as they were often referred to, used two-person crews, with a motorman at the front and a conductor stationed just in front of the middle doors. Passengers entered from the front doors and could either sit in the front half of the car, and pay the conductor as they left, or travel to the rear of the car, paying the conductor as they passed. This helped improve passenger flow and fare collection compared to the arrangement in earlier cars, which involved a conductor roving through the car like a railway conductor collecting fares.[4]

These new steel-bodied cars were far more advanced and comfortable than the wood-constructed streetcars of the old TRC. They were also wider, meaning that virtually the entire streetcar network had to be rebuilt to accommodate the new cars. The space between the two sets of track (known as the devilstrip) was too narrow for the new Peter Witts, so almost all of the network was dug up and the rails were moved further apart. This is not to be confused with the track gauge (the width between one set of rails). It remained at Toronto's unique measurement of 4′ 10 7/8″ (1,495 mm), slightly wider than standard railway gauge (4′ 8½″ or 1,435 mm). This oddity dates back to the horsecar era in the nineteenth century, when city politicians dictated that the street railways should be operated with a different gauge from steam railways, in order to prevent freight trains from running down city streets.

The improvements made by the TTC to track, other infrastructure, and rolling stock in the 1920s were no small accomplishment for what was one of North America's largest streetcar networks. The streetcar system reached its peak in 1928. Not only was the entire city now fully connected and integrated, but on busy routes such as Yonge, modern Peter Witts pulling trailers became the norm as passenger numbers soared. This municipal investment in the 1920s, amounting to some $30 million in 1921 dollars ($435,319,149 in 2019 dollars), would set the stage for further investments in the coming decades, which would make the TTC the envy of most North American cities.[5]

The 1930s was a tough decade for the streetcar industry. Competition from automobiles and buses was siphoning off passengers. The Depression also led to a drop in ridership and revenue. It was in this context that the leaders of most of the major North American transit companies came together to create a new streetcar to lure passengers back. They formed the Electric Railway Presidents' Conference Committee to coordinate investment in, and the development of, a new vehicle. After several years of research, the PCC car, named after the committee, debuted in Brooklyn in 1936. Two manufacturers built the cars for the North

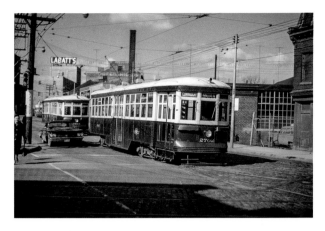

Figure 6.1 | A Peter Witt streetcar purchased by the newly formed TTC in the early 1920s. It is shown here forty years later, towards the end of its service life. Photographer: unknown, Brian and Michael Doucet collection.

American market – the Pullman-Standard Company from Worcester, Massachusetts, and the St. Louis Car Company. Cities tended to favour one over the other. In Toronto's case, the St. Louis firm won out, with some of the assembly carried out by Canadian Car and Foundry in Montreal, which allowed the TTC to avoid payment of import duties in an era well before free trade. In 1938, the TTC purchased the largest order to date for 140 cars at a cost of $3 million ($53,842,105 in 2019 dollars). PCCs 4001 and 4002 went on display at the 1938 Canadian National Exhibition, and the cars entered service on the St. Clair line later that year.[6]

The new PCCs were deemed a success. As war broke out and passenger numbers rose due to rationing of gasoline and increased war-related employment, the TTC purchased another 150 cars spread over four orders. As wartime public transit ridership skyrocketed across North America, several cars intended for Toronto were diverted by the Canadian government to Montreal and Vancouver, two cities also in need of new streetcars. These first orders of PCCs were known as air-electrics, because of their air-operated doors, windshield wipers, and brakes. Visually, they can be distinguished by their larger passenger windows and lack of standee windows (figure 6.2).

After World War II, the PCC streetcar builders introduced an all-electric vehicle as an improvement to the prewar air-electric cars. Their biggest change in appearance was the addition of small standee windows along the sides of the cars, included so that

Figure 6.2 | An example of the air-electric PCCs purchased by the TTC between 1938 and 1945. Photo taken on 20 January 1968 on the Queensway, just west of Roncesvalles. The Roncesvalles carhouse is to the left. Photographer: unknown, probably John F. Bromley, Brian and Michael Doucet collection.

standing passengers could see out of the streetcar, and an increased slope to the front window, a modification designed to reduce glare for the driver. Toronto continued to modernize its streetcar fleet with the purchase of 250 postwar all-electric PCCs, spread over three orders between 1947 and 1951. One hundred of these streetcars had couplers and multiple-unit controls, allowing them to run in two-car trains on the busy crosstown Bloor line. In total, Toronto purchased 540 new PCCs, the third most in North America, behind only Chicago and Pittsburgh.

In the early 1950s, the TTC was still interested in buying new streetcars. However, the cost of a new PCC had increased considerably. At the same time, many American cities that had purchased PCCs were abandoning their streetcar networks in favour of buses. In many cities, ridership was dropping fast, even as new PCCs were introduced. Rising auto use, combined with new suburban house construction far beyond the ends of the trolley lines, was taking away thousands of riders each year. Amid the growing racial tensions in many US cities, many white residents were moving from the city to the suburbs (often referred to as "white flight").[7] In the United States, owners of predominantly private transit companies decided that buses could provide cheaper and more efficient service. Hundreds of relatively new second-hand PCCs flooded the trolley market as cities such as Cincinnati, Cleveland, Chicago, Detroit, and Minneapolis abandoned their streetcar networks in the 1950s.[8]

The TTC was a major buyer of second-hand PCCs, purchasing 52 cars from Cincinnati in 1950 (figure 6.3), half of which were less than three years old, 75 cars from Cleveland in 1952, 48 from Birmingham, Alabama, later that year, and 30 from Kansas City in 1957. Unlike the rest of the TTC's PCC fleet, the cars from Cleveland and Birmingham were built by Pullman-Standard. In total, 205 second-hand PCCs were

Figure 6.3 | A PCC running in Cincinnati, Ohio. It only ran for fifteen years in the Queen City before being sold to Toronto; newer, postwar all-electric PCCs ran for less than three years in their original city. 5th and Broadway, Cincinnati. Photographer: unknown, Brian and Michael Doucet collection.

purchased from American cities, bringing the total PCC fleet to 745 streetcars, the largest number of any North American city. They brought with them unique aspects that reflected their cities of origin. Along with tinted glass in the standee windows to protect riders from the harsh Alabama sun, Birmingham cars had Jim Crow–era signs indicating separate doors for Blacks and whites – which were obviously removed upon arrival in Toronto. And despite being postwar all-electric PCCs, the Kansas City cars had no standee windows at all, reflecting the head of Kansas City Public Services' disdain for, as he put it, "those little apertures."[9] The second-hand cars allowed for the retirement of the very last of the old Toronto Railway Company wood-bodied streetcars in 1951.

After the war, Yonge was the city's busiest streetcar line. Therefore, it was decided to replace it with Canada's first subway line, which opened on 30 March 1954, from Union to Eglinton. Electrically powered trolleybuses took over the northern portion of the streetcar line, from Eglinton to the Glen Echo loop at Yonge Boulevard.

Yonge was the first major streetcar line to be abandoned in Toronto and, in contrast to other North American cities, transit capacity was increased with the arrival of the subway, rather than decreased with a replacement bus. With the abandonment of the Yonge streetcar line, most of the Peter Witts bought by the TTC in the early 1920s were scrapped. The subway brought with it new developments and increased density. North Toronto, which was developed in the 1920s as a typical streetcar suburb, was transformed with major retail, commercial, and apartment developments, particularly around the intersection of Yonge and Eglinton. This meant more transit riders, which fuelled expansion of the system, leading in turn to more development, taller buildings, and more demand for transit. This process is still going on. Today, one can clearly see the spikes of development around subway stations, particularly along Yonge Street.[10]

However, it was not just development around the subway stations that led to high ridership volumes. The TTC was a pioneer in creating a grid of frequent suburban buses that extended into the post-1945 suburbs, many of which fed directly into the subway system.[11] Eglinton was the first station to feature this type of transfer – providing bus connections along Avenue Road, Mount Pleasant, Eglinton, Lawrence, and through Leaside – but it became standard practice for TTC subway growth, particularly as the network expanded into Etobicoke, Scarborough, and North York in the late 1960s and early 1970s; Finch, Islington, and Warden are good examples of stations that support this type of feeder network. In most cases, the buses entered into loops that were already in the fare-paid portion of the station, meaning convenient connections to the subway without the need for a paper transfer.[12] Today, as a result of this frequent grid of buses, many of North America's busiest bus routes can be found running through postwar Toronto neighbourhoods.[13] As the subway network was being expanded in the 1960s and 1970s, transit planners from across the world came to study Toronto's efficient and modern system. The TTC regularly ranked among the best transit operators in North America. Australian transportation planner Paul Mees cited Toronto's model as being a key example of the concept of a "network effect," which he describes as "a network of routes [that is] is provided, allowing passengers to travel between all parts of a city by transferring from one route, or line, to another, just as motorists navigate a road system by turning at intersections."[14]

When it came to public transit, Toronto was clearly Canada's, and possibly North America's, most progressive city in the early postwar decades. In the 1950s, while Toronto opened its first subway line, the last streetcars would run in Ottawa, Montreal, Vancouver, Edmonton, and Calgary. Each city would have to wait decades for rail transit to return, with the Montreal Metro opening in 1966, Edmonton's pioneering LRT (the first new light rail line in North America) commencing operation in 1978, followed shortly after by the C-Train in Calgary in 1981. Vancouver opened its first SkyTrain line, using technology similar to Toronto's Scarborough RT, in 1985, and Ottawa began operation of its O-Train, a very short diesel-operated line, in 2001.[15] Its electrically powered Confederation Line, opened in 2019, the same year as the ION LRT opened in Kitchener-Waterloo (where the last electric streetcars ran in 1946).

In the 1960s, even more American cities were abandoning their streetcar networks and more second-hand cars came onto the market. But by this time there were very few cities left with streetcars and even Toronto was no longer looking for additional cars. Hundreds of PCCs were scrapped as cities such as Baltimore, Washington, St. Louis, and Los Angeles abandoned their networks, and systems in Boston, Philadelphia, and Pittsburgh were reduced to a fraction of their former sizes.

The 1960s was a time of streetcar abandonment in Toronto, but again, this came as a result of subway expansion. The short University subway, from Union to St. George, marked the end of both the Dupont line and the Peter Witt streetcars in 1963. But the big abandonments would come three years later when the east-west Bloor-Danforth subway opened in February 1966. The Bloor, Danforth, Coxwell, Parliament, Fort (a line from Bathurst and St. Clair to the CNE), and Harbord lines were all abandoned on 26 February 1966, the day the subway opened. Streetcar shuttles continued to serve the outer ends of Bloor Street West and Danforth Avenue, beyond the subway terminals at Keele and Woodbine, respectively, for two more years until subway extensions made these routes redundant. For the first time since the arrival of the Peter Witt cars in the early 1920s, the TTC had a surplus of streetcars. Most of the air-electric cars built before 1945 were put up for sale; with no Canadian or American cities in the market for used streetcars, 127 PCCs were purchased by Alexandria, Egypt, in 1968 and 10 by Tampico, Mexico, in 1971 (figure 6.4).

In the postwar decades, the Toronto Transit Commission was a very progressive and innovative agency. Its subway system was expanding to meet the needs of the growing city and region. Between 1951 and 1971,

Figure 6.4 | An ex-TTC, ex-Cincinnati Street Railway PCC running in its third home, Tampico, Mexico. Photographer: unknown, Brian and Michael Doucet collection.

Metro Toronto's population grew by over 75 per cent, reaching over two million, with most of that growth taking place outside the old City of Toronto. In 1968, the Bloor-Danforth subway line was extended beyond the old termini of the streetcar line at Jane in the west and Luttrell (between Main and Victoria Park) in the east, stretching into Etobicoke and Scarborough. Plans were then under way to extend the Yonge line north into North York.

However, at this time, progress did not include the streetcars. In 1966, when the Bloor-Danforth subway opened, an official streetcar abandonment policy was put in place, and the TTC envisioned the gradual elimination of the streetcars by the early 1980s. A subway under Queen Street was planned to replace the downtown streetcar lines. Across North America, streetcars were seen as outdated and obsolete. By the early 1970s, fewer than ten North American cities had streetcars; most that did, including Pittsburgh, Boston, Philadelphia, and San Francisco, only kept them because they either did not have the funds to replace them or because they used tunnels or subways on parts of their lines, thereby providing faster service than buses. Toronto was unique in that it was the only large city to retain a complete street-running urban network, without tunnels or large sections of dedicated, private rights-of-way. But for all the investment in the new cars, the TTC had built very little new track since the 1920s; plans for underground streetcar tunnels downtown never went very far. Apart from the short private right-of-way along the Queensway, the track

Figure 6.5 | In 1975, Philadelphia's transit system suffered a disastrous carhouse fire, where several dozen of their PCCs were destroyed. Desperate for equipment, they purchased streetcars from Toronto, including cars the TTC had purchased from Birmingham, Alabama (pictured), and Kansas City. They kept their TTC paint schemes, although Philadelphia gave them new fleet numbers. This image is from 5th Street and Snyder Avenue in South Philadelphia. The abandoned and decayed nature of this neighbourhood (like others in Philadelphia) stands in contrast to much of Toronto at that time. Photographer: unknown, Brian and Michael Doucet collection.

Figure 6.6 | A new prototype CLRV depicted in 1978. Photographer: Edward J. Wickson, Brian and Michael Doucet collection, used with permission.

remained much as it was when reconstructed by the TTC in the early 1920s.

Among TTC planners and managers, there was a desire to rid the city of these holdovers from a bygone era. However, in the early 1970s, a group of community activists formed Streetcars for Toronto to lobby the TTC and City Council to retain streetcars. They focused their efforts on producing a report highlighting the economic, social, engineering, and planning benefits of retaining the streetcars. The report argued that because of the high volumes of riders on most streetcar routes, and relatively low capacity of buses compared to streetcars, it would prove far costlier in the long run to operate buses on them – more buses (and more bus drivers) would be required. While they received less media attention than the group organized to stop the Spadina Expressway in the 1970s, their lobbying and campaigning worked; in November 1972, the TTC voted unanimously to retain the streetcars. Two peripheral lines – Mount Pleasant and Rogers Road[16] – were abandoned and replaced by trolleybuses. But apart from this, the streetcar system remained largely the same from the late 1960s until the early 1990s.[17]

With the decision to retain streetcars, the TTC began to think about replacing the aging PCCs. However, no new streetcars had been built in North America since 1952. Several dozen of the PCCs the TTC had purchased second-hand from American cities were resold to Philadelphia (figure 6.5), San Francisco, and Shaker Heights, Ohio (a suburb of Cleveland), three systems desperate for streetcars in the late 1970s. For these cities and for Toronto, PCCs were not a long-term solution. As a temporary measure, however, the TTC embarked on a rebuilding program to completely overhaul 173 of the PCCs it had purchased new between 1947 and 1951.

In the mid-1970s, the aircraft manufacturer Boeing-Vertol began constructing new articulated cars for Boston and San Francisco that were intended to be a standard LRV for the remaining trolley networks (and the new systems in cities such as San Diego that were already in development). But the specifications for this car, combined with its poor running qualities, caused the TTC to look elsewhere for new equipment. In the end, the Commission worked with the provincial government's Crown Corporation, the UTDC (Urban Transportation Development Corporation), to develop a new streetcar for Toronto, known as the Canadian Light Rail Vehicle (CLRV). The idea was that this vehicle would also be used on the Scarborough Rapid Transit line; however, the provincial government of the day was eager to test new technology it was developing, so these vehicles never ran there.

The first six CLRV prototypes were built by the SIG company of Switzerland, and arrived in Toronto in 1977 (figure 6.6). The remaining 190 vehicles, built in Thunder Bay, entered service between 1979 and 1981. Disappointed with their Boeing LRVs, Boston tested three CLRVs in the early 1980s, though they would eventually place an order with a Japanese firm for their new cars. While the CLRVs ran in Toronto for forty years, both Boston and San Francisco began looking for replacements for their Boeing LRVs almost immediately after they entered service, with the last vehicles retired in 2007 and 2001 respectively.[18]

The CLRVs quickly formed the backbone of Toronto's streetcar fleet. In the 1980s, they were augmented by the remaining PCCs that had been rebuilt in the early 1970s. In 1987, the first of 52 ALRVs (Articulated Light Rail Vehicles) entered service. They had a similar design, but were longer (23.2 metres versus 15.2 metres), with an articulation joint in the middle, and had six axles rather than four. The ALRVs were the mainstay of service on the 501 Queen and 511 Bathurst lines for almost three decades. The ALRVs allowed for most of the remaining PCCs to be scrapped. It was not uncommon during the 1980s to see dozens of rusting PCCs sitting in the yard of the old St. Clair Carhouse on Wychwood Avenue awaiting transport to the scrapyard, something both of us vividly remember (figure 6.7).

By the late 1980s, there was talk of expanding Toronto's streetcar network for the first time in decades. Across North America, many cities were looking to reintroduce the streetcar, in the form of modern light rail transit (LRT), in an effort to reduce automobile congestion, rein in sprawl, reduce pollution, and spur new development. The San Diego Trolley opened in 1981 and was used as a model by other cities. In the late 1980s, the TTC rebuilt 19 PCCs (with two vehicles, 4500 and 4549 restored to their original 1951 appearance). Classified A-15, they were intended for the new 604 Harbourfront Light Rail Line, running from Union Station to a loop at Spadina and Queens Quay. The line included a tunnel under Bay Street, from Union Station to Queens Quay. The rebuilt PCCs ran on Harbourfront from its opening on 22 June 1990 until October 1994, eventually proving to be too noisy for condo dwellers along the line. After that, they ran in rush hour service on the 506 Carlton, 505 Dundas, and 504 King lines until 7 December 1995,

Figure 6.7 | PCCs awaiting scrapping at Wychwood yard in the 1980s. In the 1980s, many Doucet family journeys around the city included a detour to look at these old streetcars. Photographer: Michael Doucet.

Figure 6.8 | Former Toronto PCCs at the Halton County Radial Railway in Milton, Ontario, September 2002. Photographer: Michael Doucet.

when they were withdrawn from service. These last PCCs met their end owing to a combination of budget cuts, service reductions, and decreased ridership due to the deep recession of the early 1990s. Maintaining a standardized fleet of only CLRVs and ALRVs was also highlighted by the Commission as being more efficient and cost-effective.[19]

Once again Toronto had a surplus of streetcars. Seventeen of the rebuilt cars were sold to various museums and cities across North America where many still operate today. Three of the cars found their way to the Halton County Radial Railway near Guelph, Ontario, where they joined, among others, PCC 4000, Toronto's first streamlined streetcar (figure 6.8). The

TTC retained PCCs 4500 and 4549, which, along with Peter Witt 2766, and the recent additions of CLRVs 4001 and 4089 and ALRV 4207 form the Commission's historic fleet.

While the PCC era came to an end, there were several bright spots for Toronto's streetcars. On 27 July 1997, route 510 Spadina opened, linking the Harbourfront line with Spadina subway station. Three years later, on 21 July 2000, a short extension of track along Queens Quay between Spadina and Bathurst led to the creation of the 509 Harbourfront line, running from Union Station to the Canadian National Exhibition grounds; this line provides a quick and direct link from the new condominiums built along the western waterfront to the downtown core. More recently, the 512 St. Clair line was rebuilt to run on a private right-of-way. Work started around 2005, but the rebuilt line was not fully operational until 30 June 2010. While controversial at the time, the traffic chaos and empty storefronts envisioned by its detractors never materialized. Quite the opposite in fact; the right-of-way has contributed to an enhanced quality of life along St. Clair, with gentrification of both residential and commercial spaces being a more pressing concern than abandonment and decline. Finally, on 18 June 2016, the 514 Cherry line was created. Most of this route ran along King Street, between Sumach and Dufferin, but it included a new short section of track on Sumach and Cherry Streets between King Street and the eastern entrance of the Distillery District. At its western end, the line ran along little-used track on Dufferin Street, heading south towards the western entrance to the Exhibition grounds.[20] The 514 Cherry line was short lived, however, and has been replaced by two separate sections of the 504 King line: one running from Dundas West Station, along Roncesvalles, King, and Sumach/Cherry to the Distillery loop and another running from Broadview station, along Broadview, Queen, King, and Dufferin to the Dufferin loop. This provides more frequent and reliable service along the central part of the King line, although it is no longer possible to ride continuously from Broadview to Dundas West stations.

After 1995, the TTC's streetcar fleet comprised the 196 CLRVs and 52 ALRVs (one car, CLRV 4063, was scrapped in 2009 as it was the prototype for a rebuilding program that was later cancelled). As more people have been living and working along the streetcar lines in recent decades (in areas such as King West), ridership has increased and pressure on the system has

Figure 6.9 | Toronto's new Flexity light rail vehicles were built by Bombardier and were modelled after those running in Brussels. 13 August 2012. Photographer: Brian Doucet.

grown. In 2019, the eleven streetcar lines carried more passengers than the entire GO Transit rail network.[21]

With high ridership and an aging fleet of high-floor CLRVs and ALRVs (which were inaccessible to wheelchairs and other mobility devices and difficult to board for those pushing strollers), the TTC sought a new replacement streetcar. In 2009 an order for 204 cars was placed with Bombardier. The Quebec-based company, whose rail division has since been sold to the French firm Alstom, had become an industry leader in building modern streetcars and light rail vehicles, primarily for the large European market. The TTC's Flexity Outlook cars are similar to those running in Brussels (figure 6.9). The first of these cars entered service on 31 August 2014 on the 510 Spadina line. Despite delivery delays, these vehicles rapidly became part of everyday street scenes in Toronto.[22] As the CLRVs did before them, they have dramatically altered the look and feel of Toronto since they emerged on city streets.[23] As more Flexitys entered service, the last ALRVs ran in September 2019. The final day of CLRV service on Bathurst was on 28 December 2019. The following day, a parade of six CLRVs ran along Queen Street, officially marking the end of 40 years of service from these iconic Toronto streetcars (figures 6.10 and 6.11).[24] While most of these streetcars were sent to the scrapyard, 13 CLRVs and 2 ALRVs were preserved, most by the TTC or streetcar museums. The TTC put several vehicles up for sale and one was purchased by Alex Glista, a recent planning graduate, whose family owns a farm near Priceville, Ontario. He

Figure 6.10 | The final night of CLRV operation in Toronto. Bathurst and Harbord Streets, 28 December 2019. Photographer: Brian Doucet.

Figure 6.11 | A farewell parade of CLRVs after 40 years of service in Toronto. The final run went from the Wolseley Loop at Bathurst and Queen to the Russell Carhouse near Queen Street East and Greenwood Avenue. The LRV is operating on a regular southbound run on the 511 Bathurst route. 29 December 2019. Photographer: Brian Doucet.

bid $3,400 for the vehicle and paid as much again to transport it to its new home.[25]

The new Bombardier streetcars will be a familiar site on Toronto's streets for decades to come and there have been several significant improvements to the system in recent years. The streetcar system is now entirely wheelchair accessible. Stops along the network have been modified to create curb cuts so that passengers using wheelchairs or pushing strollers can easily access the streetcar. Along a few streets, the roads have been redesigned to allow for easier boarding. On Roncesvalles Avenue, platforms have been created in the curb lane, to avoid passengers having to walk across a lane of traffic to board the streetcar. Some streets have seen the addition of bicycle lanes, a trend that has accelerated during the pandemic. A pilot project along King Street between Jarvis and Bathurst began in late 2017. It was designed to assess the impact of placing some restrictions on the flow of vehicular traffic on the performance of streetcars. It was deemed so successful that it was made permanent and streetcar service along the 504 King route – the busiest surface route in the city – has dramatically improved. A new carhouse and maintenance facility on Leslie Street opened in 2015 (figure 6.12). Despite these changes, the general tendency has been to alter streets and sidewalks very little. Historic images reveal how little change is evident throughout much of the streetcar network; track has been renewed, roads have been resurfaced, but the overall layout and design of streets and sidewalks has been remarkably consistent since the 1960s.

Other significant changes have come in the form of proof-of-payment and all-door boarding, which were introduced on all streetcar lines in 2015 and have speeded up wait times at stops, particularly on busy routes such as Spadina. However, this has raised a new set of challenges around fare enforcement, with the TTC adopting aggressive and accusatory advertising to discourage fare evasion, even going so far as to wrap two LRVs in full-length ads warning riders about the consequences of not paying their fare.[26] The TTC has the highest fines of any transit system in North America (up to $425), which, critics point out, far outweigh fines for parking violations.

In 2019, TTC fare inspectors and enforcement officers were accused of racial profiling. A 2019 *Toronto Star* investigation examined the practice of issuing TTC Field Information Cards that collect sensitive information such as a person's name, address, driver's licence number, physical appearance, and race when an inspector or enforcement officer believes a person has committed fare evasion or other offence, without issuing that person a ticket. While only 10.7 per cent of Toronto residents who commute to work by transit are Black, 19.3 per cent of the cards on which race was recorded listed the subject as Black.[27] While this affects all transit – bus, subways, and streetcars – because the streetcar network now operates on a full proof-of-payment system, on board fare inspection is more common than on a bus or subway, though some bus routes have switched to all door boarding and fare inspectors in order to speed up service.

A new and pressing challenge for transit around the world is how to deal with the dramatic reductions in ridership as a result of the COVID-19 pandemic. Despite this, the streetcar system, in the form of modern light rail, is being expanded. The 19-kilometre Crosstown LRT now being constructed along Eglinton will bring fast, frequent, and reliable transit to parts of the city that have never enjoyed such service. One of the intentions behind the construction of the new LRT lines in Toronto's inner suburbs is to enhance the quality of life there. This is likely to happen, though in some neighbourhoods along the line, particularly around Weston Road, there is concern that this enhanced mobility will also bring about gentrification and displacement.[28] The Finch West LRT is also under construction, running from the Finch West subway station to the north campus of Humber College in northwest Etobicoke. Metrolinx is also advancing plans for the Ontario Line, which is proposed to run between Ontario Place and the Ontario Science Centre. The $11 billion, 15-kilometre line, a mix of subways, elevated guideways, and at-grade tracks, is an adaptation of proposals that date back decades to provide relief to the overcrowding on the Yonge subway line. The Ontario Line is not expected to open before 2030. Plans for the East Bayfront, or Waterfront East LRT, are also advancing to connect Queens Quay East and the Port Lands with Union Station.

In some ways, the streetcar network, like the city it runs in, has suffered from its own success. Unlike in previous eras, when the TTC had a surplus of vehicles, ridership increases on King and other routes has

Figure 6.12 | Three generations of Toronto streetcars, seen at the new Leslie Barns, 4 September 2016. Photographer: Michael Doucet.

meant the 204 Bombardier LRVs are now insufficient to provide streetcar service on all routes, even with reduced service because of the pandemic. In 2019 and 2020, track work and construction meant that some routes had to be operated by buses. However, for the foreseeable future, there will always be one or two streetcar routes operated by buses owing to this shortage of vehicles. In October 2020, the TTC board approved a plan to purchase 13 additional Bombardier streetcars (with an option for 47 more, which was exercised in 2021). While these vehicles are necessary for the long term, it is unlikely that ridership will quickly return to pre-pandemic levels. While service has been reduced during the pandemic, the TTC has avoided some of the more dramatic cuts to service that have been seen in many American cities, partly because both the provincial and federal governments provided emergency funding to cover operational costs. With the challenges of the COVID-19 pandemic, and the provincial government's attitude towards transit in general, and streetcars specifically, there remains much uncertainty about new transit investment in Toronto and across Ontario. In December 2019, the provincial government abruptly cancelled Hamilton's LRT; it would take an injection of federal money in 2021 in order to revive the project.

PHOTO PORTFOLIOS

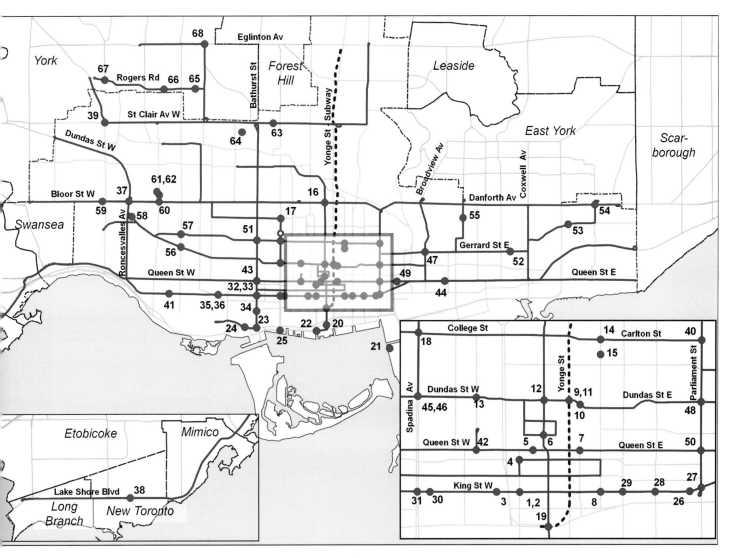

Photo locations | The city of Toronto and the streetcar and subway network in 1960.

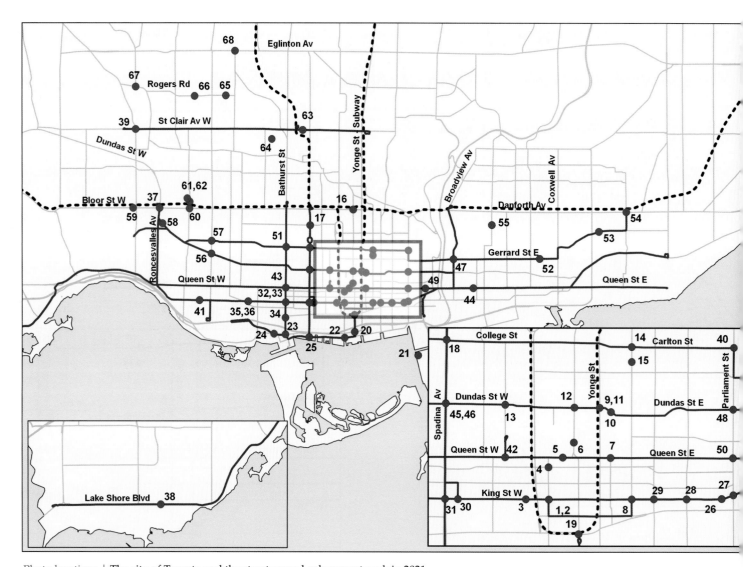

Photo locations | The city of Toronto and the streetcar and subway network in 2021.

PORTFOLIO 1: DOWNTOWN

Downtown is where Toronto's transition towards a major global city is most striking. This is vividly evident in the changing skyline, particularly around the Financial District. The landscapes in and around downtown shown in the photos in this portfolio clearly point to the emergence of Toronto as a major centre of global finance, as well as the tremendous boom in condominium construction.

But going deeper into these photos reveals other trends. Not only are the bank towers growing in height, but the financial district is also spreading outward as well. In the 1960s, going just a block or two away from King and Bay transported you to another type of economic activity entirely: newspaper offices, warehouses, parking lots, mid-range hotels. In the intervening decades, these have all but disappeared as Toronto's transition to a global financial centre has extended the finance, insurance, and real estate clusters over a much larger area.

Beyond the Financial District, surface parking lots, which once dotted the downtown, have become endangered species, often replaced by new high-rise residential and office towers. Unlike many other Great Lakes cities, Toronto's downtown retail sector has thrived and expanded. Megaprojects of the 1960s and 1970s swallowed up entire blocks to create the Eaton Centre, the Atrium on Bay, and new buildings around Yonge and Dundas, meaning that large parts of the downtown are completely unrecognizable today compared with the 1960s. Old buildings that remain have been renovated and restored, often by changing their function from factories or warehouses to shops, residences, and offices. In some cases, only the façades of historic buildings have been preserved and incorporated into the podiums of large office or condo towers.

1 | King Street West and York Street. We start our visual tour downtown in what used to be known as Newspaper Row – both the *Globe and Mail* and *Toronto Star* had their offices on the north side of King, between York and Bay Streets (on the left side of the photo). The Toronto Star building was demolished in 1972 to make way for First Canadian Place. Commerce Court North, completed in 1931, dominates the scene; it was the tallest building in the British Empire until the 1960s. 22 February 1966. Photographer: John F. Bromley © All rights reserved, Brian and Michael Doucet collection, used with permission.

In the more than fifty years since the first photo was taken, we can clearly see the spatial expansion of Toronto's Financial District. Commerce Court North remains, but virtually everything else, including the city's newspaper offices, are gone. The *Globe and Mail* building has been replaced by the Exchange Tower (left), opened in 1981. 10 September 2019. Photographer: Brian Doucet.

2 | King Street West and York Street. Staying at the same intersection, we now turn to the northwest corner, which at the time of this photo was the location of the Lord Simcoe Hotel. Its existence was brief, opening in 1956 and closing in 1979. 14 June 1970. Photographer: Robert D. McMann, Brian and Michael Doucet collection.

As the Financial District grew outward from King and Bay, the Lord Simcoe was replaced by the Sun Life Centre, built in 1983–4, stretching the entire block between York Street and University Avenue. It became the corporate head office of the insurer after they left Montreal due to Quebec's strict language laws. However, more recently, Sun Life has left the Financial District and moved its offices to One York Street in the emerging South Core Financial district. 5 April 2018. Photographer: Brian Doucet.

3 | King Street West and Simcoe Street. It is easy to forget the number of surface parking lots that existed in downtown Toronto in the 1960s, 1970s, and 1980s. The one visible behind the PCC streetcar occupied the northwest corner of King Street West and University Avenue, right on top of the St. Andrew subway station. The Lord Simcoe Hotel is visible in the background. 3 September 1973. Photographer: Edward J Wickson, Brian and Michael Doucet collection, used with permission.

In the downtown core in general, and the Financial District in particular, there are very few surface parking lots left. Real estate is too valuable and demand for office space too great. 28 July 2020. Photographer: Brian Doucet.

4 | York Street and Richmond Street West. Looking south, the streetcar visible in this photo is turning from Adelaide Street to head north on York Street. Toronto's financial district has long been centred around the intersection of King and Bay, three blocks away. However, this 1967 view looks more like the heart of an industrial and warehousing district rather than the complex financial services cluster it would become. 25 June 1967. Photographer: unknown, Brian and Michael Doucet collection.

All traces of the old buildings and the economy that went with them are gone, replaced by the continued outward and upward spread of the Financial District; only the streetcar track remains although the southbound track has been removed on this one-way street. 10 September 2019. Photographer: Brian Doucet.

5 | Queen Street West, across from Nathan Phillips Square. 14 November, 1964. Photographer: John F Bromley © All rights reserved, Brian and Michael Doucet collection, used with permission.

Queen Street West, between Bay and York Streets, is a prime example of the large urban redevelopment megaprojects that dominated downtown planning in the 1960s and 1970s. The Sheraton Centre, opened in 1972, consumes the entire block. Its development led to the demolition of all the buildings on the south side of Queen, visible in the previous view. Behind the photographer is New City Hall and Nathan Phillips Square, another example of an urban renewal project that swept everything away to create a blank slate for its construction. Normally, this is a very busy location, but as a result of the COVID-19 pandemic, there were few of the hot dog carts, tour buses, or crowds of people that can usually be found along this part of Queen Street West. 3 April 2020. Photographer: Brian Doucet.

6 | Albert Street and Bay Street. Until the Bloor-Danforth subway opened in 1966, Dundas streetcars only operated west of downtown, terminating at an on-street loop situated just behind Old City Hall; Harbord streetcars ran along Dundas Street East. After the subway opened, Dundas streetcars ran all the way to Broadview station, as they do today; however, some cars were short turned at the City Hall loop until it was closed for the redevelopment of the Eaton Centre in the mid-1970s. On the right-hand side of the photo is the rear of Old City Hall; on the left, Eaton's Annex occupies the majority of the block. In this view, Albert Street continues all the way over to Yonge Street. 14 May 1967. Photographer: unknown, Brian and Michael Doucet collection.

Only a small portion of Albert Street remains. To the west of Bay (behind the photographer), the street was removed to create New City Hall and Nathan Phillips Square; the Eaton Centre now occupies the space between James and Yonge Streets. The original plans for the redevelopment of this area would have seen Old City Hall demolished (with its tower preserved). However, growing opposition to megaprojects, along with an interest in historic preservation, led to changes in planning that have resulted in many buildings (or at least their façades) being preserved as new development takes place around them. There has also been a much more concerted effort since the 1970s to maintain the existing urban fabric, rather than continue to create superblocks, such as the Eaton Centre, that dominated downtown development for decades. 14 August 2020. Photographer: Michael Doucet.

7 | Queen Street East and Victoria Street. This is a view from a multi-storey parking garage that still exists today. The red band at the top of the photo contains signage for a Woolworth store in the Jamieson Building (1895) at Yonge Street. 5 May 1965. Photographer: John F. Bromley © All rights reserved, Brian and Michael Doucet collection, used with permission.

The Jamieson Building at Queen and Yonge has been preserved while the rest of the block was demolished to make way for the Eaton Centre. Only the façade of the Bank of Montreal branch (1909–10) on the northeast corner of Queen and Yonge remains. The rest of the block, over to Victoria, is a modern office building. The parking lot on the northwest corner of Queen and Victoria is long gone. 30 July 2020. Photographer: Brian Doucet.

8 | King Street East and Church Street. This series of four views of the northwest corner of this intersection depicts a familiar trend in downtown Toronto: older, lower-rise buildings demolished and replaced by parking, only to be later replaced by condo towers. 6 September 1964. Photographer: Robert D. McMann, Brian and Michael Doucet collection.

16 August 1969, Photographer: Robert D. McMann, Brian and Michael Doucet collection.

The condo on the corner, King Plaza, was completed in 1991. 7 August 2014. Photographer: Brian Doucet.

During the COVID-19 pandemic, King and Church was an unusually quiet intersection, with non-essential businesses shut and downtown office workers working from home. 3 April 2020. Photographer: Brian Doucet.

9 | Yonge Street and Dundas Street East. The Brown Derby Tavern opened in 1949. It was a mainstay at Yonge and Dundas for decades, closing the same year this photo was taken. 20 April 1974. Photographer: Robert D. McMann, Robert Lubinski collection.

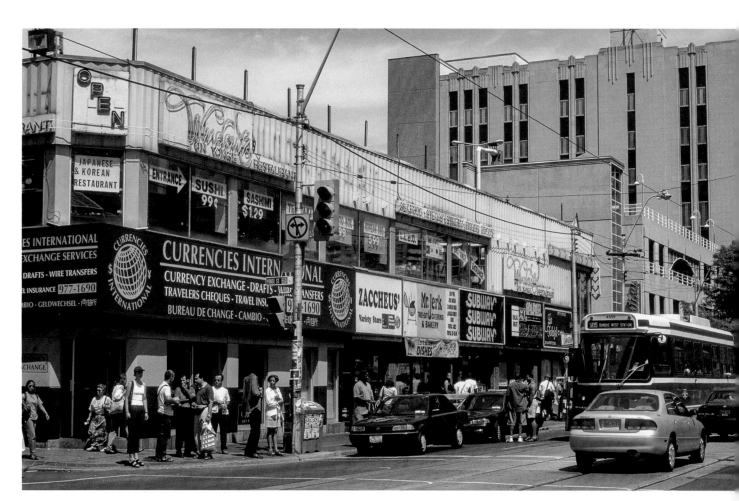

Yonge and Dundas by the 1990s was both edgy and seedy. Behind the scenes, the city of Toronto was busy purchasing properties on the east side of the street in order to redevelop the intersection as the focal point for both tourism and consumption. The low-rise building was a transitional structure that was demolished in 1999. 15 August 1998. Photographer: Michael Doucet.

This new complex at 10 Dundas East was constructed as part of the city's development strategy. Gone were the arcades, seedy bars, and strip clubs of Yonge Street downtown; in their place came chain retail, corporate restaurants, and a multiplex cinema. Some of its theatres also serve as lecture halls for Ryerson University. 31 July 2014. Photographer: Michael Doucet.

A quiet scene at a normally busy intersection during the first summer of the global pandemic. Typical of Toronto during the pandemic, many pedestrians in this photo are wearing masks. 14 August 2020. Photographer: Michael Doucet.

10 | Dundas Street East and Victoria Street Lane. The Eaton Centre had been completed by 1977, running all the way from Dundas to Queen, but on the east side of Yonge Street (behind the streetcar), the mix of low-rise buildings remains. May 1977. Photographer: Harvey Naylor, Brian and Michael Doucet collection.

Yonge-Dundas Square was the new focal point of this reimagined downtown experience. To make way for this new public space, all buildings in the block bounded by Victoria, Dundas, Yonge, and Dundas Square were demolished. Large advertising billboards, reminiscent of Times Square, are visible in the background. While Yonge-Dundas Square gives all the outward appearances of a genuinely public space, it is maintained through a public-private partnership which limits activities that can occur there. Loitering, busking, and panhandling are prohibited, for example. Although intended to make an area safe and enjoyable for consumption, these spaces come at the cost of limiting people's rights within them. 30 July 2020, Photographer: Brian Doucet.

11 | Dundas Street West and Yonge Street. Another example of where virtually an entire block of buildings was demolished in order to make way for a large redevelopment can be found on the north side of Dundas, between Yonge and Bay Streets. The Ford Hotel was built in 1928. The clothing boutique in the former TTC streetcar in the parking lot was called Desire, presumably a nod to Tennessee Williams' play. August 1973. Photographer: John F. Bromley © All rights reserved, Brian and Michael Doucet collection, used with permission.

The Atrium on Bay opened in 1981. Like many other projects of its time, it contains retail, underground parking, and offices, as well as direct connections to the PATH (an underground pedestrian walkway), the subway, and the former inter-city bus station, since moved adjacent to Union Station. 30 July 2020. Photographer: Brian Doucet.

12 | Dundas Street West and Bay Street. October 1963. Photographer: unknown, Brian and Michael Doucet collection.

While the planning ideas that produced the Eaton Centre, the Atrium on Bay, and other large megaprojects have fallen out of favour, a few recent examples of this type of redevelopment can still be found today. Across the street from the Atrium on Bay, no traces of the buildings in the previous photo can be found. The eastern end of the block became part of the Eaton Centre complex; at the western end, the mixed-use building on the corner is a more recent addition to downtown. It houses big-box type stores and the Ted Rogers School of Management, part of Ryerson University. 17 May 2016. Photographer: Michael Doucet.

13 | McCaul Street and Dundas Street West. The only structure clearly visible in both photos is the Main Building of OCAD University, formerly the Ontario College of Art and Design, which opened in 1921. It was the first building in Canada to be built for art education. The lineup of streetcars on McCaul must have been due to a diversion caused by an accident, road closure, or track work. 18 November 1967. Photographer: unknown, Brian and Michael Doucet collection.

The corner of Dundas and McCaul that once housed a branch of the Toronto Dominion Bank is now home to the Art Gallery of Ontario, the third-largest art museum in Canada (behind the Montreal Museum of Fine Arts and the National Gallery of Canada in Ottawa). The museum has been expanded and renovated several times since it opened in 1913. In 2008, Toronto-born architect Frank Gehry's new redevelopment, Transformation AGO, opened; it was his first Canadian project. OCAD has also seen the arrival of a signature building, the award-winning Sharp Centre for Design, which opened in 2004. It was designed by British architect Will Alsop in partnership with Robbie/Young + Wright Architects, based in Toronto. In 2013, *Toronto Star* architecture critic Christopher Hume named it one of the five most influential buildings in Toronto. 10 April 2015. Photographer: Brian Doucet.

14 | Church Street and McGill Street. Maple Leaf Gardens is visible in the background, at the nearby intersection of Carlton and Church Streets. Three months before this photo was taken, the Toronto Maple Leafs won their last Stanley Cup, defeating the Montreal Canadiens 4-2. 13 August, 1967. Photographer: unknown, Brian and Michael Doucet collection.

The round apartment building that looms over the Gardens in both pictures is at 50 Alexander Street (28 storeys, built in 1965), a distinct departure from the slab apartment buildings of that period. As of 2020, the surface parking lot on the corner of Church and McGill Streets is one of the few that remain in this part of the city. However, along much of the rest of this portion of Church Street, new condo and apartment towers have been erected since 2015. 6 May 2014. Photographer: Michael Doucet.

15 | Carlton Street and Church Street. The Toronto Maple Leafs finished in third place (out of six teams) in 1966, the year this photo was taken; Montreal won the Stanley Cup, defeating the Detroit Red Wings. 30 June 1966. Photographer: Robert D. McMann, Brian and Michael Doucet collection.

Maple Leaf Gardens closed as a hockey arena in 1999 and the Toronto Maple Leafs moved to their new home near Union Station (originally called the Air Canada Centre, now Scotiabank Arena). In 2004, Maple Leaf Gardens was purchased by Loblaws for $12 million. In one of the more creative uses for an old building in Toronto, the company partnered with Ryerson University to convert the former hockey arena into a grocery store and athletic centre, which opened in 2009. A new hockey arena, featuring 2,800 seats, was constructed above the retail space. On the ground floor, the original location for centre ice is now visible in one of the aisles in the supermarket and some of the original seats decorate part of a wall. 12 December 2014. Photographer: Brian Doucet.

16 | Bloor Street West and Yonge Street. One of the earliest photos in our collection, this view shows Bloor and Yonge in 1953, one year before the Yonge subway opened. Passengers getting off these Bloor-Danforth streetcars still needed to transfer to a Yonge streetcar for the rest of the trip downtown. 1953. Photographer: unknown, Brian and Michael Doucet collection.

Today the intersection of Yonge and Bloor is one of the busiest spots in the city, with new condo towers on the south side, although passengers changing between subway lines do so underground. Very little transit infrastructure is visible in these photos – even the station entrances are found within the basements of the large office and retail structures that dominate the intersection. 13 November 2020. Photographer: Michael Doucet

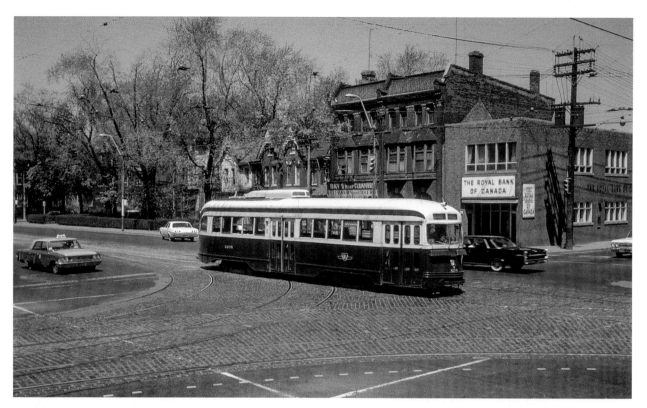

17 | Spadina Avenue and Harbord Street. The streetcar tracks to the left were part of the Harbord streetcar line, which was abandoned when the Bloor-Danforth subway opened in 1966. In this view, we are looking towards the northeast corner of the intersection. This streetcar is probably entering service, or is on diversion, as no streetcars had operated in regular service north of Harbord since the Spadina line was abandoned in 1948. 15 May 1965. Photographer: John F. Bromley © All rights reserved, Brian and Michael Doucet collection, used with permission.

After the Harbord line was abandoned, streetcar tracks on Spadina were removed between Bloor and College Streets. In the late 1990s, they would be put back for the new 510 Spadina route. In August 2014, the first new Bombardier Flexity Outlook LRVs debuted on the Spadina route. This photograph clearly shows the growth of the University of Toronto; Robarts Library is visible in the background. It opened in 1973 and is one of the best examples of Brutalist architecture in the city. The prominent building in the foreground is a residence for graduate students. 6 June 2015. Photographer: Michael Doucet.

18 | College Street and Spadina Avenue. While today Canada is dominated by the "Big Five" banks, in the 1960s, there were many more banking institutions, with mergers and acquisitions reducing this number in the intervening decades. Consequently, many major intersections in Toronto had one or more bank branches prominently situated on the corner. 8 August 1964. Photographer: John F Bromley © All rights reserved, Brian and Michael Doucet collection, used with permission.

The Canadian Imperial Bank of Commerce still stands on the northwest corner of College and Spadina; however, the construction crane behind it suggests increased density is coming nearby. While many downtown scenes are beyond recognition today when compared to how they appeared in the 1960s, it is remarkable how much of the original urban form of two- or three-storey commercial buildings still remains within the downtown core. This photo was taken during the final summer of CLRV operation in Toronto; these streetcars ran on the College line until late November 2019. 12 July 2019. Photographer: Brian Doucet.

19 | Union Station and Bay Street. Tearing down or rebuilding the Gardiner Expressway is a frequent subject of political and planning discussions today, but back in 1964 it was a new urban highway under construction. The bents, visible in the background, have been erected, but the highway they support is still under construction. It was completed later that year. This view was taken from the railway viaduct over Bay Street, looking south towards Lake Ontario. 7 June 1964. Photographer: John F. Bromley © All rights reserved, Brian and Michael Doucet collection, used with permission.

An exact replication of the previous photograph is no longer possible. In 1964, John F. Bromley was able to stand by the railings across the tracks to get an unobstructed view south along Bay Street. This view was taken from one of the passenger platforms of Union Station. To fully show the scale of change, it was also necessary to zoom outward and upward from his original image in order to make visible the incredible growth that has taken place south of Union Station, an area now known as South Core. This image lies at the heart of a downtown whose population is expected to grow to grow by 80 per cent to over 130,000 between 2013 and 2030. 28 November 2018. Photographer: Brian Doucet.

20 | Bay Street and Harbour Street. Looking north on Bay Street towards Union Station and the downtown, this view shows very few tall buildings. In 1965, the skyline was dominated by the Royal York Hotel and what is now called Commerce Court North. 1 July 1965. Photographer: unknown, Brian and Michael Doucet collection.

The Royal York Hotel, once so prominent, has been dwarfed by the dozens of buildings that have been constructed in the intervening decades. The original Commerce Court North is not visible in this view; it has been eclipsed by many other, more recent, skyscrapers. 4 April 2018. Photographer: Brian Doucet.

21 | View from Polson Pier. As Toronto's streetcar network declined after the opening of the Bloor-Danforth subway in 1966, surplus PCC streetcars, including 4591, were sold to Alexandria, Egypt. In this photo, it is waiting on the quayside for loading onto a ship bound for Egypt. This PCC was itself a second-hand streetcar, purchased by the TTC from the Cincinnati Street Railway in the early 1950s. Image date 1966. Photographer: John F. Bromley © All rights reserved, Brian and Michael Doucet collection, used with permission.

As late as 2015, ships came into the harbour at Polson Pier, as is evidenced by this photo, with the boat docked at the same location as in the 1966 view. 9 May 2015. Photographer: Michael Doucet.

Even in the short period of time between 2015 and 2020, the changes to the city's skyline have been striking. This view zooms out from the first image to show the complete downtown skyline. Many of the new additions have been condominium towers. Somewhat surprisingly, the Royal York Hotel is visible in all three views. 30 July 2020. Photographer: Brian Doucet.

PORTFOLIO 2: (DE)INDUSTRIALIZATION

In the 1960s, Toronto, particularly its urban core, was dominated by industry. The streetcar network ran through many industrial sections of the city. Factories, warehouses, and railway yards were not only clustered along the waterfront and the railway lines that radiated from the downtown core; they were also scattered throughout all but the wealthiest neighbourhoods. While the city lacked the kind of mega factories that could be seen in Hamilton, Detroit, or Buffalo, the landscape was dominated by a mix of industries producing all sorts of products. Unlike Detroit, with its focus on automobiles, Toronto's industry was extremely diverse, as can be seen in this portfolio. One of the largest factories was the Massey Ferguson complex on King Street West, which produced farm equipment. In the northwest corner of the city, the stockyards district at St. Clair and Keele was home to a large cluster of slaughterhouses and meat-packing plants.

What is remarkable when looking at these locations today is the extent to which these manufacturing activities (and to a lesser extent their industrial buildings) have disappeared from the city's urban core. Today none of the manufacturing activity shown in the earlier views remains. As discussed earlier, despite the deindustrialization we see in the urban core, the Toronto region remains Canada's largest and most important industrial centre, but most of the industrial activity is now concentrated in the suburbs or further afield.

What has replaced this industry within the Streetcar City? While specific activities vary, they are all centred on aspects of the post-industrial economy. Many of Toronto's formerly industrial spaces have been transformed into new residential developments, primarily in the form of high-rise condominiums. When people speak of the condominiumization of Toronto, it is in these old industrial spaces, rather than in adjacent residential neighbourhoods, where this trend is most visible. New condos are more likely to replace factories than single-family homes and entire neighbourhoods of 20-, 30-, and 40-storey towers have emerged on spaces that were previously factories, warehouses, and railyards. In previous chapters, we connected this to specific policies at all levels of government. But, as we will discuss later on in the book, not only are these condos home to a growing number of professionals working downtown, but they are also major sites of capital accumulation, speculation, and investment, something that is not always immediately or inherently visible in photographs.

22 | Queens Quay West at York Street. The Terminal Warehouse Building (1927) is still functioning as a warehouse and cold storage facility in this photo; most of the waterfront is industrial space. July 1961. Photographer: unknown, Brian and Michael Doucet collection.

The Terminal Warehouse was renovated in 1983 and was the city's first major example of an adaptive reuse of a formerly industrial building. It now contains shops, offices, a dance theatre, and condominium units, the latter built to extend above the roof of the original structure. 20 July 2015. Photographer: Michael Doucet.

23 | Fleet Street and Bathurst Street. 22 October 1966. Photographer: John F. Bromley © All rights reserved, Brian and Michael Doucet collection, used with permission.

The industry has gone along Toronto's downtown waterfront and has been largely replaced by condominium towers. 11 April 2015. Photographer: Brian Doucet.

24 | Fleet Street near Fort York Boulevard. Molson had a large brewery along Fleet Street, which was characteristic of the industrial nature of the land between the railway tracks and the waterfront. 13 July 1969. Photographer: unknown, likely John F. Bromley, Brian and Michael Doucet collection.

A slightly different angle was needed for this view, which gives perspective to the previous image by emphasizing the restored Queen's Wharf lighthouse (1861). The brewery was demolished in the mid-2000s and was replaced by condos in the years that followed. The 509 Harbourfront and 511 Bathurst streetcars use this section of Fleet Street, and stops adjacent to these new buildings are busy. 24 April 2015. Photographer: Michael Doucet.

25 | Queens Quay West and Spadina Avenue. One of the newest images to comprise the first view in a photo set, we are at the very southern tip of Spadina Avenue. By the time the Harbourfront LRT line opened in 1990, most of the industry along the waterfront had disappeared. But around this intersection, as well as to the west and north, there was still plenty of space for development, a landscape that Ken Greenberg called a "terrain of availability." 24 September 1993. Photographer: Edward J. Wickson, Brian and Michael Doucet collection, used with permission.

In a more recent view, it is clear that that "terrain of availability" has largely been developed. Behind the Gardiner, the tall towers are part of CityPlace, built by the Vancouver-based Concord Adex corporation on former railway yards. Originally situated between Spadina and the Rogers Centre (hidden behind the condominium on the corner), it now stretches all the way to Bathurst Street. 17 October 2018. Photographer: Brian Doucet.

26 | The Two Kings (I): King and Parliament.
King Street East at Berkeley. The relaxing of planning restrictions at the Two Kings in the 1990s (see chapter 3) led to extensive adaptive reuse of old industrial buildings as well as the construction of new, primarily mixed-use, developments, and has been the template for other underutilized spaces within the core of Toronto. Some change predated this policy, however: the grey building on Berkeley Street (extreme left) was part of a row built in the early 1870s and renovated by Joan Burt in 1969, making it one of the first examples of gentrification in Toronto. In this view, the structure on the corner has not yet seen similar investment. 11 September 1977. Photographer: Robert D. McMann, Brian and Michael Doucet collection.

15 June 2018. Photographer: Michael Doucet.

27 | The Two Kings (I): King and Parliament.
Parliament Street and King Street East. 19 November 1966. Photographer: John F. Bromley © All rights reserved, Brian and Michael Doucet collection, used with permission.

The Smith Brothers and Snack Bar signs are long gone, as is the nineteenth-century structure on the northwest corner. The remaining industrial buildings around this intersection are now lofts and offices. 7 August 2014. Photographer: Brian Doucet.

28 | The Two Kings (I): King and Parliament.
King Street East and Sherbourne Street. The CIBC bank branch on the northeast corner was built in 1908 for the Imperial Bank of Canada. This view is typical of much of King Street, both east and west of downtown: low-rise commercial and industrial structures punctured by surface parking lots. 17 May 1979. Photographer: Harvey Naylor, Brian and Michael Doucet collection.

This view shows a contemporary Toronto trend in urban redevelopment: façadism. The façade of the bank has been preserved as an entrance to a new condominium tower, while the rest of the building has been demolished. This specific example is called Kings Court, which opened in 2005. It is worth noting the red-brick industrial-looking building immediately to the right of the former bank; it is a new structure that filled in a gap previously left by a surface parking lot. On commercial streets throughout the Streetcar City, the surface parking lots that emerged in the second half of the twentieth-century as older buildings were demolished have largely been redeveloped in the twenty-first century. 15 June 2018. Photographer: Michael Doucet.

29 | The Two Kings (I): King and Parliament.
King Street East and Jarvis Street. The changes in zoning around King and Parliament stretched all the way to Jarvis Street, where, in this photo, the building on the northeast corner is looking rather rundown. Its main commercial occupant is an army surplus store. 11 September 1977. Photographer: unknown, Rob Hutchinson collection.

The building on the corner dates from 1850 and replaced structures destroyed in Toronto's fire of 1849. Neighbouring structures were demolished and replaced by larger developments. 17 October 2018. Photographer: Brian Doucet.

30 | The Two Kings (II): King and Spadina.
King Street West and Charlotte Street. 20 August 1966. Photographer: John F. Bromley © All rights reserved, Brian and Michael Doucet collection, used with permission.

The mixed use and density changes central to the Two Kings policy are evident in this image. Also evident is the lack of mega-buildings that characterized much of the downtown redevelopment in the 1970s and 1980s. Lot sizes are small, leading to a variety of buildings, designs, and uses, albeit with affordability being a major concern. 8 August 2014. Photographer: Brian Doucet.

The final image in this series was taken a few metres further west than the previous two images in order to show more clearly the public space changes that were implemented as part of the King Street Pilot, which was introduced as an experiment in November 2017. In April 2019, City Council voted to make the changes permanent and it was subsequently renamed the King Street Transit Priority Corridor. This photo shows both the parking restrictions and public space enhancements that have been part of the redesign of King Street. 12 July 2019. Photographer: Brian Doucet.

31 | The Two Kings (II): King and Spadina.
King Street West and Spadina Avenue. There had been a hotel on the northwest corner of King and Spadina since 1875. The premises went through several iterations in its early days but almost a century later, when this image was taken, it was known as the Spadina Hotel. While the hotel was a part of the Queen Street West music scene, and its beach-themed Cabana Room featured performances by the Tragically Hip, Leonard Cohen, and the Rolling Stones, it had a reputation of being a rather seedy hotel. 27 July 1973. Photographer: Robert D. McMann, Brian and Michael Doucet collection.

In 1997, the building became the Global Village Backpackers hostel, the largest youth hostel in the city. The hostel closed three months before this photo was taken, although the signage was still in place. 11 April 2014. Photographer: Brian Doucet.

The building was bought by the Konrad Group and redeveloped into an innovation hub. Its main occupant is BrainStation, which is a digital skills training company. 12 July 2019. Photographer: Brian Doucet.

32 | The Two Kings (II): King and Spadina.
King Street West and Bathurst Street. The scene, looking back towards downtown, is similar to set 28, seen earlier in this portfolio. It is a mix of older and newer industrial buildings, a bank on the corner (in this case the Toronto Dominion Bank) and several gap sites used as parking lots. 18 July 1968. Photographer: unknown, Brian and Michael Doucet collection.

One of the few planning restrictions in the Two Kings policy was that buildings had to retain the original fabric of the area (i.e., not occupy an entire city block). The combination of reusing old buildings and new construction on King Street West is clearly visible in this image. 12 November 2019. Photographer: Brian Doucet.

33 | King Street West and Bathurst Street. 6 September 1965. Photographer: John F. Bromley © All rights reserved, Brian and Michael Doucet collection, used with permission.

While the Two Kings policy spurred a condo and development boom that has been relatively unbroken for two decades, some redevelopment of former industrial lands predates this policy. These condos on King Street West at Bathurst Street are some of the oldest downtown condos in the city. For example, the building with the Second Cup and McDonald's, known as the Westside Lofts, opened in 1987. The one just beyond it, 720 King Street West, dates from 1976. 11 April 2015. Photographer: Brian Doucet.

34 | Bathurst Street Bridge. The industrial landscapes around the iconic Bathurst Street Bridge, officially known as the Sir Isaac Brock Bridge, are on full display as work equipment car W-28 runs over it. Built in 1903 to carry railway tracks over the Humber River, it was disassembled and moved to Bathurst Street in 1916 (and realigned in 1931). In 2020, the bridge was completely overhauled. W-28 was originally constructed as a wood-bodied Toronto Civic Railways streetcar and is now preserved at the Halton County Radial Railway. 18 July 1964. Photographer: Robert D. McMann, Brian and Michael Doucet collection.

Given the scale of new residential development that has taken place on both sides of the bridge over the past few decades, it is clear to see why rebuilding it was necessary. The large building on the opposite side of the bridge is the Minto Westside condo. A few weeks after this photo was taken, the city's iconic CLRV streetcars were all retired from service. Bathurst was the last route where they operated. 12 December 2019. Photographer: Brian Doucet.

35 | Shaw Street and King Street West. This view shows the Massey-Ferguson factory complex, which was centred at this intersection. The company made tractors and other farm equipment and was a major employer on the city's west side. 15 July 1967. Photographer: John F. Bromley © All rights reserved, Brian and Michael Doucet collection, used with permission.

In this view, all traces of the former factories are gone, though the company's office building to the east of the intersection remains. Shaw Street has been extended south of King, to incorporate new residential developments on the old Massey Ferguson site. The tracks on Shaw Street still connect up to Queen Street, and are used for diversions and short turns on both the 501 Queen and 504 King lines. 11 April 2015. Photographer: Brian Doucet.

36 | King Street West and Shaw Street. Another building in the Massey Ferguson complex is visible in this view facing west along King Street towards the railway underpass. 4 September 1967. Photographer: unknown, Brian and Michael Doucet collection.

Ground floor retail, with condominiums above, has been characteristic of the mixed-use nature of the redevelopment of many former industrial sites, particularly along major streets such as King. 11 April 2015. Photographer: Brian Doucet.

37 | Bloor Street West and Dundas Street West. Situated on the north side of Bloor between Dundas and the Canadian National/Canadian Pacific railway tracks to the east, the ToastMaster bread factory takes up much of the block. This nondescript industrial building was typical of many throughout the city that were found adjacent to railway lines. The large sign was visible to passing motorists driving along Bloor Street. 19 February 1966. Photographer: John F. Bromley © All rights reserved, Brian and Michael Doucet collection, used with permission.

ToastMaster was replaced in 1974 by the Crossways Centre, which is comprised of shops, offices, and two 29-storey apartment buildings. To the right of this complex is the Bloor Station for GO and UP Express trains. A direct link to the Dundas West subway station (to the left of the photo) has not yet been built, so passengers connecting between the two stations need to walk along this stretch of Bloor Street. 9 May 2016. Photographer: Brian Doucet.

38 | Lakeshore Boulevard and 11th Street. The former municipalities of Mimico, New Toronto, and Long Branch, became part of Etobicoke in 1967. They were part of a ring of early twentieth-century industrial suburbs that also included Weston and Leaside. Most of this industry, however, was located north of Lakeshore Boulevard, closer to the railway tracks. Goodyear was one of the factories that reached as far south as the streetcar tracks of the Long Branch line. 11 February 1987. Photographer Edward J. Wickson, Brian and Michael Doucet collection, used with permission.

The former factory has been replaced with townhouses and apartment buildings of modest height and generous open space. In contrast to many of the formerly industrial spaces depicted in this portfolio, many of the developments visible in this photograph are in non-market forms of tenure, such as co-ops. Downtown, the St. Lawrence neighbourhood adopted similar principles to housing, both in terms of tenure and design. 16 May 2016. Photographer: Michael Doucet.

39 | St. Clair Avenue West and Keele Street. The intersection of St. Clair and Keele Street was the heart of Toronto's stockyards, with numerous slaughterhouses and meat-processing plants located in the vicinity. Its proximity to both the Canadian Pacific and Canadian National railway lines made it an ideal location for these activities; livestock arrived by train from across Canada, and processed meats could be easily shipped to markets throughout the country. Looking west towards Keele Street, we see Swift's Packers, one of the largest meat-packing firms in the area, with the streetcar descending into the railway underpass, 10 August 1969. Photographer: Robert D. McMann, Brian and Michael Doucet collection.

Urban geographers often refer to cities shifting from sites of production to sites of consumption. The site of the former Swift's facility has now become the Stockyards Centre, a physical transformation that exemplifies this shift. The Stockyards Centre opened in 2014. Also visible in these views is the growth of new housing on formerly industrial land; the townhouses to the right (on the east side of Keele) were constructed in the 2000s. 28 July 2020. Photographer: Brian Doucet.

PORTFOLIO 3: NEIGHBOURHOODS

If physical change to the built environment is the dominating visual theme of formerly industrial areas, in residential neighbourhoods the theme is one of continuity. In much of the city, it is remarkable how little the urban form and the built environment have changed since the 1960s.

In Toronto's urban neighbourhoods, main streets still consist of predominantly two- and three-storey buildings, with shops and retail on the ground floor, and apartments above. On the surrounding residential streets the same houses remain. In this portfolio, views of Queen Street are a good example of the former, while Gerrard Street East represents the latter. There is a commonly held idea that these spaces constitute "stable residential neighbourhoods." This "stability" is evident through a lack of physical change in land use, density, and urban morphology.

However, this stability in the built environment belies the dramatic social changes taking place *within* these buildings. Within the Streetcar City, the dominant trend is gentrification. On average, it has become considerably wealthier over the past fifty years, especially compared to the city as a whole. Houses that were occupied by the factory workers employed in the adjacent industrial districts, or were the first entry points for newcomers to Canada, are today, increasingly home to affluent, well-educated, and professional households. Amenities, likewise, have also gone upmarket, focusing more on leisure and entertainment (i.e., cafés and restaurants) than the necessities of daily life.

These changes to the visual landscapes are more subtle than what we saw in portfolios 1 and 2: houses are maintained, renovated, and enlarged. Occasionally there are small pockets of new investment along major streets, such as an infill development or small-scale condos. However, on residential streets, time – as measured through the built environment – has largely stood still. The political pressure to maintain the character of these neighbourhoods of detached and semi-detached houses is a major contributor towards the glacial pace of physical change that limits opportunities for intensification or increased density. As a result, in large parts of the Streetcar City, the population is decreasing, despite being at the heart of Canada's largest city.

There are, of course, still some small pockets of ungentrified housing within the Streetcar City, such as parts of Regent Park, and the 1960s and 1970s apartment buildings that dot the inner-city, though these are some of the new frontiers of gentrification. While statistical analysis shows the core of Toronto becoming wealthier, a visual analysis – in which small details and subtle differences are rendered visible – reveals the more variegated and fine-grained processes occurring within its changing neighbourhoods. Photography is, therefore, well positioned to provide "a ground-up representation of inner-city change to counterbalance more distant and large-scale accounts."[*]

---

[*] Geoffrey DeVerteuil, "The Changing Landscapes of Southwest Montreal: A Visual Account," *Canadian Geographer* 48, no. 1 (2004): 82.

40 | Carlton Street and Parliament Street. Cabbagetown was one of the first neighbourhoods in Toronto to gentrify. However, at the time this photo was taken, much of the housing stock on the surrounding streets was in poor condition and the area housed many low-income residents, with old Victorian houses often subdivided into small apartments. There were still debates about demolishing the neighbourhood and rebuilding it in a style similar to that of nearby Regent Park. Within a decade, however, the first pockets of upgrading were emerging and Victorian houses covered in decades of dirt and soot were painted white, an early indication of gentrification in Toronto. 6 July 1967. Photographer: unknown, Brian and Michael Doucet collection.

Although the built environment in this view has remained static over more than fifty years, much has changed economically and socially in Cabbagetown. Gentrification has made it one of the most exclusive parts of the city and retail upgrading has also dramatically reshaped the economic and social character of the neighbourhood and its main streets. 12 December 2014. Photographer: Brian Doucet.

41 | King Street West and Spencer Avenue. Small neighbourhood grocery stores such as this Dominion in Parkdale, with its classic Vitrolite sign, characterized many streets. Daily needs were within walking distance and the streetcar ran through the heart of the neighbourhood. While today there is talk of a "15-minute city," where all services and amenities are within a quarter of an hour's walk, this view shows that was a common feature of the Streetcar City in the 1960s. August 1965. Photographer: Robert D. McMann, Brian and Michael Doucet collection.

In this view, the urban form remains, but the types of businesses are now very different. Many neighbourhoods have lost their local grocery stores, meaning a long walk or drive to a large supermarket (with abundant parking). As neighbourhoods have downgraded, other have businesses arrived; the Parkdale Village Coin Laundry was an essential service to many low-income residents who lacked washing machines and dryers in their apartments and rooming houses. 8 August 2014. Photographer: Brian Doucet.

Much has changed in Parkdale since the second view in 2014, although the urban form of this block has remained static. The contrast between the static physical form and the rapid gentrification of Parkdale is evident in this third view. Parkdale was skipped over by earlier waves of gentrification, but in recent years it has been a major frontier of capital investment. Along the small commercial strip on the north side of the street, all the businesses seen in 2014 have disappeared apart from the convenience store. The entire block has been renovated. The coin laundry, so vital for low-income residents, has now become a sleek espresso and wine bar. Beside it is an indoor golf simulator. On surrounding streets, real estate investment trusts (REITs) have purchased and renovated many apartment buildings, displacing low-income residents. 28 July 2020. Photographer: Brian Doucet.

42 | Queen Street West and St. Patrick Street. This section of Queen Street West was one of the first retail streets to gentrify, beginning its gentrification trajectory as an edgy, alternative strip in the 1970s. 1 July 1979.
Photographer: Robert D. McMann, Brian and Michael Doucet collection.

The alternative vibe of Queen West has largely been replaced by high-end shops and chain stores, as rising prices and rents have displaced many smaller, independent, or niche stores. Visible in the background, one of the city's first adaptive reuses of an old industrial building was the television studio at King and John, long home to Citytv and its famous Speakers Corner. Built in 1913 as the Wesley Buildings, for many years it was home to book publisher Ryerson Press. It is now home to a number of TV operations owned by Bell Media. 28 November 2018. Photographer: Brian Doucet.

43 | Queen Street West and Bathurst Street. 22 August 1964. Photographer: John F. Bromley © All rights reserved, Brian and Michael Doucet collection, used with permission.

Queen Street West near Bathurst. A bank, a restaurant, an independent fruit market, and a chain wine shop typify much of the retail landscape in the Streetcar City. 28 November 2018. Photographer: Brian Doucet.

44 | Queen Street East and Booth Avenue. 12 September 1982. Photographer: Robert D. McMann, Brian and Michael Doucet collection.

Restaurants, cafés, and bistros are a classic sign of commercial gentrification, as is evident in Leslieville. Rather than catering to the daily needs of local residents, retail streets such as Queen have become "destinations" for people from across the city. While this urban form offers the potential for a "15-minute city," the lack of grocery stores, hardware stores, and other essential businesses means that local residents now need to drive (or rely on online deliveries) to shop for many of life's necessities. In this photo, we zoomed out from the original 1982 view in order to show the construction taking place across the street during the summer of 2018. It is now a seven-storey mixed-use building containing ground floor retail and residential units above. 30 August 2018. Photographer: Michael Doucet.

45 | Dundas Street West and Spadina Avenue. On the northeast corner of this busy intersection, the Victory Burlesque Theatre opened in 1961 in the former Standard Theatre (1921). It closed in 1975. 20 September 1968. Photographer: Robert D. McMann, Brian and Michael Doucet collection.

Dundas and Spadina has become the heart of the downtown core's largest Chinatown, and many Chinese-owned businesses can be seen to the east of the former Victory Burlesque building. The former theatre became a bank, and today is a chain drug store, with signage in both English and Chinese. The marquee is long gone, and the roof is full of cell phone towers. 3 September 2014. Photographer: Michael Doucet.

One of Toronto's busiest intersections lies quiet during the first wave of the pandemic, as most nearby shops, restaurants, and businesses were shut and office workers transitioned to working from home. 3 April 2020. Photographer: Brian Doucet.

46 | Spadina Avenue and Dundas Street West. While the last Spadina streetcar ran in 1948, Harbord streetcars continued to use this section of track until 1966, after which it was used for detours and short turns. 20 May 1967. Note the angled parking on Spadina Avenue, which was a characteristic of this wide street until it was rebuilt in the 1990s for the 510 Spadina streetcar line. Photographer: John F. Bromley © All rights reserved, Brian and Michael Doucet collection, used with permission.

Regular streetcar service returned to Spadina Avenue in 1997 with the construction of the 510 Spadina line, which uses a private right-of-way for streetcars. This is one of the few routes in the streetcar network where the design of the road itself has changed; in addition to the streetcar right-of-way, there are left turn lanes for cars and the angled parking has been removed. However, this redesign has not given true priority to the streetcars and they still need to wait for cars turning left, meaning that delays and large gaps in service still exist on this very busy line. 10 December 2014. Photographer: Brian Doucet.

47 | Broadview Avenue and Gerrard Street East. This series consists of four views of the southeast corner of this busy intersection, where three streetcar lines converge. August 1973. Photographer: unknown, Brian and Michael Doucet collection.

In the 1980s a large Chinese, and later Vietnamese, cluster of businesses emerged around this east-end intersection. 4 January 2005. Photographer: Brian Doucet.

The building on the corner was destroyed by fire on 13 July 2013. 10 April 2014. Photographer: Michael Doucet.

A newly built structure keeps the same urban form as its predecessor. It houses a chain fast food restaurant that today coexists beside the remaining Chinese and Vietnamese retailers. 30 August 2018. Photographer: Michael Doucet.

48 | Dundas Street East and Parliament Street. Regent Park, bounded by Dundas, Parliament, Gerrard, and River Streets, was Canada's largest public housing project when it was built in the late 1940s. The entire area became one large superblock, with internal streets eliminated or severely curtailed. The apartments depicted in this photo were envisioned as a slum clearance project, replacing an earlier neighbourhood of Victorian-era houses. This full-scale demolition and renewal of the neighbourhood served as a template for several other urban renewal projects in the postwar decades. 4 April 1966. Photographer: John F. Bromley © All rights reserved, Brian and Michael Doucet collection, used with permission.

At present, Regent Park is being completely rebuilt as a mixed-use, mixed-income neighbourhood. The original street network is being put back in place and new retail, including grocery stores, restaurants, and a bank (all of which were missing from the old Regent Park) line Dundas and Parliament Streets. As with the old Regent Park 75 years ago, this new design, based around a public-private partnership, is acting as a template to restructure other large housing projects built in the postwar decades. However, a major concern is that many low-income tenants have been unable to return to the new Regent Park, despite an increase in the overall number of housing units. 6 April 2014. Photographer: Brian Doucet.

49 | Queen Street East and Sumach Street. The Regent Park–style approach to redevelopment and slum clearance continued throughout the 1950s and 1960s. By the time this photo was taken, Trefann Court (bounded by Queen, Parliament, Shuter, and River Streets) was planned for a similar type of redevelopment. Had it gone ahead, all the buildings in this photo would have been demolished. 12 March 1967. Photographer: John F. Bromley © All rights reserved, Brian and Michael Doucet collection, used with permission.

Local residents, under the leadership of future Toronto mayor John Sewell and others, devised their own plan for Trefann Court that involved rehabilitating existing buildings when possible and constructing new structures that were in keeping with the existing design of the neighbourhood. Social housing played a key role in keeping the neighbourhood affordable. These design and planning principles are evident today; most of the structures visible in 1967 are still standing. Mid-block, the modest three-storey public housing structure is not out of character with the overall physical and social characteristics of the community. 28 July 2020. Photographer: Brian Doucet.

50 | Queen Street East and Parliament Street. Plans for Trefann Court – developed around the time this photograph was taken – would have resulted in the demolition of all the buildings visible behind the streetcar. Regent Park, Don Mount Court, and Alexandra Park followed this approach to urban renewal. 28 August 1965. Photographer: John F. Bromley © All rights reserved, Brian and Michael Doucet collection, used with permission.

By the late 1960s, that type of redevelopment approach had fallen out of favour. Like those shown in the previous set of images, the existing structures were preserved and new affordable housing was added on gap sites or where a building had deteriorated to such an extent that it had to be demolished. 28 July 2020. Photographer: Brian Doucet.

51 | College Street West and Bathurst Street. 26 May 1969. Photographer: unknown, Brian and Michael Doucet collection.

In 1989, the College Street United Church (1885) was mostly demolished and replaced with the eight-storey Channel Club condominium (1990), which kept part of the original church structure, including the tower. 12 July 2019. Photographer: Brian Doucet.

52 | Gerrard Street East and Highfield Road. October 1967. Photographer: John F. Bromley © All rights reserved, Brian and Michael Doucet collection, used with permission.

While most buildings along this part of Gerrard Street East remain, this area is now known as the Gerrard India Bazaar, which extends east from here to Coxwell Avenue, and includes several blocks of South Asian–oriented businesses, which first emerged in the 1970s after the opening of a cinema showing Bollywood movies. Most Toronto ethnic retail strips were situated in close proximity to their communities; however, the residential streets around the Gerrard India Bazaar never housed a large South Asian community. 7 August 2014. Photographer: Brian Doucet.

53 | Gerrard Street East and Golfview Avenue. This view of modest semi-detached and detached houses, closely spaced on narrow plots, is typical of the Streetcar City neighbourhoods constructed between 1910 and 1930, although few residential streets like this ever had streetcars running along them. Every house in this photo sports a television antenna. 12 February 1967. Photographer: John F. Bromley © All rights reserved, Brian and Michael Doucet collection, used with permission.

The stability of the built environment of the Streetcar City is clearly evident in the almost fifty years between these two views. However, while "stable residential neighbourhood" is a popular term to describe much of the city, areas such as Upper Gerrard and the "Upper Beaches" have undergone tremendous demographic, social, and economic change. They became frontiers of gentrification after nearby neighbourhoods such as The Beach or Riverdale were gentrified and became unaffordable. Not a single television antenna remains. 14 April 2014. Photographer: Brian Doucet.

54 | Main Street and Danforth Avenue. The Hope United Church (1930) occupies the northwest corner. Two months before this photo was taken, the Bloor-Danforth subway opened as far east as Woodbine. Streetcar shuttles continued east to Luttrell for another two years. 1 May 1966. Photographer: unknown, Brian and Michael Doucet collection.

The Hope United Church is still active today. Note the phone boxes on the corner. On the three corners of Main and Danforth visible in this photograph, it is difficult to tell that there has been a subway stop here for more than fifty years! On the southeast corner, to the right of the photographer, however, a large apartment complex, Main Square Apartments, opened in 1972. Despite the presence of the subway, there are few dense developments along the Danforth. 7 August 2014. Photographer: Brian Doucet.

55 | Pape Avenue and Strathcona Avenue. The Harbord streetcar line started at Pape and Danforth and weaved its way downtown and then out to the northwest part of the city. It was discontinued in conjunction with the opening of the Bloor-Danforth subway shortly after this photograph was taken. February 1966. Photographer: unknown, Brian and Michael Doucet collection.

This is another view where the static nature of the built form of the Streetcar City is evident. This view is remarkable because Pape subway station is a mere 500 metres away. Homes such as these, originally constructed for factory workers, now sell for well in excess of $1 million. Also of note is the 30km/hr sign; Pape Avenue is one example of where the city has lowered speed limits to promote safety, albeit without any changes to the road design. 30 July 2020. Photographer: Brian Doucet.

(Opposite page) 56 | Dundas Street West and Dufferin Street. Three generations of Toronto streetcars form the focal point of this spread. Looking west along Dundas, we see a typical landscape in the Streetcar City: a thriving retail strip anchoring a neighbourhood, with a streetcar running down the middle of the neighbourhood's main street. Most buildings in the photo date from the late nineteenth and early twentieth centuries. 14 August, 1965. Photographer: John F. Bromley © All rights reserved, Brian and Michael Doucet collection, used with permission.

This urban form has proven remarkably durable and enduring across the core of Toronto. While businesses have changed hands and gentrification has led to more restaurants, cafés, and high-end shops along Dundas, the urban form remained frozen in time between these two views. 8 August 2014. Photographer: Brian Doucet.

The only change in urban form since 2014 has been the arrival of a mid-rise development one block west of Dufferin. Buildings such as this represent a new trend along main streets such as Dundas, as we also saw on Queen Street East. Slowly there are new developments that are increasing the density of two- and three-storey structures that dominated for more than a century. Like their predecessors, they contain ground-floor retail, with residential units above. This intensification is in keeping with the city's Official Plan for greater density and mixed uses along main corridors. To date, however, these ideas have not spread to surrounding residential streets. 28 July 2020. Photographer: Brian Doucet.

57 | College Street and Dufferin Street. A combination of rain and snow did not stop this unknown streetcar enthusiast from heading outside to take this photograph! Like the previous photo set, this image captures the essence of the Streetcar City's urban form. 17 April 1965. Photographer: unknown, Brian and Michael Doucet collection.

The 506 streetcar was replaced by buses for much of 2020 in order to facilitate track reconstruction. The biggest physical change in this image is the presence of trees along the sidewalk. While they somewhat obstruct the view of the buildings behind them, as elsewhere in the Streetcar City, very little physical change has taken place. 28 July 2020. Photographer: Brian Doucet.

58 | Dundas Street West and Ritchie Avenue. A typical scene from Toronto's industrial past, where factories were a part of many of the city's neighbourhoods. The buildings along Dundas Street West were a combination of ground floor retail, with apartments above, interspersed with manufacturing. The factory on the corner was constructed in 1911 for the Toronto Feather and Down Company, which made pillows. 16 October 1965. Photographer: unknown.

The factory building remains but is now a loft-style condominium known as the Feather Factory Lofts, which opened in 2007. This building can also be seen in figure 3.5. 28 July 2020. Photographer: Brian Doucet.

59 | Bloor Street West and Mountview Avenue. This image looks east along Bloor Street towards the intersection of Keele/Parkside and beyond. When the Bloor-Danforth subway opened in February 1966, it ran only as far west as Keele. Until the extension to Islington opened two years later, streetcars operated a shuttle service between the intersection of Jane and Bloor and Keele station; the track into the loop can be seen behind the westbound streetcar. Bloor Street was also a major automobile thoroughfare, evidenced by the numerous advertising billboards on the north side of Bloor Street, aimed at catching passing motorists' attention, as well as the many auto-related businesses (gas stations, repair shops) visible in the photo. The ToastMaster billboard from set 37 is visible in the distance. 5 August 1967. Photographer: John F. Bromley © All rights reserved, Brian and Michael Doucet collection, used with permission.

This is one of the few parts of the Bloor-Danforth subway line that has seen intensification around its stations; the first major development being the Quebec/Gothic site near High Park Station. More recent intensification is also evident in this photo. However, what is striking is the lack of tall buildings in the background along Bloor Street. 19 May 2016. Photographer: Michael Doucet.

60 | Bloor Street West and Lansdowne Avenue. August 1965. Photographer: John F. Bromley © All rights reserved, Brian and Michael Doucet collection, used with permission.

At Lansdowne, two subway stops east of Keele, there are few tall buildings. A historically working-class part of the city, situated near several railway lines and numerous factories, Bloor and Lansdowne was skipped over by previous rounds of gentrification. In recent years, however, it has become a major gentrification frontier, with new hip cafés and restaurants sitting side by side with older businesses catering to immigrants and less affluent residents. Also of note in this view are the new bicycle lanes that have been progressively extended along Bloor Street and Danforth Avenue. This stretch was installed in the summer of 2020. 6 September 2020. Photographer: Michael Doucet.

61 | Lansdowne Avenue and Jenet Avenue. In this view, we are 100 metres from the entrance to the Lansdowne subway station, which will open to the public in less than two months. This track was used for accessing the Lansdowne carhouse; trolleybuses operated on the Lansdowne route from 1947, as evidenced by the double set of overhead wires. 1 January 1966. Photographer: John F. Bromley © All rights reserved, Brian and Michael Doucet collection, used with permission.

In the more than fifty years since the Bloor-Danforth subway opened, this view of Lansdowne Avenue, north of Bloor, has seen no changes to its built form, apart from some modifications to the houses, despite being steps from the subway. However, two 30-storey towers have been proposed around the corner on Bloor Street. 3 December 2018. Photographer: Brian Doucet.

62 | Paton Road (Lansdowne carhouse). 3 April 1965. The Lansdowne carhouse was one of two that were used by Bloor-Danforth streetcars, which until 1966 served the city's busiest route. Originally built by the Toronto Railway Company, it housed streetcars until the Bloor and Harbord lines were abandoned to coincide with the opening of the subway, less than a year after this photo was taken. Photographer: Robert D. McMann, Brian and Michael Doucet collection.

The Lansdowne carhouse became a trolleybus and bus garage in 1966, housing the trolleybuses that served the west end of the city until 1993, as well as various diesel bus routes. Budget cuts led to its closure by the TTC in 1996. The building remained vacant until it was demolished in 2003. The lot remained empty for many years; borrowing Ken Greenberg's evocative phrase, it was a "terrain of availability." In 2021, part of the site was proposed as the location of the Magellan Centre, a seven-storey affordable rental and senior assisted living centre, primarily aimed at the Portuguese Canadian community. 16 September 2020. Photographer: Brian Doucet.

63 | St. Clair Avenue West and Tweedsmuir Avenue. In this image, a PCC streetcar is exiting the ramp from the new St. Clair West subway station, which opened earlier that year. 28 May 1978. Photographer: unknown, Brian and Michael Doucet collection.

A larger Loblaws supermarket, atop the subway station, and condominium towers along St. Clair and Bathurst are now visible. The section of St. Clair from here to Yonge Street is one of the few parts of the streetcar network (outside the downtown core) where high-rise developments are common and most of the original structures lining the street have been replaced. 17 October 2018. Photographer: Brian Doucet.

64 | St. Clair Carhouse – Artscape Wychwood Barns. An extended series of photographs of the St. Clair Carhouse, which was originally constructed by the Toronto Civic Railways in 1913 and was used as an active storage and maintenance yard by the TTC from 1921 until the 1970s. It was one of the most important facilities of the TTC, and stored streetcars that ran on the St. Clair, Dupont, Rogers Road, Earlscourt, and Bathurst lines. 28 August 1965. Photographer: John F. Bromley © All rights reserved, Brian and Michael Doucet collection, used with permission.

25 February 1966. Photographer: unknown, Brian and Michael Doucet collection.

Inside the carhouse. 12 February 1966. Photographer: John F. Bromley © All rights reserved, Brian and Michael Doucet collection, used with permission.

By the late 1970s, most of the routes that once ran out of the St. Clair Division had been abandoned. To consolidate operations, the TTC decided to close the carhouse and move the remaining streetcars to Roncesvalles. The facility closed as an active storage yard in 1978. 28 August 1978. Photographer: unknown, Brian and Michael Doucet collection.

The property remained important after its closure; the new CLRV streetcars were tested here in the late 1970s and early 1980s. Throughout the 1980s, it was also used as a dead storage facility for old PCCs awaiting shipment to the scrap yard. Dozens of old PCCs could be spotted here at any one time, all in various states of disrepair, something both authors vividly remember. October 1987. Photographer: Michael Doucet.

Dead storage of old streetcars and trolleybuses continued until 1992. By the time this photo was taken 10 years later, the property was in sad shape. November 2002. Photographer: Brian Doucet.

While adaptive reuses of older buildings can be found within the Streetcar City, the former St. Clair Carhouse is an example of a streetcar facility that has been reinvented as a park, arts space, and community hub. Now known as Artscape Wychwood Barns, ownership of the former TTC property was transferred to the City of Toronto in 1996. City councillor Joe Mihevc and architect Joe Lobko worked to transform the site into a multipurpose venue and park, which opened in 2008. 11 May 2015. Photographer: Michael Doucet.

The site is leased to Toronto Artscape for $1/ year for fifty years. Much of the carhouse has been preserved and now functions as a multipurpose space. The outdoor storage tracks have been replaced by a park. 11 May 2015. Photographer: Michael Doucet.

Joe Mihevc, the city councillor who spearheaded the redevelopment of Wychwood, pictured inside the former carbarn. Photo courtesy of Joe Mihevc.

A busy outdoor market. Photographer: George Matthews. Photo courtesy of Wychwood Artscape.

65 | Rogers Road and Glenholme Avenue. 19 February 1966. Photographer: John F. Bromley © All rights reserved, Robert Lubinski collection, used with permission.

The next few sets depict Rogers Road, which was one of the last streetcar lines to be abandoned. Today, it is a frontier of gentrification, though the process is in its early phases, and therefore not always immediately evident in visual landscapes. Like in other photo sets, pictures along Rogers Road reveal a fine-grain detail of a mix of uses, activities, and social classes that is typical of many gentrification frontiers. 30 July 2020. Photographer: Brian Doucet.

66 | Rogers Road and Nairn Avenue. 19 July 1974. Photographer: Robert D. McMann, Robert Lubinski collection.

28 July 2020. Photographer: Brian Doucet.

67 | Rogers Road and Keele Street. 19 July 1974. Photographer: Robert D. McMann, Robert Lubinski collection.

30 July 2020. Photographer: Brian Doucet.

68 | Eglinton Avenue West and Oakwood Avenue. It is difficult to find good quality images of the Oakwood line, which was abandoned on New Year's Day, 1960. The line ran from Oakwood and St. Clair, north to Eglinton, then turned west along Eglinton to a loop at Gilbert Street. Despite the poor quality of this image, we use it to end our repeat photography portfolios. This stretch of Eglinton Avenue will see dramatic transformations in the coming years as new streetcars, in the form of modern light rail transit, will soon be operating beneath the street. 1 January 1960. Photographer: unknown, Brian and Michael Doucet collection.

It has been more than 60 years since streetcars ran on this part of Eglinton Avenue West. But the new Eglinton Crosstown LRT, Line 5, is scheduled to open in 2022 and will feature an underground station at this intersection. The construction site at the right of the photograph will be an entrance. The busy Eglinton West bus will be replaced by a new higher-order transit mode connecting large parts of the city. Like many of the neighbourhoods within the Streetcar City, the built form and urban morphology along the former Oakwood line have changed very little since they were built. Social changes are far more evident than physical ones; gentrification is starting to reshape the economic and social character of the neighbourhood, a trend that will undoubtedly continue once the area is connected to the rapid transit network. Currently, this part of Eglinton is known as the International Market. It is part of Little Jamaica and has many businesses catering to the area's large West Indian population. Just to the east of the intersection is Reggae Lane. 3 December 2018. Photographer: Brian Doucet.

7

INTERPRETING VISUAL CHANGE IN A DIVIDED CITY

On 31 May 2019, the Toronto Raptors appeared in their first NBA Finals game. Before the game started, Sportsnet, which broadcasted the Finals across Canada, played a short introductory video featuring images of Raptors players, as well as skyline views of Toronto and scenes of people playing basketball in front of tall apartment buildings.[1] Among these images – and only visible for a split second – was a shot of something rather peculiar for a basketball video: a CLRV streetcar. To many Canadians, however, this was immediately recognizable, something inextricably linked to Toronto.

Toronto's streetcars are indeed a symbol of the city; after the CN Tower, they are the city's most recognizable icon.[2] Just as the PCCs were decades ago, the CLRVs became "a symbol of Toronto that adorns everything from postcards to t-shirts."[3] The Spacing Store – the retail outlet of the eponymous urban affairs magazine – stocks a variety of streetcar-themed products, from magnets to candles, and even has a sign explaining to its customers that the dense network of streetcars seen in downtown Toronto is something unique in North America. In neighbourhoods throughout the city, more than two dozen murals pay homage to these transit vehicles. Even in areas such as Weston, where they have not run for more than seventy years, they are visual evidence that streetcars are a much-loved part of the character of Toronto. A local craft beer, Downtown Brown Ale by Amsterdam Brewery, features the image of a PCC streetcar on its cans and bottles.

Streetcars have also played crucial roles in works of fiction set in Toronto.[4] New Orleans has its famous green streetcars and San Francisco has its historic cable cars. While both help to define their cities, neither have the scale and presence of Toronto's streetcars, not to mention the volume of riders that makes them not only iconic but a vital piece of transportation infrastructure. The red and white streetcars are instantly identified with Toronto, and are part of the city's global brand. After New York and Los Angeles, Toronto is the most important production centre for English-language television shows and motion pictures. Usually, Toronto plays some other place, but when it is allowed to portray itself, streetcars can be found. For example, three notable Canadian television series – *Street Legal* (CBC television, 1987–94 and 1999), *Private Eyes* (Global television, 2016–21), and *Kim's Convenience* (CBC television, 2016–21) – each had streetcars in its opening credits. Interestingly, all three series were rooted in premises located on Queen Street (figure 7.2).

But, of course, streetcars do not serve the entire city, one that is quickly approaching 3 million inhabitants, let alone the sprawling metropolitan area that can be defined as having 6.4 million people (the Greater Toronto Area [GTA]), 7 million people (Greater Toronto and Hamilton Area [GTHA]), or upwards of 9.2 million (the Greater Golden Horseshoe [GGH]). Furthermore, both the city and region are increasingly divided along economic, social, ethnic, and political lines. Some of these divisions can be visualized through the repeat photography we have just presented, whereas others revolve around the boundaries of the Streetcar City (built up before World War II) and the post-1945 city (primarily built around the automobile). These require further explanation and elaboration. These fault lines raise an important question: if streetcars are an icon of Toronto, and Toronto is an increasingly

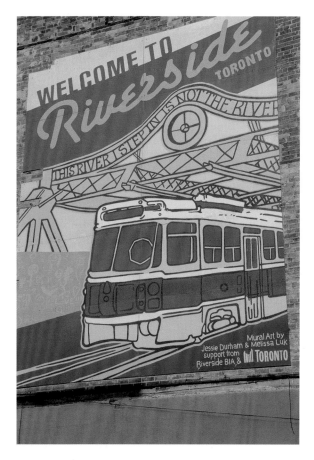 

Figure 7.1 | Two ways of viewing Toronto's streetcars: celebrating a much-loved icon of the city or sitting in traffic stuck behind one while trying to get out of the city. Photographer: Michael Doucet.

Figure 7.2 | Kim's Convenience and an eastbound 501 streetcar. Photographer: Michael Doucet.

divided city, what Toronto are they an icon of? The remainder of the book will explore this question, drawing both on key studies of the city and on our own visual analysis.

While many scholarly studies have examined long-term changes within Toronto, ours is the first major project to use visual analysis to interpret these divisions. Photography has a powerful ability to start new conversations (or continue existing ones) about cities and urban space. Visual analysis therefore plays a key role in the three pillars of critical urban planning outlined by Peter Marcuse:[5] exposing injustices and inequities in urban space, politicizing the geographies of the everyday city, and proposing a more socially just and equitable future. Photography contributes towards this critical urban scholarship by rendering visible what is often invisible, by revealing fine-grained detail, and in sparking new questions and ways of looking at the city. Photographic analysis also helps to unify what are usually separate debates about transit, land use, housing, economic development, sustainability, racism, and equity into the same conversation about challenges and opportunities of contemporary cities. Repeat photography's role in this is to bring historical images into a conversation with these contemporary debates by grounding them in a much clearer understanding of the histories, path dependencies, and trajectories of cities. In the final two chapters, we will explore what insights this visual analysis can bring to present-day debates about Toronto and future challenges for cities more broadly. To do this, we will focus both on what we have seen in these photographs and on how the changes depicted in our repeat photography relate to the rest of the city and region. In this chapter, we start by exploring the extent to which the boundaries of the Streetcar City serve as one of the key spatial fault lines within the city; we then turn to an internal examination of the various forms of change within the Streetcar City itself.

Economic, Social, and Political Fault Lines

Jason Hackworth, a geography professor from the University of Toronto, argues that there are three processes that define the current socio-spatial divisions within the Toronto region: rapid suburban growth on the fringes of the region, decline and disinvestment in the inner-suburbs, and reinvestment, in the form of gentrification, in the inner city.[6] In this sense, the streetcar is an icon of an increasingly affluent and gentrified piece of Toronto. This is no coincidence; across North America, gentrification tends to first arrive in neighbourhoods with a dense and walkable urban fabric, close to downtowns, and with good transit. In other words, in neighbourhoods that developed during the streetcar era.

While these are often the most visible parts of a city, they need to be contextualized within a wider regional perspective. In the 2016 census, the Streetcar City (as we defined it in chapter 3) had 980,155 inhabitants, out of total of 2,731,571 in City of Toronto. The Toronto Census Metropolitan Area (CMA) counted 5,928,040 inhabitants, though the GTA and GTHA are both larger than this statistical boundary tallied in the census. Within the city, the old, pre-amalgamation City of Toronto is growing at a rapid rate, increasing by 17 per cent between 2006 and 2016, far faster than the city as a whole. Visually, this manifests itself in the condominium towers you see while driving into the downtown core along the Gardiner Expressway, or in the changes we have depicted along King Street and Queens Quay in particular. While this provides a dramatic visual impression, the bulk of the Toronto region's growth actually takes place at its fringes, rather than in its core. The population of the Streetcar City grew by 8.5 per cent (or 76,541 residents) between 2011 and 2016, primarily through the construction of new high-rise buildings clustered in a handful of areas. Outside of the Streetcar City, the rest of the Toronto CMA saw its population grow by 268,435 people (an increase of 5.7 per cent). Most of this growth took place beyond the city of Toronto, in low-density, automobile-dependent suburbs. Growth within the city of Toronto, but outside of the Streetcar City (that is, the inner suburbs) was modest: 39,970 people, or an increase of only 2.3 per cent.

Growth in the urban core and on the periphery can look and feel very different. The experiences of a new sprawling subdivision, with its car dependency, provide very different lenses from which to experience urban space than new condo towers that are situated beside streetcar stops, bike lanes, and wide sidewalks. These different built environments and the mobility options (or lack thereof) present within them also impact how people relate to each other and their communities.[7]

Understanding how people get around in different parts of the city is important to interpreting the divisions that exist within it. Aggregating data for the Streetcar City and the automobile city (the rest of the city of Toronto) for journey to work modes from the 2016 census reveals some insightful trends. Across the city as a whole, a slim majority – 50.6 per cent – of people drove to work. In the Streetcar City, however, only 37 per cent of people drove. In the rest of the city, this figure was much higher, at 59.6 per cent. Transit use, however, was rather similar in the two parts of the city: 39.2 per cent in the Streetcar city and 35.5 per cent in the automobile city. For the city as a whole, it was 37.0 per cent. Part of this high level of suburban transit use is due to Toronto's unique grid of extremely frequent buses within the first ring of postwar suburbs.[8] As with all modes, however, there are large internal variations, with transit use highest along the subway lines and in high-density, low-income neighbourhoods in the inner suburbs, where many bus routes remained crowded during the COVID-19 pandemic. The neighbourhoods of Liberty Village and Parkdale also stand out as having high transit mode share within the Streetcar City.

While transit usage is relatively consistent across the different built forms, cycling and walking (as a primary journey to work mode), are almost non-existent in the automobile city. Across the CMA, these figures were 0.6 and 3.4 per cent in 2016, compared with 6 and 16.5 per cent in the Streetcar City.[9] The figures for the city as a whole were 2.7 and 8.6 per cent. Within the CMA, there were thirteen census tracts – two in Etobicoke, one in North York, three in Scarborough, three in York Region, and four in Peel Region – where no one indicated that they walked to work. This should not be confused with an absence of walking or cycling, however. For transit users, there is almost always a walk to get to a stop, even if this is not recorded as the main mode of transport. For cycling, 370 of the CMA's 1146 census tracts recorded no one who cycled to work in 2016. Fewer than ten of these were in the Streetcar City. However, as Julian Agyeman, professor of urban and environmental policy and planning at Tufts University, reminds us, many people who cycle, particularly low-income and visible minority residents, rarely get counted in official statistics. Partly as a result of this official data, peripheral neighbourhoods where a growing proportion of low-income residents reside see far fewer investments in walking and cycling infrastructure like good sidewalks and dedicated bike lanes. Though Agyeman and others stress that good infrastructure alone is insufficient to create equitable cycling and that changes in political, economic, and cultural factors (such as policing) are essential for the benefits of cycling to be enjoyed by everyone.[10]

Conversely, the highest percentage of people who walk and cycle to work can be found in neighbourhoods within the Streetcar City, which, as we have stressed, have become wealthier since the 1960s and are less ethnically diverse than neighbourhoods built after 1945. In areas just beyond walking distance to downtown, such as Cabbagetown and neighbourhoods west of Bathurst and south of Bloor, around one in five people cycled to work in 2016. The highest concentrations of walkers can be found in the heart of the downtown core itself, where seven census tracts recorded more than half of all respondents walking to work. In the area bounded by Queen, Simcoe, Front, and Bathurst Streets, home to many new condominium towers constructed on formerly industrial land, around six out of every ten people walked to work in 2016!

Residents of the Streetcar City therefore enjoy a variety of different transport modes. While many own cars, they are not dependent on them. Indeed, it is quite common to find that the new condominiums in the Streetcar City provide more parking spaces for bicycles than they do for motor vehicles (figure 7.3). The Streetcar City has a dense and frequent network of public transit, and the majority of the city's subway stations and many of its busiest surface routes are located within it. While networks and infrastructure for walking and cycling are not as good as they could and should be, they are far better than in many suburban locations.

Provisions for cycling have improved as a result of the pandemic. In June 2020, City Council approved an ActiveTO plan to accelerate the development of new bike lanes. It approved 40 kilometres of new lanes, the largest one-year increase in the city's history and they were quickly installed throughout the rest of the year. However, the geography of this new infrastructure reinforces the mobility disparities found in different parts of the city. Of the eight projects, five are on, or south of, Bloor Street. On top of subway Line 2, there is now a relatively seamless separated bike lane from High Park in the west, all the way to Dawes Road between the Main Street and Victoria Park subway stations. Our most recent view at Bloor and Lansdowne clearly shows this new bike lane (set 60). Only two projects were in Scarborough, one was in North York, and none were in Etobicoke.[11] In 2021, new bike lanes were installed along Yonge Street between Davisville and Bloor, replicating the mobility choices found along large portions of subway line 2.

Economic divisions are also evident within Toronto. The images in our book show a part of Toronto that has been almost completely deindustrialized, with condos, townhomes, shops, offices, and entertainment facilities replacing factories, warehouses, and railway yards. The Financial District has grown both vertically and spatially, a clear visual representation of Toronto's

Figure 7.3 | Development proposal sign for 2 Bloor Street West showing 256 parking spaces for cars and 1,579 for bikes for a three-building complex containing 1,507 housing units. Photographer: Michael Doucet.

ascendancy of the global urban hierarchy as it has gained more control over key functions in the financialized economy. However, these spectacular and recognizable visual transformations are only part of Toronto's economic story. The boundaries of the Streetcar City largely align with the economic fault lines outlined by University of Toronto geography professor Alan Walks, with financial services dominating the core of the city and industry (specifically the automotive sector) central to the economy of large parts of the suburbs.[12] Financial services and automotive production are the region's two biggest economic sectors; while images of old factories that have been demolished to make way for condo towers may give the impression that manufacturing no longer matters, what we do not see in these streetcar photos is that the GTA remains home to Canada's automotive production industry. Ford, GM, Chrysler, Toyota, and Honda all have assembly plants in the Toronto region, and there are also many auto parts suppliers, as well as other industries, warehouses, and logistical firms that are central to the region's economy. While in the 1960s, much of this was visible in the core of the city, today it is found almost exclusively outside of the Streetcar City, and, increasingly, out of the city of Toronto altogether.

These economic and spatial divisions have also been evident during the pandemic. To start, low-income Canadians, including those working in retail, hospitality, and tourism, were more likely to lose their jobs as a result of the economic fallout from the pandemic than well-educated professionals. For those who retained their jobs, there were also starkly different trajectories. While office workers were able to transition to working from home, those employed in warehouses, logistics firms, factories, sorting facilities, grocery, and drug stores, and healthcare facilities were still required to show up for work each day. According to Statistics Canada, four in ten Canadians have jobs that can be done from home. However, these tend to be better paid, office-based jobs requiring higher-levels of education.[13] Lower-income Canadians are more likely to have a job that requires them to go to a workplace, thereby exposing them, and others in their household, to more risk of infection. As we have explained, low-income residents are increasingly residing in the inner suburbs of large Canadian cities, rather than in their downtown cores, as was common fifty years ago.

In Toronto, race and ethnicity are closely related to income, economics, and geography. Data released in the summer of 2020 also showed that Black, Indigenous, and People of Colour (BIPOC) residents constituted 83 per cent of all reported COVID-19 cases in Toronto, despite constituting only half the city's population. The city's Black community in particular was disproportionately affected by COVID-19 infections: accounting for 21 per cent of all cases, while only representing 9 per cent of the population. The city's white population, in comparison, only saw 17 per cent of reported cases, despite comprising 48 per cent of the population. Toronto was one of the few cities to collect race-based data on COVID-19 infections; while an analysis of this data made for sobering reading, it came as little surprise to many experts.[14]

An investigation by the *Toronto Star* demonstrated how these ethnic, racial, and social inequities translated to spatial differences in infection rates across the city. During the early days of the pandemic, rates of infection rose relatively evenly in both Toronto's wealthiest and poorest neighbourhoods, as well as in neighbourhoods that were predominantly white and those inhabited predominantly by visible minorities. However, after the provincial government ordered all non-essential workplaces to close, on 23 March 2020, the curve for the whitest and wealthiest neighbourhoods flattened almost instantly. It kept rising in neighbourhoods with the greatest share of both low-income and visible minority inhabitants, peaking almost two months later.[15] In the North Toronto neighbourhood where Michael lives and Brian grew up, the rate of infection was low, with only 13 cases of COVID-19 as of June 2020. In contrast, neighbourhoods near Jane and Finch, a mere 15 kilometres away saw infection rates ten times higher.[16]

During the early stages of the pandemic, it was common to hear slogans of "we're all in this together" and COVID-19 being the great equalizer. However, this quickly gave way to a greater understanding of how COVID-19 amplifies existing inequities in society, finds its weakest links, and works in tandem with forces such as racism and discrimination to produce very different trajectories, outcomes, and experiences, depending on who you are and where you live in the city. Kwame McKenzie, the CEO of the Wellesley Institute and a professor of psychiatry at the University of Toronto, stated, "COVID-19 is not a great equalizer – it discriminates."[17] These disparities can be found in planning and policy responses to the pandemic, particularly with regard to how cities were altering their public spaces, including streets. As several writers have noted, many of the high-profile responses to the pandemic such as creating new bike lanes, closing streets to cars, and allowing more outdoor dining spaces have been implemented primarily in gentrified urban cores, which have far fewer cases of COVID-19, rather than in the spatial and societal peripheries where the virus is highly prevalent. Pandemic responses, such as the rapid implementation of sidewalk patios, have largely been aimed at addressing challenges for established businesses and middle-class consumers, rather than marginalized communities and individuals, such as the homeless or street vendors, who also use that same space.[18]

This raises the question of how to visualize the ways the pandemic has impacted different parts of the city and their residents. During the pandemic, there were many photographers who took pictures of empty and deserted streets and contrasted them with bustling pre-pandemic views. However, to understand the inequities of the pandemic, we have chosen two views of different locations, emphasizing the different nature of work during lockdown. One shows a deserted King Street; the other, a nondescript suburban industrial facility near Toronto, its in-person employment evident by the vehicles in its parking lot (figure 7.4).[19]

The Politics of the Streetcar and the Streetcar City

Returning to the broader divisions within the city, economic shifts and differences in urban form are also important in interpreting political divisions. In general, progressive, left-leaning voters tend to be found in urban areas, while more conservative and right-wing support can be found in lower-density, automobile-oriented suburbs, as well as in smaller towns and rural communities. Former Toronto mayor John Sewell has argued that car-dependent suburbs promote a sense of isolation and the idea that residents are in competition with one another, something manifested by drivers, who, alone in their vehicles, jostle with each other to get an edge in rush hour traffic. By contrast, the denser built form of the city means that people from different socioeconomic and ethnic backgrounds interact more (such as riding a streetcar together) and have a greater sense of being part of a larger community.[20]

Several scholars have tried to empirically test the relationship between urban form and wider political patterns and voting behaviour. Alan Walks conducted several studies on the spatial distribution of voting in Canadian cities. He explicitly posed the question of the role the boundaries of the pre-1945 and post-1945 city played in voting behaviour by analysing the results from every federal election between 1945 and 2000. He laid out two competing theories about the role that these spatial divisions could play in explaining differences across an urban region. The first is a jurisdictional explanation, which has to do with municipal boundaries; the second, a morphological interpretation, which is centred on differences in urban form. Jurisdictional differences are far more common in the United States, where there is greater unevenness in taxation levels, income, and racial compositions between different municipalities in a metropolitan area, for example, between the core city and suburban municipalities than in Canada (with much less regional cooperation to even out these differences). The latter theory tends to better explain voting patterns in Canadian

Figure 7.4 | Contrasting images of the pandemic city. King Street West, Toronto, 3 April 2020, and Otanabee Drive, Kitchener, 29 April 2020. Photographer: Brian Doucet.

and European cities, where these jurisdictional differences in taxation are less extreme. The morphological explanation is more closely related to lifestyle differences that exist between different urban forms: the dense, mixed-use, and transit-oriented neighbourhoods constructed before World War II, and the lower-density, automobile-oriented suburbs built after 1945.[21]

In his analysis, Walks found that morphological factors, specifically the boundaries of pre- and post-1945 neighbourhoods, were the most powerful explanation of voting patterns. He concluded that "urban spatial form plays a somewhat more direct role in the mediation of political culture, behaviour, values and ideology, instead of merely acting as an empty container for processes occurring at other scales of analysis."[22] Assuming that there is some degree of self-selection of where people want to live (particularly with middle- and upper-class residents), he went on to add, "if the primary mechanisms producing zonal polarisation are self-selection based on the aesthetical qualities of inner-city and suburban neighbourhoods, and local experiences are themselves depending upon the built form, then this posits urban space and place not only as a location where social relations are negotiated, but as a central element in the establishment and maintenance of class habitus, consciousness and distinction, particularly among fractions of the (new) middle classes."[23]

Walks would delve deeper into the causes of city-suburban political divides through a detailed examination of Toronto's Beaches East-York riding, which contains both older, urban neighbourhoods and different types of postwar environments, including apartment clusters and low-density-single family districts. While this research painted a more complex picture, the same divisions between pre-1945 and post-1945 neighbourhoods could be found. He argued that the middle-class search for community could best be found in the aesthetic qualities of the inner city, thereby contributing to its gentrification. The urban spaces visualized within this book therefore constitute a positional good that helps to shape the social distinctions and identities that separate them out from the suburbs, and act as a form of spatial political sorting between the left and the right.[24]

Alan Walks also studied the Toronto mayoral election of 2010 that saw Rob Ford become mayor, defeating his main opponent, George Smitherman, an openly gay former Liberal cabinet minister from downtown Toronto. In this election, streetcars themselves, rather than just the urban form they produced, played a central role in the campaign. Rob Ford's disdain for streetcars was well known, and he would regularly complain about getting stuck behind 505 Dundas streetcars on his way to City Hall. To experience this first hand, *Globe and Mail* reporter Marcus Gee drove the same route as Ford, from leafy, postwar Etobicoke, down Dundas Street – the most direct route – to his own late nineteenth-century neighbourhood around Dundas and Dufferin. This trip, as Gee explained, took him from "the big lawn and sprawling apartment plazas of Etobicoke, to the cafes, art galleries and Portuguese sports bars of [his] own neighbourhood. It takes no more than 20 minutes in light midday traffic, but it can feel like a journey between separate worlds."[25]

These different worlds have different priorities when it comes to mobility. Ford promoted many ideas that appealed to suburban voters, who felt alienated from the city's booming downtown and lived in a built environment where the automobile is the dominant means of transportation. His campaign was built around a narrative of "stopping the war on cars"; this "war" included an increased vehicle registration tax, as well as new bike lanes and transit projects, both of which took space away from cars. His disdain for streetcars played a major role in his campaign and he proposed replacing all the city's streetcars with buses and subways. He also promised to cancel the Transit City project, an ambitious light rail transit (LRT) plan developed by the previous mayor, David Miller, a left-of-centre politician from the west end of the old city of Toronto who was able to garner support during his two election victories across all parts of the city.

Transit City was to be a series of LRT lines running throughout the inner suburbs – much of David Hulchanski's City 3 – primarily above ground, along their own private rights-of-way on major arterial roads such as Eglinton, Finch, Sheppard, Don Mills, and Jane. This was intended to provide better transit to some of the most underserved parts of the city. The $8.15 billion project would have been almost entirely funded by the provincial government. Had it gone ahead, its seven new LRT lines would be up and running by now.

After seven years of planning Transit City, and construction having already started on Sheppard East, Ford, on his first day in office, triumphantly declared, "Ladies and gentlemen, the war on the car stops today … Transit City is over … We will not build any more rail tracks down the middle of our streets … all new subway expansion is going underground."[26] These comments spoke directly to the automobility of the suburbs and the ways in which streetcars and light rail transit – their modern adaptation – are perceived by many people as an obstacle to free-flowing traffic and the freedom to drive. Throughout his tenure as mayor, Ford continued to press for extending subways into the suburbs, which tapped into a sentiment that residents of areas such as Scarborough deserved their fair share of infrastructure. Because building subways requires huge investments of time and money (he suggested that the private sector would finance them, without evidence to support this), the likelihood of them ever being built was slim.[27]

Economic divisions also played a role in Ford's populist success. He promised to cut taxes and reduce waste, and stop the so-called gravy train, appealing to many conservative voters. His election victory also coincided with the economic and financial crisis of the late 2000s, which had a profound impact on manufacturing in Ontario (manufacturing, as we outlined earlier, is today almost exclusively found outside of the Streetcar City). In 2009 alone, 121,000 manufacturing jobs were lost in the province. But, as Walks argues, at the same time, the banks and other financial institutions (headquartered downtown) received support from the federal government that helped them continue to offer mortgages which fostered more downtown gentrification, leading to resentment about downtown "elites."[28]

In the 2010 mayoral election, George Smitherman obtained only 36 per cent of the vote, although he won a plurality in the old City of Toronto and contiguous parts of Hulchanski's City 1. By contrast, Rob Ford won the biggest share of votes in every single neighbourhood in Etobicoke, North York, and Scarborough, winning 47 per cent of the total vote across the city. According to Walks' analysis, at the neighbourhood level, the number one variable linked to voting for Rob Ford was the proportion of people in a neighbourhood who drove to work. Housing stock built after 1945, was the second most important variable. Another model that he ran looked at the three main candidates (including Olivia Chow) and confirmed that the strongest variable to explain voter preference was the proportion of the labour force employed in manufacturing.[29]

These same political trends can be seen in more recent elections as well. Rob Ford's brother Doug, who became the premier of Ontario in 2018, was elected largely with rural and suburban votes. While transit was a less critical issue in this election, Doug Ford campaigned heavily on reducing the gas tax, and therefore the cost of driving, thereby appealing to voters in automobile-centred communities. In the run-up to the election, his views on transit were summarized in the *Globe and Mail*:

> Ford is an avowed motorist who seems to see anti-car conspiracy everywhere. In 2012, he called a proposed lightrail line evidence of the "war on the car" and he has cited the recent King Street pilot as another example. For him, bike lanes and streetcar projects are attacks on regular people, irresponsible experiments cooked up by the lefty, looney, bicycle-wielding downtown elite. To many observers, Ford's "war on the car" rhetoric is proof of his small-minded, dangerous, right-wing thinking.[30]

In December 2019, the Ford government abruptly cancelled Hamilton's LRT, despite years of planning and Metrolinx having already spent more than $165 million on the project.[31] Only the guarantee of federal money has since revived the project.

Doug Ford's 2018 election victory highlighted the political fault lines between older, core urban areas built around streetcars, and the automobile-oriented suburbs constructed since 1945. While Doug Ford's Progressive Conservative (PC) party won 11 of Toronto's 25 seats, representing North York, Etobicoke, and Scarborough, they won none fully within the Streetcar City.[32] By contrast, the PCs took every seat in suburban York Region to the north, Mississauga and Halton Region to the west of Toronto, and every riding in Durham Region to the east of Toronto, apart from Oshawa. The left-leaning New

Democratic Party (NDP) took every seat fully within the Streetcar City.³³

The same trends can be seen across the province, with the governing Progressive Conservatives shut out of all predominantly pre-1945 urban ridings in Ottawa, Kitchener, Hamilton, and London (all cities which had their own streetcar systems in the early twentieth century). They are also absent from Windsor, St. Catharines, Niagara Falls, Guelph, Kingston, and Waterloo.³⁴ While many parts of these cities share similar spatial patterns and urban morphologies, they are economically and socially very different from each other. Like his brother before him, Doug Ford attempted to win over suburban voters with the promise of a 10 cents per litre reduction in the gas tax (much of which goes to funding transit), as well as extending Toronto's subway into Richmond Hill to the north and Pickering to the east (again without providing specific details as to how this would be funded).

Zach Taylor, a political scientist from Western University, has found that since 1979, conservative parties have been more successful at winning ridings in the suburbs, while the NDP has received more votes in the urban core, with the Liberals switching back and forth. He explained this with reference to the automobility of the suburbs such that "the individualist experience of detached-home ownership and automobile commuting has been correlated with lower political support for redistribution and collective benefits … The characterization of the suburbs as politically conservative derives in part from the lifestyles generated by physical environment and associated mobility systems."³⁵ He argued that when ridings reach a density of more than 50 people per hectare – typical of an urban area – a shift in voting patterns away from the Conservatives, and towards the NDP, can be seen. However, this shift in density does not always form the boundary of the Streetcar City, as many dense clusters of apartment buildings exist in Toronto's inner suburbs.³⁶

This dependency on the automobile has been noted by Pierre Filion, emeritus professor of planning from the University of Waterloo. He argues that within the automobile city, there remains a fundamentally suburban dimension to the ways that people live their lives that transcends the growing economic, social, and ethnic diversity found in Canada's suburbs. This commonality revolves around their heavy reliance on the car as the primary mode of transport.³⁷ This, in turn, influences and is influenced by land-use patterns which are the antithesis of the city: abundant public and private spaces, a spatial separation of functions, and facilities to support a car-dominated environment, such as parking.

Filion states that this gives rise to a culture in the suburbs which is centred around the automobile and other aspects of daily life that the car enables: large single-family homes, shopping centres, mass consumption, as well as the experience of spending large amounts of time in a car. Over the past few decades, planning and policy initiatives have attempted to bring urban elements to the suburbs, such as better transit, increased density, and mixed-use development. This tends to be done under the mantra of environmental sustainability. Filion argues that the political backlash against these types of "transgressions" is due to their perceived threat to the automobility and the car-dominated culture of the suburbs. This helps to explain the popularity of both Doug and Rob Ford and their anti-streetcar, pro-automobile policies. While the suburbs may be economically and ethnically diversifying, their common element is the car, and therefore, this automobility-induced environment is a "force of durability" in the suburbs, which, Filion argues, will endure for decades to come.

In the Streetcar City, daily life is very different; many residents forgo owning a car entirely, as parking costs are very expensive and there is far less need to own a car simply to get around. In Toronto, the streetcars are therefore also a flashpoint between the hundreds of thousands of people who rely on them each day to get around (at least before the pandemic) and others for whom they are seen as an impediment to automobility. Being stuck behind them while driving downtown is a common complaint. The streetcars are often full, but most suburban residents never have a need or reason to ride them. Like the throngs of pedestrians and weaving cyclists, streetcars are seen as an impediment to driving downtown; some suburban politicians have called for them to be abandoned rather than look for ways to improve their speed and reliability (which invariably means taking space away from cars).³⁸

Therefore, the streetcars themselves, and not just the urban form they enabled, are a point of division within the city. They are a much-loved and cherished icon, a distinct part of the city's increasingly affluent downtown core, perhaps best visually represented by the Riverside neighbourhood mural depicted at the start of this chapter. They contribute to the authenticity and uniqueness of the city and constitute part of what distinguishes the urban experience from the suburban one. While the CN Tower is an icon of Toronto, the streetcars are the most iconic image of gentrified Toronto. However, there is another common, but less endearing, image of the streetcar, the only one that millions of people in Toronto, the GTA, and the Greater Golden Horseshoe ever encounter: the rear of a streetcar slowly creeping along a downtown street viewed through a car windshield as one tries to get onto the Gardiner or Don Valley Parkway in order to get out of the city as quickly as possible (see figure 7.1).

Density, Divisions, and Transit Improvements

To say that the solution would be to simply nudge, encourage, or entice people who live beyond the Streetcar City out of their cars and into transit is to ignore the culture of the suburbs, its entrenched automobility produced by decades of car-centred development, the challenges of density and sprawl, and the state of public transit, particularly once you enter the 905 region around Toronto. Even for those who would like to take transit downtown, or within their own communities, the options, particularly in outer parts of the region, are slow, infrequent, and expensive, so, like it or not, driving remains the default option.[39] While streetcars and subways arrive every few minutes, detailed maps of the GO Transit system are, in the words of York University's Roger Keil, largely a "a work of fiction, a utopian representation of regional mobility that anyone who knows reality will recognize as such."[40] In other words, these maps may look nice on paper, but when a bus journey takes three times longer than by car and when there are no more trains to Toronto after 8 am, they provide no alternative to driving.

In this sense, the streetcar is both a luxury and a threat. They are a luxury (though we do not say this too loudly, in case you are reading this on an overcrowded streetcar) because they, and the urban environment they created, give people choice and the possibility to live without the need to be reliant on a car, while still enjoying a high quality of life in the city. The urban form, morphology, land-use patterns, and transportation options mean that residents of the Streetcar City are not dependent on their automobiles. However, most residents of the Greater Toronto Area reside in car-dependent communities, where driving is the only reasonable option. Therefore, to many others, streetcars are perceived as a threat, because they impede the ability to drive around the city quickly.

While Pierre Filion is right to suggest a common culture and suburban experience based around the car, and that this is fundamentally different from an urban experience, and that for millions of people (particularly many new immigrant groups) the suburbs, and the lifestyle associated with them, represent their aspirational living environment, two points are worth mentioning. First, not everyone living in the suburbs is able to own and operate a car. This is particularly true in Toronto's inner suburbs, which contain both the city's largest concentration of low-income residents, and some of North America's busiest bus routes. Many people live without a car beyond the Streetcar City because they simply cannot afford one. For those who depend on transit, walking, or cycling in the suburbs, the experience is one of navigating hostile environments that are designed for others: sidewalks are absent in many places and the wide arterial roads can be impenetrable or dangerous to cross on foot. This becomes highly problematic for transit users who need to cross these busy roads mid-block simply to access a bus stop because the nearest safe crossing is hundreds of metres away. Beyond the city limits, bus service, particularly in the 905, is rarely fast and frequent. A consequence of this environment, as noted by the University of Toronto's Paul Hess, is that many suburban low-income residents see purchasing a car, rather than better transit, as their mobility solution.[41]

The second point worth mentioning is that in today's Toronto, many people have no housing options other than those found in the inner suburbs. The clusters of apartment towers constructed in the decades after World War II that dot the inner suburbs are often among the least desirable places in the city, but because they are comparatively affordable, they are home to people who have few other options. The suburbanization of poverty has been gradual, yet persistent; the images of the Streetcar City from the 1960s depict a largely working-class, immigrant, and low-income city; recent views show more signs of affluence and gentrification.

Visual methodologies are well-positioned to contribute new insights to understand housing challenges in the Streetcar City and beyond. Part of the appeal of living in the Streetcar City is economic proximity: downtown Toronto has the largest concentration of jobs in the region. The aesthetics and class identity of professional and creative individuals are also inherently connected to this urban form and morphology, and, in turn, to the different mobility options of the Streetcar City. All of these factors are putting pressure on housing prices. Hence the recent trend of gentrification in the outer fringes of the Streetcar City: 1920s-era nondescript, formerly working-class neighbourhoods in York, Weston, and East York. These are some of the last pockets of "affordability," though this is a decidedly middle-class interpretation of affordability. For those on low incomes, there are even fewer options. Small, kit-built houses along the former Rogers Road streetcar line – one of the last streetcar suburbs to be constructed in Toronto – now fetch in excess of a million dollars (figure 7.5). Like elsewhere in the city, some are bought purely to be demolished and replaced with newer and larger ones.

While the rows of older houses and mixed-use main streets, complete with their streetcar lines, are both enduring features of the Streetcar City and part of its appeal, one of the biggest physical changes has been the arrival of tall condominium towers that have created an entirely new residential skyline. But rather than replacing this urban form these new clusters of towers are largely concentrated in a small number of formerly industrial areas, such as the waterfront, downtown, or the central spine of Yonge Street. In Portfolio 2, many views from the 1960s are completely unrecognizable today, both economically and physically. These extremely dense areas of development, such as Liberty Village or CityPlace, are where Toronto's Official Plan dictates that much of the city's growth will take place. Across the entire city, these growth areas comprise roughly 5 per cent of the city's landmass.

In the other 95 per cent, dense new developments are largely prohibited. This is done in order to protect employment zones, green space, or a neighbourhood's "residential character." The preservation of this last element is evident throughout many of the photographs we presented in Portfolio 3, where the urban form of the Streetcar City's many residential neighbourhoods has remained largely static. Socially and economically, however, they are anything but static: gentrification is evident through the physical upgrading of old houses and changes in retail. The visual cues include the early practice of painting houses white (today replaced by sandblasting back to the original brick), renovations, additions, new, historic-looking doors, spelling out the address on the front of the house, removing elements of ethnic detail such as certain types of cladding, and changes to the exterior landscaping. Another aspect of gentrification in Toronto is the demolition of older houses and the construction of larger ones. Collectively, these constitute changes that maintain the same density, urban form, and land use, thereby restricting a neighbourhood's ability to grow in population. Toronto's ascendancy to Canada's largest city and its transition from a provincial, industrial, and predominantly white city to a global, financial, multicultural metropolis has occurred without significantly altering the basic urban morphology of the city's best-positioned residential neighbourhoods. Thus our repeat photography along College, Gerrard, Bloor, Danforth, Pape, and most of the Streetcar City reveals layers of inequality baked into the sediment of a built environment where, in the city's much celebrated "stable" urban neighbourhoods, physical change moves at a geological pace. The photographs are a visual manifestation of a planning system that works, above all else, to protect the visual "character" of these neighbourhoods from development and intensification.[42]

Changes in zoning, to allow for mid-rise, townhouses, or other, marginally denser, forms of housing on Toronto's residential streets are therefore challenging, if not impossible, to achieve, not least because their proponents (progressive planners, anti-poverty advocates, renters, and low- and, increasingly, middle-income residents advocating for more affordable housing) are often drowned out by larger forces of property-owning NIMBYism. One of the best-known examples of this opposition to intensification in a residential neighbourhood occurred in 2017 when Margaret Atwood and other wealthy neighbours in the Annex opposed an eight-storey building on Davenport Road that would have housed 16 luxury residential units.[43] Another extreme example was in early 2021, when some East York residents objected to turning

Figure 7.5 | Rogers Road, one of the last frontiers of gentrification in the Streetcar City. Visible elements of gentrification are not always immediately apparent, but these small, nondescript homes constructed in the early twentieth-century now regularly sell for a million dollars or more. Rogers Road and Dufferin Street, 15 July 1974 and 30 July 2020. Photographers: Robert D. McMann and Brian Doucet.

a parking lot into 64 units of affordable housing for some of the city's homeless population. In defence of preserving the parking lot, one resident stated, "This parking lot is the hub, it's the heart of the community. It provides everybody an opportunity to partake in everything that's here."[44]

It is not just NIMBY homeowners, however, that work to limit intensification and changes to the city's residential neighbourhoods. The Toronto-based planner Gil Meslin has pointed out that an underlying rationale within the planning system is the protection of single-family residential neighbourhoods against even modest increases in density. This is due to the threat that new multi-family housing is *perceived* to pose to adjacent property values.[45] However, he is quick to debunk this perception by highlighting the fact that many older neighbourhoods where small, walk-up apartments intermix with single-family homes (including Atwood's Annex) are some of the most desirable areas of the city. Meslin dubbed large parts of the city "yellowbelts," owing to the protection they enshrine on the built environment, a play on the term greenbelt, which preserves the countryside. By some estimates, 38 per cent of the city's landmass, and 60 per cent of its residential areas are zoned in a way that makes it impossible to add density.[46] While intensification is challenging in Streetcar City neighbourhoods, zoning rules are even more restrictive in areas constructed after 1945.

The stability of the visual landscapes is a product of official planning documents and policies that date back over one hundred years. In the 1960s – a decade of tremendous change in Toronto and the starting point for our visual research – planners enshrined earlier bans on apartments and walk-ups in existing residential neighbourhoods, designating such neighbourhoods as "stable," at least as defined by their urban form. A draft of the 1966 Official Plan stated, "When specific areas of stability have been established, the area should be regarded as inviolate, and a firm commitment made that no basic change in zoning in character would be allowed for a period of ten years."[47]

Subsequent plans utilized similar language to protect the physical character of single-family neighbourhoods by prohibiting any increase in density. The city's current Official Plan states that "neighbourhoods are considered physically stable areas ... development in established neighbourhoods will respect and reinforce the existing physical character of the neighbourhood."[48] Amendment 320 to the Official Plan, which passed in 2016 with little public scrutiny, enshrines the architecturally and spatially ambiguous terms of "prevailing character" and "prevailing building type" as central to the approval of new developments in residential neighbourhoods. Under this amendment, should a developer or homeowner attempt to build something out of character (defined in vague terms) – such as a triplex on a street with single-family homes – local residents, neighbourhood associations, and even the city have grounds to object, effectively prohibiting even modest density increases.[49] Both the static physical nature of many Toronto neighbourhoods and the dramatic additions to the skyline in a small number of areas (both evident in our repeat photography) are products of these planning regulations; the combination of a rapidly growing population and most of the city being off-limits to that growth means that where new housing can be added, it is done in extremely dense forms.

While physical change has moved at a glacial speed (ironic because much of the early repeat photography measured glacial and geological changes), socially, however, there have been many changes within Toronto's neighbourhoods; some are clearly visible, while others require further explanation. Notwithstanding their physical appearance, these neighbourhoods have, in fact, been anything but static. Gentrification and class and ethnic transformations are, by far, the dominant trends. Working-class neighbourhoods have become middle- and upper-class areas, and immigrant reception neighbourhoods have shifted to the suburbs. In the Streetcar City, once home to the main immigrant reception areas, only 34.2 per cent of the population self-identified as a visible minority in the 2016 census. Across the city the figure was 51.5 per cent. Finally, while the basic urban form of retail streets has remained the same, the businesses occupying the ground floor premises have shifted away from small, and often independent, shops catering to the daily needs of local (lower-income) populations, towards chain stores, cafés, restaurants, and high-end destination retail.

However, the photographs we have presented also indicate that these processes are far from complete within the Streetcar City. There are many examples where capital investment and social change are evident within streetscapes, houses, and retail buildings, but other instances where these are not yet visible. Gentrification is the dominant trend, but a visual analysis reveals many ungentrified spaces as well. While statistics reveal patterns and trends aggregated to the level of a neighbourhood, photography enables us to see the fine-grained detail and variations within a neighbourhood.

Visual analysis also enables us to clearly interpret differences between the massive new developments in formerly industrial areas and the static nature of the built form in many residential neighbourhoods. This allows us to better understand one of the biggest trends within the Streetcar City, albeit one that rarely receives as much attention as the construction of new condo towers: substantial decreases in population within many urban neighbourhoods. While the population of the Streetcar City increased from 893,315 in 1971 to 980,115 in 2016, this growth has largely occurred in those small pockets of formerly industrial land or in the downtown core, as dictated in the Official Plan. In her recent work, Anna Kramer, a planning professor based at McGill University, showed that in some neighbourhoods considered to be "stable residential" areas, particularly in the west end of the Streetcar City, populations have declined by as many as 200 people/hectare since the 1970s. Kramer's analysis is featured in the book *House Divided: How the Missing Middle Can Solve Toronto's Affordability Crisis*; in the same book, Alex Bozikovic noted that in his own neighbourhood of Seaton Village (Bathurst to Christie, Bloor to Dupont), the population dropped from around 9,200 to 5,600 since 1971.[50] Similar decreases in population between 1971 and 2016 can be found in Little Italy (the census tract bordered by Bathurst, College, Grace, and Harbord), where the population decreased from 6,740 to 3,980; Bloor and Dufferin (bounded by Bloor, Dovercourt, College, and Brock), which saw a drop from 7,345 to 5,803; the west side Roncesvalles Village (west to High Park), with a drop from 7,290 to 5,399; and Bloor West Village (bounded by Bloor, Jane, Annette, and Runnymede), with a decrease from 7,365 to 5,525 inhabitants over this time period. These declines are due to a combination of factors, including gentrification leading to many small apartments and rooming houses being converted back to single-family homes, and older couples staying in their houses after their children leave. Having little or no new development to increase the housing supply is also a major factor, especially when combined with these demographic and socio-economic trends. Despite these neighbourhoods' stability in urban form, their population decline has resulted in a highly unstable social situation, as schools are threatened with closure, shops are shut due to a lack of local customers, and many amenities disappear.

But demand for urban living remains high, even during the pandemic. While many people in the Toronto region dream of a quiet life in the suburbs, with ample parking and a big house, many others end up dealing with long and cumbersome commutes, driving further out along the 401, because they cannot qualify for a mortgage to live where they want to: in a walkable neighbourhood, close to good transit, amenities, and their jobs in the city. Even during the pandemic, demand for detached housing in Toronto showed no signs of slowing down, as prices jumped by 25 per cent between July 2019 and July 2020.[51]

However, even with large parts of the city off-limits to intensification, new housing supply is being added. Toronto is North America's condominium construction capital; at any one moment, there are dozens of high-rise residential towers going up across the city. Shouldn't this mean that there is enough housing for everyone? In our final chapter, we will reflect on where housing is built and who it is built for, as well as the relationship between good mobility and affordable housing. We will also discuss the role that visual analysis can play in interpreting these challenges, as well as proposing some alternatives and solutions for a more socially just and equitable city.

8

NEIGHBOURHOOD CHANGE, MOBILITY, AND SOCIALLY JUST SOLUTIONS

Good photography documents what is in a scene. Great photography prompts new questions, both about what is visible within a photograph and what is not. Visual analysis is therefore well-positioned to play a leading role in enhancing conversations about cities. Repeat photography's specific contributions include a deeper understanding of how places change over time, and an examination of both fine-grained detail and broad patterns taking place across the city. Urban conversations are becoming increasingly complex; questions about who cities are for, and how to build a more socially just and equitable distribution of wealth within them, require placing issues that have long been siloed – land use, housing, economic development, climate change, design, equity, and inequality – within the same conversation. In her writings and public speaking, urban planner and cycling advocate Tamika Butler emphasizes this point with a quote from Audre Lorde: "There is no such thing as a single-issue struggle because we don't lead single-issue lives."[1] Photography enables us to bring these issues together in order to see how they interact with the major forces of change shaping urban space.

As our photographic analysis has shown, the Streetcar City in Toronto is largely characterized by new 30-, 40-, or even 50-storey condominium towers that are juxtaposed beside time capsules of charming Victorian and Edwardian neighbourhoods, where, despite sitting in the centre of Canada's largest city, populations are falling. It is an unaffordable place for most people to live, particularly if one requires enough space to raise a family. Congestion throughout the Toronto region is getting worse. The city's much-loved streetcars, while appearing every few minutes, are stuck in the same traffic as everyone else. Economic shifts have also changed the core of Toronto from one of Canada's leading manufacturing centres in the 1960s, to one of the world's major financial centres, its professional workforce gentrifying formerly working-class neighbourhoods nearby. Toronto is also one of the world's most multicultural cities, but while the city and region have become more ethnically diverse and non-white since the 1960s, this is far less evident in the Streetcar City. Across the city, there are growing challenges, particularly pertaining to how people get around and the cost of housing, that are putting serious strains on Toronto and its residents. In this final chapter, we will explore some solutions to these challenges, drawing on ideas that have emerged from our photographic analysis as well as the wider research that has accompanied it.

A major transportation and mobility challenge in Toronto is the slow speed of streetcars and buses. Toronto's streetcar system is one of the few in North America that operates in much the same way that it did one hundred years ago. In transit terms, it is often referred to as a "legacy" system, one of the few North American streetcar networks dating from the nineteenth century that remain in operation today. Looking at the photos we have presented, one of the most striking features is that the actual streets themselves have remained remarkably similar. Yet the streetcars are not vintage relics; they carry more people each day than all the GO trains put together and the network is one of North America's busiest light rail systems. Some rights-of-way have been constructed on new lines such as 510 Spadina and 509 Harbourfront, or through rebuilding the tracks along 512 St. Clair. However, these are exceptions; throughout most of the streetcar network, the road layout and design have remained virtually static for decades. Today, most of the streetcar system is still plagued by the problems of running in mixed traffic on narrow streets, where a single automobile turning left can delay a few hundred people on streetcars for several minutes. Even where rights-of-way can be found, streetcars rarely enjoy true priority either; when the Spadina route was constructed in the mid-1990s, traffic engineers from the Metro Toronto government insisted that there be two lanes of vehicular traffic along its entire length, as well as left turn lanes (with signal priority) at most intersections. These demands also resulted in the narrowing of some of the city's busiest sidewalks, particularly around Chinatown.[2]

While it is always tricky to look to Europe for answers to North American problems, the volume of people and density of Toronto's downtown streetcars have more similarities with tram systems in Amsterdam, Brussels, or Zurich than they do with many of the new LRT systems that have been built on this side of the Atlantic since the late 1970s.[3] In many European cities, trams used to run in mixed traffic with cars and trucks, albeit on even narrower streets. However, European city leaders have taken conscious and deliberate steps to prioritize transit (as well as walking and cycling) by giving trams their own lane, priority signalling at intersections and limiting – or in some cases eliminating – access for cars. This provides an opportunity to improve the quality of the streets and public spaces as well (figure 8.1). Closer to home, Waterloo Region's new LRT system has followed some of these principles, including redesigning King Street – the region's main thoroughfare – from four lanes to two, with a central right-of-way for the LRT. This, combined with signal priority at intersections, means that the LRT is able to maintain a consistent schedule that is often quicker than driving (figure 8.2).

A promising initiative in Toronto, and one of the few spaces where the road itself has been redesigned, is the highly successful King Street Pilot, which was launched in November 2017, and made permanent in August 2019. Now known as the King Street Transit Priority Corridor, it runs between Bathurst and Jarvis Streets, connecting the Two Kings and the Financial District (figure 8.2). It is a remarkably basic and low-cost solution to enhancing the reliability of the 504 King streetcars, and is premised on the idea of prohibiting automobiles from driving through intersections. At virtually every traffic light, drivers must turn right.[4] This has effectively stopped King Street from being a through street for cars, while streetcars and bicycles are allowed to go straight through an intersection. It has demonstrated how, with minimal effort, dramatic improvements in travel time and huge growth in ridership are possible when transit is given priority. The 504 King streetcar is the TTC's busiest surface line, but prior to these improvements, travelling between Liberty Village and downtown was often faster on foot than by streetcar. While streetcars were scheduled to arrive every few minutes, it was not uncommon to wait fifteen minutes or more, only to then have three or four arrive at the same time, most of which were already full.

Despite rather patchy enforcement of these rules for drivers, and congestion late in the evening, when taxis are allowed to drive through intersections, the reliability of the 504 King streetcar has increased dramatically. Ridership soared from an average of 65,000 to an astonishing 84,000 riders per weekday within the first year of the pilot! Pre-pandemic, this streetcar line carried more passengers than any of the GO Transit commuter lines, the combined six trolley routes that still run in Philadelphia, entire LRT systems in Houston, Seattle, Salt Lake City, full metros in Baltimore and Miami, and Chicago's Green line "L." There have also been placemaking initiatives along King, through the installation of seating, planters, and some public art, and better cycling provisions (see photo set 30). As with St. Clair a decade earlier, fears expressed by some vocal opponents that the street would turn into a ghost town have not materialized. If the question is how to improve service on Toronto's streetcar lines, and consequently quality of life for the people who live nearby, then the King Street pilot is a cost-effective way to improve transit reliability and enhance the quality of the urban environment.

But as we stated at the outset of this chapter, we cannot look at transit in a vacuum. Investments in and improvements to transit are spatially limited and a scarce resource inherently linked to issues of housing, equity, and inequality. What if improving the reliability of the streetcars in the downtown core

Figure 8.1 | Two images of Amsterdam showing distinct changes to the built environment to shift focus away from cars towards pedestrians, bikes, and transit. Leidsestraat (top) is one of the busiest shopping streets in the city. It is pedestrianized and three tram lines run along this narrow street. Cars have been prohibited from driving on the Leidsestraat since 1971. Plantage Middenlaan (bottom) was made car-free in 2015. Trees were planted and wide bike lanes, painted red, were also added. Photographer: Brian Doucet

Figure 8.2 | Waterloo Region's new ION LRT. The signal priority, reserved rights of way and intense development around new stations are evident in this photo. Frederick and Duke Streets, Kitchener, 27 June 2019. Photographer: Brian Doucet.

serves to further increase the spatial divisions within the city by making the Streetcar City an even more attractive (and consequently expensive) place to live? This paradox of quality-of-life improvements is one of the key contradictions of contemporary planning: enhancements to the urban environment, such as better transportation, new amenities, parks, placemaking, public spaces, improved facilities, and so forth, serve to make an area more attractive and desirable to live in, and, consequently, without measures to mitigate against this, more expensive as well.[5] The transformation of the abandoned St. Clair Carhouse into a park, community centre, and arts space is also part of this paradox (see photo set 60). Artscape Wychwood Barns enhances quality of life by providing green space and new urban amenities and, therefore, makes the area around it more attractive. Apart from the most die-hard streetcar enthusiasts, most people would prefer to live beside a park than a derelict and abandoned trolley barn. While not the only reason this part of St. Clair has gentrified, this type of redevelopment certainly plays a role. A similar process can now be seen in Weston, with the new Artscape Weston Common. Part of the problem is that these new amenities, such as bike lanes, parks, and cultural facilities, contribute to making areas more expensive. Another is that planning and development rarely include the voices of low-income or visible minority residents, many of whom end up being directly, or indirectly displaced as an area becomes more attractive through these quality-of-life improvements.[6] It is therefore essential to look beyond a single issue, plan, or site to understand the challenges, conflicts, and contractions inherent to cities. For example, Julian Agyeman from Tufts University and planners Tamika Butler and Destiny Thomas speak of the need to talk about race and racism when discussing cycling infrastructure, especially as bike lanes have been a popular planning and policy response during the COVID-19 pandemic.[7] They specifically focus on the lack of low-income and racialized voices in the planning process, as well as the differences in how public space is experienced depending on the colour of your skin, something that also applies to riding a bicycle.

This is not to say that improvements to the urban environment should not be made. Safe and convenient cycling brings many new opportunities, as do parks, and green spaces. Measures to make transit run more smoothly and reliably are urgently needed across the city. In dense urban environments, higher-order transit, such as streetcars, are far more efficient at moving people around than either buses (at least three of which are required to meet the capacity of one modern streetcar) or private automobiles, which are among the most inefficient users of urban space, especially when parking is taken into account. The success of the King Street Pilot could easily be replicated along other routes where streetcars still operate in mixed traffic: Queen, Dundas, College, and Bathurst. Two other modes of transport that are perfectly suited to the landscapes and geographies of an urban environment – walking and cycling – would also benefit from a reorientation of our largest public spaces – our roads – away from cars. However, doing this primarily in the core of the city, which is already highly gentrified and desirable, or only focusing on one corridor in the suburbs will further contribute to the uneven geographies of the city and increase the divisions we discussed in the previous chapter.

During the pandemic, the bike lanes implemented by ActiveTO also served to reinforce these divisions (see chapter 7): neighbourhoods in the city's gentrified core that already had comparably good cycling infrastructure got more, while neighbourhoods in the inner suburbs, which increasingly contain the city's low-income and racialized populations, got very little. To make matters worse, these areas have some of the city's highest COVID-19 infection rates and levels of

Figure 8.3 | The transformation of King Street as a result of the King Street Pilot (now known as the King Street Transit Priority Corridor), date unknown and 2020. King Street West near Simcoe Street. Photographers: unknown and Brian Doucet.

overcrowding, as well as greater percentages of the workforce who cannot work from home. Yet they have seen very few of the public space and mobility improvements designed to mitigate the health, social, and economic impacts of the pandemic, including new bicycle lanes, wider sidewalks, outdoor dining, and traffic-calming measures. Additionally, implementing measures to improve transit efficiency, such as proof of payment and all-door boarding, without thinking about how interactions and experiences with fare inspectors vary greatly depending on class, gender, and race, is unlikely to produce equitable outcomes and a safe and welcoming space for everyone.

Equitable and just responses to the pandemic, as well as planning the post-pandemic city, need to treat housing, mobility, racism, development, and land use as interconnected parts of the same planning, policy, and political conversations. This means that ideas about transit need to be about more than how many kilometres of subways are built, and discussions about housing need to be about more than the number of new condos or changes in zoning. In the next sections, we discuss ideas on making fast and reliable transit a reality across the city, and how genuinely addressing housing affordability means both emphasizing housing as a human right, and working to decouple profit and speculation from new and existing housing supply.

On Fast Transit, Housing, and Inequality

Delivering fast, frequent, and reliable transit to all parts of a city as large and complex as Toronto is a major challenge. The Transit City project would have come close to that, though it was cancelled in 2010 and most of its LRT lines are unlikely to ever be built. While the urban form of the automobile city may have been dictated by the car, and John Sewell talks about the "rational suburban resident us[ing] a private car for transportation needs"[8] (see chapter 7), it is simply not true that everyone living there is completely car dependent. Some of the TTC's busiest surface routes operate along streets like Finch, Jane, Don Mills, and Eglinton, and each carry more than 40,000 riders per day. They also run through many of the city's poorest neighbourhoods with the highest concentrations of visible minorities, neighbourhoods that have seen very little investment in better or faster transit, or in better provisions for walking and cycling.

Many of the TTC's suburban bus routes actually enjoy high frequencies throughout the day. As we pointed out, transit ridership as a mode-share for journey-to-work is almost the same in the Streetcar City and the automobile city. Jonathan English, a recent PhD graduate in planning from Columbia University and an expert on post-war suburban transit expansion, is keen to remind planners and policymakers that under well-established formulas based around land use and density that determine frequency of service, many suburban TTC routes would only have one bus per hour, as is the case in many American transit systems. But route 36 Finch West, for example, has a bus scheduled every 90 seconds in rush hour, and a five-minute headway for most of the rest of the day – more frequent service than many streetcar lines! It, and other routes like it, also require a lower subsidy per rider than streetcar lines. The grid system of suburban buses that the TTC developed in the early 1960s established a frequent network of service meaning that wait times were low and transferring between bus routes (or into the subway) was relatively quick and easy. As Jonathan English, Paul Mees, and others have argued, this makes many more journeys possible by transit, meaning that even if the built form favours automobiles, viable alternatives are possible. The coverage, span, and frequency of the TTC's suburban buses are unrivalled in North America and even sit in stark contrast to transit in the 905 region, despite its remarkably similar urban form and morphology.[9]

This extensive grid of bus routes that connect to the subway also lessens the need for extremely high densities around subway stations, because buses can efficiently deliver riders to the subway from well beyond the immediate station area. Jonathan English compared Bethesda station, just outside of Washington, DC, with Warden station in Scarborough, on the TTC's Line 2. In Bethesda, there is a lot of office and condo development around the station, but there are poor bus connections. Warden has very little development in its immediate vicinity, yet is served by nine separate bus routes that are integrated into the same flat-rate fare. The result: Bethesda station has around 18,000 boardings a day, Warden almost 40,000. The key, according to English and others, is frequency of service on bus routes that either connect with each other, as part of a grid, or to a higher order transit line, such as a subway. Because frequency greatly reduces the least enjoyable aspect of a transit journey – waiting – Jarrett Walker, a consultant in public transit design, is keen to stress that a frequent service is of vital importance to a good transit system. In his words, "frequency is freedom."[10] While transit in much of the 905 is less frequent than in Toronto, Brampton offers some lessons and insights on how to improve suburban bus systems. In 2009 the city switched to a grid system for its buses, with high frequency service all day. Despite its predominantly automobile-oriented urban form, the city saw a 40 per cent increase in ridership as a result.

Fast and frequent bus routes could help provide better connections to suburban GO stations as well, offering a competitive alternative to driving for many more journeys to downtown Toronto or points in between. While most subway riders walk or take a bus to the station, at present, most GO stations are surrounded by large parking lots, as 62 per cent of Toronto-bound GO train riders drive to the station (with a further 15 per cent being dropped off by car). Metrolinx is working to reduce the dependency on driving to GO stations. With large increases in ridership projected, especially as the system moves away from catering solely to downtown office workers and shifts towards providing all-day service in both directions, this is a necessity. There is simply not enough space near stations to build enough new parking, which can cost $40,000 per spot to build in a new multi-storey garage. As a result, Metrolinx is hoping to shift today's ratio of one parking space for every 1.4 to riders to 2.5 riders for every parking space by 2031. With over 70,000 parking spaces across the network, Metrolinx is not only the largest commuter train operator in Canada but the largest provider of parking, the vast majority of it being free. But around one-third of GO train riders live under two kilometres from the station.[11]

Frequent and reliable buses that provide quick and easy connections to a growing number of peak and, particularly, off-peak trains, combined with better walking and cycling infrastructure to get to stations could mean that coming downtown for shopping, the theatre, or a baseball game from Woodbridge, Georgetown, or Mississauga could be almost as easy as taking the TTC. It would also open up more options to travel in the opposite direction as well, in addition to more local journeys within the suburbs. This would not force people to stop driving, but it would give them a viable choice of how they can get around Canada's largest urban region. Unfortunately, in early 2020, the provincial government cancelled the fare discount of $1.50 when switching between GO Transit and the TTC, making it less attractive to connect between the two biggest transit providers in the region. (The fare discount was also cancelled for the UP Express.) Several experts have noted that this type of fare integration should not be reduced, but rather expanded, especially to ease parking requirements at suburban GO stations.[12] However, the discount was considered to be too expensive precisely because it was so popular; more riders were using it than initially projected.

In Toronto, while more attention is paid to streetcars, its heavily used buses also run in mixed traffic and are subject to delays, bunching, and congestion. Like the streetcars, they can also be unreliable, with large gaps between buses, or unscheduled short turns a regular occurrence despite their high frequency. Improving these already busy bus routes by giving them dedicated bus lanes and signal priority at intersections would be cost-effective ways to improve the reliability, speed, and, consequently, attractiveness of these bus lines. A small start has been made with the addition of a series of express bus routes (numbered in the 900s) along a number of busy corridors. While they are faster because they do not stop at every bus stop, there have been no improvements to the transit infrastructure for these express buses, such as priority signalling or queue jumping, meaning they are still subject to the same traffic as the regular buses.

As the King Street Pilot has shown, improvements do not need to be very expensive. An advantage of making improvements for bus systems is that they do not need to be done along the entire route, as for rail-based systems, which require, at the very least, track from one end of the line to the other. At bottlenecks or congested areas, a dedicated bus lane or priority signal could be installed to skip ahead of automobile traffic, while in quieter areas, the bus could operate on the normal street in mixed traffic. Because they are inexpensive, these improvements can also be highly flexible to respond to changing traffic patterns.[13]

Such cost-effective approaches to improving transit could be rolled out across the city. Implementing improvements is less about finding the money than it is about finding the political will to reorient road space away from private automobiles towards public transit. Prioritized transit across the city is necessary because it is becoming increasingly clear that higher-order transit (subways, light rail) favours gentrification; due to its scarcity it privileges some neighbourhoods over others, and those with more money are willing and able to pay higher prices for the luxury of living nearby. Also the new rail-based transit projects are based on a political economy that emphasizes planning and policy goals of economic development.[14] The light rail line in Waterloo Region is one such example where growth management and intensification were as important to the project's goals as moving people. While gentrification is rarely explicitly stated as a goal, it is often the result of this economic development, investment, and intensification. Most studies on transit-induced gentrification use statistical analysis to assess whether gentrification is occurring close to new transit stations. The general consensus is that gentrification does occur, although it is context-dependent and less evident in weaker markets at both regional and neighbourhood levels. The literature on whether this leads to displacement is less conclusive, largely because many of the metrics used to statistically measure displacement are not capable of capturing all its various elements and interpretations. Studies that employ a variety of methods, including interviews and field observations along with statistical data, however, find more instances of both gentrification and displacement than would automatically show up in conventional statistical analyses.[15] In Toronto, the Eglinton Crosstown LRT project has raised considerable fears about gentrification and displacement, particularly around Little Jamaica, where many businesses have closed, and there are genuine concerns about the future viability of the community, especially as intensification and new condominium development are central to the planning framework along the new line.

To mitigate the negative consequences of transit-induced gentrification, blanketing the city with a grid of transit-priority routes would negate the premium paid to live close to one or two lines. High-quality, fast, frequent, and reliable transit, whether it be bus- or rail-based, would therefore go from being a spatially scarce luxury, to a standard feature of neighbourhoods across the city. A similar approach to bike lanes (ensuring every street was safe and convenient to cycle on) would also diminish their role in promoting gentrification and instead lead them to become a mundane and ordinary part of the city, in a way that several European countries have managed to achieve.[16]

Many of the TTC's suburban bus routes already enjoy very high frequencies; if given genuine priority on the streets where they run, such an approach could be a cost-effective way of achieving greater transit equity, without leading to gentrification and displacement. There is far less evidence to indicate that bus lanes and bus rapid transit contribute to gentrification and displacement compared with rail-based transit. New bus lanes, particularly in the inner suburbs, might not appeal to everyone, though they would greatly enhance the lives of the 35.5 per cent of the population of the automobile city who take transit to work, and would provide competitive alternatives to those who currently drive. Much of the Streetcar City's appeal lies in its aesthetics and morphology, elements that are difficult, if not impossible, to replicate in the automobile city. In other words, better bus lanes are unlikely to lead to people moving from Roncesvalles to Rexdale. What it will do, however, is bring a much greater degree of transit *equity* across the city and work towards ensuring that everyone, regardless of where they live, has the kind of viable and competitive *options* to get around that are currently primarily enjoyed by residents of the gentrified Streetcar City.

Based on the success of the King Street Pilot, in the fall of 2019, the TTC proposed introducing dedicated bus lanes on five busy suburban corridors: Eglinton East, Jane, Dufferin, Steeles West, and Finch East. Ironically, these are some of the same streets that were included in the Transit City project a decade earlier. Even simply allowing all-door boarding has the potential to speed-up journey times on busy bus routes,[17] as has been done on the streetcar network. However, this raises issues of fare enforcement, in particular the racial profiling of riders by fare enforcement officers (see chapter 6).[18] As Jonathan English is keen to stress, however, simply having a dedicated bus lane with low frequencies is not enough; much of the bus rapid transit in York Region has a bus scheduled only every

20 minutes, hardly the kind of frequency that allows for spontaneous journeys, and easy transfers between routes, or is competitive with driving. However, given the frequency of service on Toronto's proposed corridors, this should not be an issue. As Tricia Wood, geography professor and regular *Spacing* columnist stated, "If the buses ran unencumbered by traffic, they could be like an alternate subway."[19]

While these bus lanes were proposed before the pandemic, it was because of challenges during the pandemic that one of them was implemented. In March 2020, transit ridership quickly switched from being the mode of choice for many riders to something of an essential service for people who needed to travel but had no other option. In the Streetcar City, where office workers were able to work from home and adopt other transportation options such as walking and cycling, most transit routes were very quiet. While ridership across the TTC dramatically dropped in March and April 2020, some routes remained busy. Sean Marshall examined when and where ridership levels were the busiest and found that crowding on morning rush hour routes was still evident on ten, primarily suburban bus lines, including 29 Dufferin, 41 Keele, 44 Kipling South, 102 Markham Road, 123 Sherway, and 165 Weston Road North. What these routes all have in common is that they run through either low-income, racialized neighbourhoods, or serve warehousing, logistics, and manufacturing districts whose businesses kept operating during the pandemic, and whose employees could not work from home.[20] As a result, the TTC had to increase service on some morning rush hour bus routes, despite cutting back service throughout much of the city.

This new reality accelerated pressure to implement better infrastructure for buses along busy suburban routes. In July 2020, the Toronto City Council approved a motion to authorize the implementation of one of the five proposed bus lanes, on the Eglinton East corridor, between Kennedy station and the University of Toronto's Scarborough campus. It was operational by the fall of 2020, its bright red paint indicating a bus-only lane (figure 8.4). As many commentators and experts have articulated, such "quick wins" need to be part of an equitable recovery from the pandemic, especially given that women, low-income households, and visible minorities have been hit hardest by its economic

Figure 8.4 | Dedicated bus lane in the Eglinton East corridor. The bus shown is serving the 116 Morningside route and is northbound on Morningside Avenue, not far from the entrance to the University of Toronto's Scarborough campus. 8 April 2021. Photographer: Michael Doucet.

consequences – all segments of the population that rely more on transit.[21] A report published in July 2021 found that peak travel times decreased by five to six minutes and reliability increased 12 per cent. It also found no discernible impact for automobile drivers.[22] This is good; however, these bus lanes are only strips of paint. They also suffer from the same design flaws as many painted bike lanes: the lane ends at intersections to allow cars to turn right. True bus rapid transit, as can be found in many Latin American cities, with seamless dedicated lanes, queue jumping, and priority at intersections, has yet to arrive in Toronto.

Whether the other bus lanes get installed remains unknown. Two LRT lines are under construction, including the Eglinton Crosstown LRT, line 5. Construction of the Finch West LRT – Line 6 – has begun, although it was mired in controversy because Metrolinx had initially planned to donate land to the city to build a new arts and community hub at Jane and Finch. Instead, the public agency decided to sell this parcel on the open market, to the highest bidder, after construction of the line was completed. This decision came without properly consulting local communities and despite discussions between local representatives and Metrolinx that had given the community reason to believe it would receive this land. However, the decision was not surprising to many, who saw it as part of a longer pattern of government underinvestment in their poor and racialized community.[23] Michelle

Dagnino, executive director of the Jane Finch Centre, a local non-profit, stated, "Metrolinx going back on their pledge would be a terrible betrayal to this community, especially at a time when this neighbourhood is still reeling from the impacts of COVID-19."[24] Metrolinx, for its part, cited provincial government realty directives on the selling of surplus property. Estimates placed the value of the land at between $7 and $9 million. The timing of this decision came at an already fractious moment for race relations; it came only a few months after the murder of George Floyd by members of the Minneapolis Police Department, an event that galvanized widespread anger against anti-Black racism around the world and raised difficult questions about institutionalized racism within Toronto and across Canada. Facing intense pressure from community leaders in Jane-Finch as well as residents from across the province, in September 2020, Metrolinx reversed their decision to sell the land on the open market and agreed to set it aside for a community hub.

The Eglinton Crosstown LRT line is predicted to be operational in 2022. Along Eglinton Avenue West in particular, we see the paradox of quality-of-life improvements in effect; as neighbourhoods along the line get better connected, they become more attractive and consequently more expensive. In what were traditionally working-class, immigrant neighbourhoods, property listings are touting the proximity of the new LRT, as narrow two-bedroom 1920s and 1930s houses are listed for around a million dollars. This is why we ended our photo portfolios with images from Eglinton and Oakwood. Because of the Crosstown LRT (as well as the relatively high-frequency UP Express, which intersects with the Line 5 near Weston Road), the Eglinton West corridor is about to see tremendous changes in the coming years, changes that are already driving up housing costs and attracting speculative investment that will displace many of its current low-income residents and the businesses that cater to them.[25] Black-owned businesses in Little Jamaica already felt marginalized in the official Business Improvement Area (BIA); severely disrupted during the LRT construction, they are concerned about the future gentrification of their community. A report written by Black Urbanists Toronto noted the loss of 140 businesses in Little Jamaica since the LRT construction began in 2011. The authors of the report also argued that the official Eglinton Connects planning study, spearheaded by the city, failed to acknowledge the unique strengths and challenges of their community.[26] While Eglinton West, Mount Dennis, and Weston will be better connected to downtown, the airport, and other parts of the city, the resulting waves of new investment and development may not necessarily benefit existing residents who call these neighbourhoods home. Without adequate measures to ensure communities have the right to stay put, low-income and visible minority communities are likely to be displaced from this gentrification frontier.

No discussion of transit equity is complete without an examination of steps being taken to ensure that those with mobility and economic challenges have access to the system. Over the years, considerable progress has been made, but the TTC cannot yet be considered to be fully accessible. For those with mobility challenges, a separate Wheel-Trans service was established in 1975. It uses small buses to take registered users from their place of residence to a destination and back. Such trips are booked in advance, and regular TTC fares are charged. At present, the service has 42,000 registered users and provides about 4.2 million trips per year. Longer-term, the TTC wants to provide a fully accessible transit system. The bus fleet became accessible to those using mobility devices and for those pushing baby strollers in 2015, with the streetcar fleet following suit in late 2019. It is on the subway system where problems remain. While all new subway stations are equipped with elevators, many older stations on both Line 1 and Line 2 have none. At the time of writing, 22 stations fell into that category, with the elevator-installation project expected to be completed in 2025.[27] With the introduction of the Presto card fare system, the TTC was able to replace turnstiles in subway stations with automated fare gates. Combined with the installation of automated doors, this has created barrier-free stations, making them easier to navigate for those with mobility challenges.[28]

As we saw in chapter 1, the TTC is one of the most poorly funded and expensive transit systems in North America. Simply put, it is too expensive for Toronto's poorest residents to use. Recently, three steps have been taken to make using the system less financially challenging. As part of its poverty reduction strategy, in 2018 the City of Toronto adopted a Fair Pass transit option that provides discounted TTC fares for Toronto residents receiving certain types of social assistance.[29] Also in 2018, the TTC introduced a two-hour transfer program which meant that riders could, for example, go shopping or attend a medical appointment, and return home for a single fare, so long as the trip fell within the permitted temporal window.[30] While not aimed directly at low-income Torontonians, the 2015 decision to allow all children aged 12 and under to ride free on the TTC must have reduced financial barriers for some low-income families.[31]

On Affordable Housing, Equity, and Mobility

While providing good transit across the city is an important first step in creating a more liveable and socially just city, major issues in the realm of affordable housing also need to be addressed. This involves thinking beyond just allowing more housing to be constructed. The images we have shown of thousands of new high-rise condominium units have not made the city any more affordable to live in. Therefore, we need to think about the policies and politics of affordable housing, which involve not only building new housing but protecting already existing affordable housing stock, and working towards protecting those in vulnerable housing situations against displacement and evictions. Questions need to be asked about who requires affordable housing, who profits from the current wave of real estate investment, and who is excluded from it.

Two interrelated approaches to housing planning and policy could help to reduce urban divisions and build a more equitable and just city. The first revolves around ownership and tenure. This means shifting debates beyond what is visible – aesthetics, height, density, setbacks, and so on – to what happens inside the building: the size of units, tenure and ownership, speculation, and what owners can and cannot do with their properties. Simply building more market-rate condo towers has not helped make the city more affordable, especially since, by some estimates, there were more than 66,000 housing units sitting empty in Toronto in 2019.[32] Major clusters of these can be found in condominium towers along the waterfront, formerly industrial land, and in neighbourhoods such as Liberty Village: many of the sites depicted in Portfolio 2. At the same time, a survey found that almost 20 per cent of GTA homeowners under the age of 35 own more than one property.[33] As Elvin Wyly and others remind us, we need to properly contextualize the images that we see and analyse;[34] what may appear to be gleaming new neighbourhoods of city-dwellers that appear on the horizon as we drive into Toronto on the Gardiner Expressway are actually investment strategies and pension nest eggs that often sit empty, or are permanently rented on short-term property websites such as Airbnb. By the best estimates, in ten downtown neighbourhoods, at least 5 per cent of rental properties were essentially full-time Airbnb units in the late 2010s,[35] not an insignificant amount when vacancy rates in the city hover around 1 per cent. Toronto is only starting to enact policies to try to restrict short-term rentals; it lags behind other cities in regulating these types of rentals. The pandemic, by temporarily halting travel and tourism, may have provided some new market-rate rental housing as owners switched from short-term to long-term rentals. However, when travel picks up again and it becomes more profitable to rent a house or condo to tourists, many owners will return to websites such as Airbnb and remove these units from the local housing supply.

Therefore, if more housing is to be built, it must be both genuinely affordable and inhabited by residents of the city. It must be built as houses to live in, not speculative investments to get rich from. Inclusionary zoning is an important start. Within Canada, Montreal has developed the most advanced policies that set the rules by which private developers construct new housing. These are based around addressing key housing challenges in the city. Therefore the city's bylaw stipulates that in many new residential developments larger than 450 square metres, 20 per cent of units must be social housing, 20 per cent must be affordable, and 20 per cent must be large enough for families.

However, there also need to be policies and programs to build affordable housing that do not rely on the actions of private developers and are not driven by motives of private profit. Higher levels of government need to help fund and build affordable housing, as they did in previous decades. We need to think about different forms of tenure, including co-ops, rent-geared-to-income housing, shared ownership schemes, and purpose-built and rent-controlled apartments. An important avenue to achieve this is to use publicly owned land to construct genuinely affordable housing that does not depend on market speculation or deal-making with private developers. Using public land for this purpose was a key election platform of Jennifer Keesmaat, former chief planner for the city, during the 2018 mayoral election.[36] Such an approach, of retaining city-owned land for the construction of affordable housing rather than selling it off to raise short-term cash, while uncommon in Canada, has been standard practice in places such as Vienna for more than a century.[37] Despite being one of the world's most liveable cities, it remains an affordable place for its inhabitants, as more than half of them live in some form of socialized or subsidized housing. In Toronto, a small, yet promising, step in this direction occurred in July 2020. City Council approved turning a municipally owned parking lot in Kensington Market into a new affordable housing development. The city will launch a competitive call for proposals to identify a non-profit developer and operator to build and manage new affordable housing on this site.[38] Public land should be kept in public ownership. If that is not possible, then there should be conditions on any housing built on formerly public land that limit speculation and ensure high percentages of genuinely affordable housing.

There are plenty of innovative solutions to the housing question that neither lead to further speculation nor rely on profit as the driving force behind affordable housing. Building new housing – something that has dominated political debate in recent years – is not the only answer. An equally important solution is maintaining and protecting existing affordable housing, much of which is under threat from gentrification and financialized landlords. Sitting tenants still enjoy some degree of rent control in Ontario; however, since the provincial government under Mike Harris introduced vacancy decontrol in 1997, landlords have free rein to charge what they want when a unit becomes vacant. As a result, there are strong and ever-growing financial incentives for landlords to evict sitting tenants so that they can dramatically raise rents in their units. Landlords can also apply for above-guideline rent increases (AGIs) when they incur costs related to capital expenditures or security. These costs can be passed on to tenants and can result in some households paying $10,000 more during the life of their tenancy. For some, this can result in displacement. Between 2012 and 2020, there was a 250 per cent increase in the number of AGI applications, most of which came from financialized and corporate landlords.[39] Closing these loopholes would help maintain affordable housing, protect tenants, and prevent rents from skyrocketing.

Planning and policy interventions such as rent control do not show up in the visual landscapes of change as easily as the construction of new housing does. However, rent control and other measures are important pieces of the strategy that all three levels of government need to incorporate. While the affordable housing crisis plays out at the city level, it is often the provincial government that, as in the case of rent control, has the power to enact policy changes. Luxury home, empty home, and second home taxes have been effective elsewhere in reining in prices and increasing housing supply, without the need to construct new dwellings. These measures are effective because they work to disincentivize using housing for non-residential purposes and encourage their owners to put them back on the market to either be sold or rented out to local residents. They also have the additional benefit of providing a source of public revenue that can directly fund new affordable housing, although the more successful these policies are, the less revenue they bring in. In July 2021, the City of Toronto approved a vacant home tax of 1 per cent of the property's current value assessment.

Strong rent control and anti-eviction regulations can work to curb steep increases in rents, which lead to displacement, and to help tenants fight against speculative practices of landlords. In the same vein, greater support for tenants facing renovictions or other no-fault evictions can help them exercise the rights that they already legally have. Municipal governments can take a more proactive role in this, and likewise, through the better enforcement of existing rules, can help deter landlords looking to circumvent them. These types of actions – to preserve and maintain the affordable housing that already exists – require efforts from all levels of government, especially the provincial government, which sets rules around rent control.

In other parts of Canada, municipalities have been more proactive in maintaining the supply of existing affordable housing. On the west coast, New Westminster, British Columbia, has enacted bylaws that fine landlords who do not relocate tenants while their units are being renovated. The intention behind this policy is to discourage landlords from evicting tenants in order to renovate their units and lease them out at higher rates to new tenants.[40] Montreal has had rules in place since 2016 that give the local government the right of first refusal of any property that comes on the market. If a private buyer makes an offer on a property, the city has 60 days to match it. This policy can help maintain existing affordable housing supply by preventing real estate investment trusts (REITs) or other financialized landlords from buying existing apartment buildings and raising rents through renoviction.

Other ideas, such as the community land trust (CLT) model can be applied to both new and existing buildings. CLTs are non-profit entities that acquire land and property and hold that land in trust in order to achieve outcomes that are of benefit to a particular community. These often involve housing, community spaces, gardens, and commercial spaces, or other assets that the private market is unwilling or unable to provide, at costs that can be met by the local community. One particularly novel initiative in the United Kingdom is a project by the East London Community Land Trust that links property prices to average wages rather than to the market.[41] This means that, in London, owning a home at one-third of the cost of market rates is possible, thereby bringing the security of home ownership within reach to people who otherwise could not afford to do so. When owners sell, however, the price remains fixed to a multiple of average earnings in the area rather than what other similar properties are selling for on the open market. While forgoing any windfall upon sale of their homes, owners get to live comfortably and affordably in some of the most accessible parts of London.

A similar approach can be found in Whistler, British Columbia, where new housing built on land originally owned by the city has restrictive covenants which cap the maximum price it can be resold for. As in London, there are some important restrictions to these Whistler Housing Authority (WHA) properties, such as that owners can own no other properties, and that they must work in Whistler. The former discourages speculation, and the latter is highly beneficial for the local community, where a major challenge is that local employees, particularly those working in the tourism industry, cannot pay market rates for housing. The result is that a WHA condominium costs around $240,000, compared to a similar unit costing around $487,000 on the open market. For a single-family home, the difference is even more striking: $735,000 for a WHA property, compared to over $2,000,000 on the open market. The scheme is proving hugely popular, and perhaps too successful, with all units occupied and over 650 people on the waiting list; in 2018, only 28 properties were resold.[42] This clearly indicates huge demand for houses that are treated as homes, rather than as commodities and sources of wealth. Restriction on what owners can sell their properties for and prices that are decoupled from the market help to ensure the long-term affordability of these houses. No one is going to get rich out of a WHA home, but that's exactly the point; these are not pension nest eggs or investment properties. Given the extreme housing challenges faced by many in Toronto, would schemes like these be appealing to its residents?

In the realm of deeply affordable housing for those on very low incomes, the Parkdale Community Land Trust (PCLT) in Toronto has been successful in acquiring an existing rooming house in a rapidly gentrifying neighbourhood. Between 2006 and 2016, at least 28 Parkdale rooming houses were converted back to single family houses, or renovated into luxury units, eroding the supply of affordable housing. The acquisition of this building by a community land trust ensures that this albeit small supply of deeply affordable housing will remain so in perpetuity, and regardless of the future gentrification surrounding it. The PCLT's aim is to acquire land and properties within the neighbourhood and then lease it to non-profit partners who can then work towards meeting the needs of Parkdale residents that the market is unable or unwilling to accommodate, such as affordable housing, spaces for social enterprises or non-profits, communal gardens, or open spaces.[43] As we outlined in chapter 3, Parkdale is currently a major frontier of gentrification, with REITs acquiring many large apartment buildings, evicting low-income tenants, refurbishing units, and renting them at much higher rates to more affluent households.

Toronto has taken dramatic steps to address major housing challenges in the past. The development of the

St. Lawrence neighbourhood involved collaboration from all levels of government, and included a mix of housing tenures, with an emphasis on genuinely affordable, subsidized housing. Today, it remains a desirable community, that is well-built and well-designed, and contains housing options for people from many different socio-economic backgrounds in a centrally located neighbourhood. Its architecture and design have stood the test of time. It is noteworthy that the federal government was directly involved in funding social housing during the time the St. Lawrence neighbourhood was being built – not assisting private developers in financing the construction of new market-rate housing as it is today. While social housing has fallen out of favour politically in Canada, could such a bold approach deliver the kind of new, and genuinely affordable, housing the city desperately needs?

If thinking about tenure and ownership is the first approach needed to address the housing crisis, the second is carefully upzoning single-family residential neighbourhoods to allow for greater density. This is particularly true close to good transit, especially since much of the Streetcar City has lost population over the past fifty years. Access to good quality transit is a scarce resource in Toronto, and will remain so for the foreseeable future, not only because very little transit has been built over the past four decades but also because, in many neighbourhoods that actually have good transit, very little new housing has been built.

Increasing densities and enabling more people to live near good transit are not just about mobility and neighbourhoods, they are of direct importance to the ongoing climate crisis. The urban form, land-use patterns, and morphology of the Streetcar City mean that the average carbon footprint in many neighbourhoods is half, or even a third of that in more automobile-oriented parts of the Toronto region.[44] If densities and populations increase in areas where carbon footprints are lower, this will not only help strengthen these neighbourhoods but also encourage a sustainable way of managing long-term population growth. These measures are also important for furthering racial and social justice; many calls to preserve "neighbourhood character," as manifested through the built environment, are, in essence, thinly veiled campaigns to prevent low-income and racialized people from residing in their communities.[45]

In recent years, there has been much talk in Toronto, and in other cities, about addressing the "missing middle" as a way to carefully increase densities.[46] As the images we presented in this book can attest, Toronto has a lot of tall towers and a lot of detached and semi-detached houses. The missing middle consists of pretty much everything in between, including townhouses, laneway housing, infill developments, small walk-up apartments, and mid-rise buildings of five or six storeys. While the latter have started to appear within the Streetcar City, they have been confined to existing commercial streets such Queen, Gerrard, Danforth, St. Clair, Kingston Road, and College, rather than on the residential streets nearby.

Things are slowly starting to change, however. Toronto recently passed a rule that gives permission for laneway houses to be constructed as of right. With 300 kilometres of laneways across the city, predominantly in the Streetcar City, there is potential to build small and more affordable units in areas where there has been very little new development or intensification. However, planning restrictions and local opposition means that it remains virtually impossible to replace single-family houses with townhomes, duplexes, or even semi-detached houses, let alone a walk-up apartment of three or four storeys. Even in neighbourhoods with very large lots, such as Long Branch, the tendency remains to replace one house with another, leading to larger and more expensive houses (a form of gentrification), but no increases in density. One hundred kilometres away in Kitchener, however, recent bylaw changes have allowed, as of right, most homeowners to have up to three units on a property. Combined with reducing minimum parking requirements for new developments, this initiative represents an attempt to stimulate the construction of more houses in residential neighbourhoods, particularly along the region's new LRT corridor.

One of the most ambitious and transformative rezoning plans in North America comes from Minneapolis, which, in 2018, approved a new housing plan to radically rezone most of its city to stimulate more housing in existing neighbourhoods. Minneapolis faces many of the same challenges as Toronto: a crisis of affordability in a city where the majority of residential areas are zoned only for single-family housing.[47] Like Toronto, the city has seen an increase in property prices at the same time as it has seen a boom in development, mostly at the luxury end of the market. The new plan rezones much of Minneapolis to allow for triplexes and heights of up to four storeys. It also eliminates parking minimums for all new developments and allows for high-density buildings around transit stops (the city opened a highly successful LRT in 2004). The goal is to increase the supply of housing, with the hope that it becomes more affordable.

The Minneapolis plan, and others that focus only on increasing supply, is largely driven by the belief that the construction of new housing will lead to the creation of more affordable housing. However, there is little empirical evidence to indicate that upzoning and intensification policies on their own will produce housing that is affordable to those on low incomes.[48] Minneapolis is seeing new housing, but much of it is in the luxury market. Design, height, and density are one thing; but planning rarely takes into account how housing is used (as homes or commodities) and by whom. Therefore, in Toronto, and elsewhere, simply relaxing zoning rules, without also considering tenure, ownership, and speculation, is unlikely to produce the types of housing that the city needs, such as deeply affordable housing for low-income residents and affordable units appropriate for families. A socially just housing approach would combine upzoning with strong rent controls, limits to speculation, and the addition of non-market forms of housing, such as co-ops. Seen from this perspective, rather than just from missing-middle density calculations, it is clear that laneway housing, for example, does little to address the city's affordable housing challenges. While they may be small, most of these units are definitely not affordable. Tucked behind single-family houses in the Streetcar City, they can rent for between $2,500 and $3,000 per month (or as much as $36,000 per year), in a city where the median household income is $65,829.[49]

At present, there is also very little stopping the owners of these laneway houses (and other forms of housing, missing middle or otherwise) from renting out their properties to tourists and visitors, notwithstanding the pandemic's impact on travel. As we have previously mentioned, we are only starting to come to grips with the scale of the short-term rental market and how this is reshaping the housing landscapes in cities around the world. But it does raise the question: What are the benefits of increasing density if the result is more luxury units and housing that ends up being rented out on Airbnb? Or phrased another way: does relaxing zoning rules to encourage intensification also require new regulations to guide that increased density towards directly addressing the city's housing challenges? If so, then limiting the use of houses as hotels is only a first, albeit important, step; the rights of some to profit from housing as an investment will need to be curtailed so that others can have the right to remain and live in the city.[50] The crux of today's housing crisis is due to its role as shelter – a basic human right – and as a financialized commodity – a source of wealth and investment. To succeed at the latter deprives many of their basic needs and rights to a home, a contradiction which current housing policies, including the federal government's National Housing Strategy, do not address.[51] It is also important to remember that while many politicians and residents speak of a housing crisis, for others, this is a market that is producing exceptional yields and huge profits.

Martine August's research into long-term care facilities has shown how their financialization can have deadly consequences: for-profit facilities own 54 per cent of beds in Ontario yet accounted for 73 per cent of deaths during the COVID-19 pandemic. It was even worse for the growing share of financialized operators (REITs); their facilities account for only 30 per cent of beds, but 48 per cent of deaths. August advocates for transformative changes in the seniors housing sector, moving away from the for-profit model that has dominated for several decades.[52] In the broader realm of housing in Toronto, a similar shift in mindset is required if the city is to become an affordable place to live.

Final Thoughts

Two of the biggest challenges facing cities in the twenty-first century are how to build and maintain affordable housing close to good transit, and how to ensure that, as areas improve, these improvements do not lead to the gentrification and subsequent displacement of low- and, increasingly, middle-income communities. Addressing these challenges is not about devising new solutions, but rather having the political courage to implement ideas that have been demonstrated to work and that have been called upon for years by many researchers and advocates. These include proactive approaches to guide new development towards directly addressing housing challenges, the construction of new, genuinely affordable, non-market housing, rent controls and regulations, curbing property speculation, banning unjust evictions, and an equitable distribution of public resources such as transit.

If we are to meet the current and future requirements for all who want, and need, to live in Toronto, good transit will have to become more widespread and distributed more equitably, and much of the new housing that will be built will have to be in genuinely affordable tenures that treat housing as places to live in rather than as speculative properties from which to extract or build wealth. This means building what is needed, rather than what is most profitable. To build a socially just city involves not just thinking about what can be visualized – the look and design of a building, its density and urban form – but also asking important, and politically challenging, questions about who owns these homes and what their owners can do with them. We need to start asking the question of who the city is for. As we have articulated throughout this book, visual analysis has a strong role to play in answering these questions. Changes (or lack thereof) are rendered visible; photography helps to ground theories, concepts, processes, and relationships within urban space. Examining Toronto's changing visual geography over a span of many decades has prompted us to ask these difficult, but necessary, questions, reinforcing Elvin Wyly's assertion that urban photography can begin conversations rather than end them.[53]

To address these challenges will take strong policies and planning, and, above all, strong political leadership. If we expect that, on its own, upzoning the city's residential neighbourhoods will magically create housing that is genuinely affordable for the majority of the population, we are sorely mistaken. The dozens of construction cranes erecting thousands of new housing units at any given moment in the city should give us pause for thought. Have they made the city more affordable? Or have they made some people much wealthier? As Annabel Vaughan poignantly states, in reference to a variety of creative forms of missing middle housing design, "Vienna's long-standing program of developing subsidized housing on publicly owned land serves as an important reminder that the planning reforms ... which are meant to encourage more missing middle housing, will not alone solve the crisis of affordability."[54]

But today's extreme housing challenges are perhaps opening the door for more transformative solutions. We have to critically ask ourselves why housing has only recently become the dominant political and planning issue that it is, while working-class and lower-income residents have been experiencing a housing crisis for decades? Are we talking about it today because of a newfound concern for the plight of the urban poor? Or is affordable housing a major political issue because it now affects middle-class professionals, including the many policymakers, planners, journalists, academics, and architects who dominate urban debates? Housing costs have become so high in Toronto and its suburbs that well-educated, professional, and comparably affluent households who never would have dreamed that they would be victims of a housing crisis now struggle to find adequate housing they can afford. Until a decade or so ago, many of these people did not really think much about a housing crisis. While one's desired neighbourhood may have been beyond reach, there was always the neighbourhood next door; the gentrification of the Annex spread westward, trendy Danforth extended eastwards, and the Upper Beaches encompassed an ever-growing section of the east end. Two professional jobs would be more than enough income to obtain a mortgage to buy one's urban "forever home."

Today, however, two professional incomes are not necessarily enough. Despite the condo boom, relatively affluent sections of the population have been shut out of the housing market completely. It is not just about buying property; today it is not uncommon for those same well-educated professionals to be evicted from the homes they rent while trying desperately to save for an ever-larger down payment needed to get a foot on the very bottom rung of the property ladder. Even if they are not evicted, the threat of eviction hangs over many renters; the longer they stay in homes at rent-controlled rates, the greater the rent gap between what their landlords currently charge and what they could charge if new tenants came in. Countless Torontonians fear what will happen if a relative of their landlord suddenly appears and wants to live in their home – a common narrative, though often merely a ploy to vacate a unit in order to raise rents. Between 2015 and the end of 2018, the city saw a 77 per cent increase in "personal use" evictions, as well as a 149 per cent increase in renovictions.[55]

Again, the pandemic has amplified these challenges. Research by the Canadian Centre for Policy Alternatives also found that, despite some high-profile stories and anecdotes of lower rental prices, on the whole, rents for a two-bedroom apartment rose by 3.5 per cent across Canada between October 2019 and October 2020. In the Toronto Region, this figure was 4.5 per cent, more than six times the rate of inflation. The authors of the report state, "the market incentive to push tenants out – due to arrears accumulated during lockdowns, via renovictions, or any other reason – is still very much present. In the absence of rent controls on vacant units, there is money to be made from evictions, and some landlords will cash that money, pandemic or not."[56]

Gentrification used to be defined as a working-class problem and a middle-class opportunity; in Toronto today, middle- and even upper-middle-income households are now victims of this process as well. Does this open up new possibilities, alliances, and partnerships centred around the decommodification of housing? If the housing market is failing a large percentage of the population, could we see this as an opportunity to look beyond the market to find inclusive and equitable solutions that address the long-standing issues for both low-income residents and middle-class residents who are increasingly priced out of the city? The challenge is finding approaches that do not pit one group against the other (where those with less income and resources will invariably lose) but instead work towards creating a more socially just and equitable city for all those who want and need to call it home. Central to this will be incorporating the knowledge and experiences of marginalized residents within the planning and policymaking that has long excluded them.

While investors and speculators purchase condo units high up in the sky, the city's waiting list for subsidized housing stands at over 95,000 people (around 8–10 years), none of whom have any options within these new towers. Even comparatively higher-income households are having to relocate to distant suburbs, or cities such as Kitchener, Peterborough, or Hamilton, because the housing market does not provide adequate housing for them in the Toronto region. Within these communities, the ripple effect is pushing households even further afield as prices skyrocket in cities such as Hamilton.[57] For those on very low incomes, there are even fewer options, and poor-quality housing, overcrowding, exploitative landlords, and the lack of subsidized housing have been major issues for decades. Decommodifying housing and disincentivizing speculation to create more affordable housing to meet a variety of housing needs are areas where working- and middle-class interests could align. Disincentivizing speculation, while ensuring the upzoning of residential neighbourhoods through a variety of tenures and ownership forms, not just different designs, offers a possibility to work towards a more inclusive and equitable city.

To conclude, it is important to reflect on how the Streetcar City we have examined in detail fits within the wider context of the city and region. Solutions will need to be specific to the areas in question: an approach that works in the gentrified inner city might not work in an impoverished suburb. It will be key to listen to affected communities and meaningfully engage with them so that they are central to the solutions and the decision-making. While upgrading and gentrification are the main trends within the Streetcar City, beyond it, in the sprawling inner suburbs built around the automobile, neighbourhood downgrading remains a dominant trend. This is the "in-between city" where "hundreds of thousands of people live in ... those unglamorous and ordinary districts that possess neither the fashionable cachet of gentrified central city neighbourhoods nor the shiny newness of exurban development."[58] While these social, spatial, and ethnic inequities were evident long before COVID-19 arrived in Toronto, it is outside the gentrified urban core that we have visually analysed where the worst effects of the pandemic have been felt, and, perversely, where comparatively few of the planning and policy measures to mitigate the economic, health, and social consequences of the pandemic have been implemented.

These are spaces that Jay Pitter, Toronto-based author and placemaker, has referred to as "forgotten densities": areas found in the city's social and spatial peripheries, where substandard and inadequate housing intersect with a lack of transportation options, over-policing, and predatory businesses.[59] Many of these spaces are very dense, but so too are the new condominium developments that constitute part of what she refers to as "dominant densities." Density, in and of itself, has been shown to play a minor role in the spread of COVID-19. Instead, it is overcrowding that is the problem; crowding occurs when accommodation is much too small for the number of people living in it. Pitter expands on this physical definition to also include psychological or perceived crowding which is about sensations or feelings related to space, which are prevalent throughout "forgotten densities."

Data collected in Toronto in the summer of 2020 show that in areas with the highest rates of household overcrowding, infection rates were 568 per 100,000 people, compared with 144 per 100,000 in areas with the lowest rates of crowding. Overcrowded housing disproportionately affects low-income and racialized residents in cities across Canada and in other countries as well. Consequently, these groups comprise a disproportionate share of COVID-19 infections. This intersection of housing, racism, transportation, and employment leads to very different experiences during the pandemic. For some Torontonians, a lack of space means having to work from home from the kitchen table of a downtown condo; for others it means riding crowded buses to get to work and the inability to self-isolate at home because there are too few bedrooms in an already overcrowded apartment.[60]

These vastly different trajectories illustrate the radically different impacts of the same forces of change in different parts of the city. This was true before COVID, and is certainly true now. As we have highlighted, however, gentrification of the Streetcar City and the downgrading of many inner-suburban neighbourhoods are two sides of the same coin. Our book has been focused on one part of the city, but our conclusions have sought to place the Streetcar City within a wider urban and suburban context. Perhaps, for our next project, we should find photos taken by bus enthusiasts, whose wider geographies within the region may enable us to uncover visual evidence of changes in the suburbs.

Our aim for this book was to continue existing conversations and begin new ones about the changing geography of Toronto, and explore how a visual analysis of long-term changes in ordinary spaces can contribute to a greater understanding of the city's current challenges. We hope that this book has inspired readers from all backgrounds both to look more carefully and critically at the city and to connect what they see (and do not immediately see) to the wider forces of change shaping urban space. We ventured far from the tourist hotspots and commonly photographed places to show ordinary spaces in an extraordinary part of Toronto, one that has shifted from an industrial centre to one of the world's major global financial hubs. By focusing on the visual changes in the city's urban core, we have shown what gentrification, in its broadest sense, looks like. Make no mistake: Toronto is a divided city and the Streetcar City we have depicted is increasingly physically, economically, socially, racially, and politically different from the one beyond the ends of the tracks.

NOTES

Introduction

1 Gillian Rose, *Visual Methodologies* (Los Angeles: Sage Publications, 2012), 23–4. The use of photographs to depict the lives of the urban poor dates back to at least 1890. See, for example, Jacob A. Riis, *How the Other Half Lives: Studies among the Tenements of New York* (New York: Dover Publications, 1971 [New York: Charles Scribner's Sons, 1890]).
2 Christopher Hume, "New Streetcars Will Rocket Ol' Hogtown into the Future," *Toronto Star*, 1 September 2014, GT1.
3 M. Klett, M. Lundgren, P.L. Fradkin, R. Solnit, and K. Breuer, *After the Ruins, 1906 and 2006: Rephotographing the San Francisco Earthquake and Fire* (Berkeley: University of California Press, 2006), 5.
4 See J.H. Rieger, "Photographing Social Change," *Visual Studies* 11, no. 1 (1996): 5–49; A.S. Metcalfe, "Repeat Photography and Educational Research," in J. Moss and B. Pini (eds.), *Visual Research Methods in Educational Research*, 153–71 (London: Palgrave Macmillan, 2016); Brian Doucet, "Repeat Photography and Urban Change," *City* 23, nos. 4–5 (2019): 411–38; G. DeVerteuil, "The Changing Landscapes of Southwest Montreal: A Visual Account," *Canadian Geographer* 48, no. 1 (2004): 76.
5 A.C. Byers, "An Assessment of Contemporary Glacier Fluctuations in Nepal's Khumbu Himal Using Repeat Photography," *Himalayan Journal of Sciences* 4, no. 6 (2008): 21–6; M.A. Crimmins and T.M. Crimmins, "Monitoring Plant Phenology Using Digital Repeat Photography," *Environmental Management* 41 no. 6 (2008): 949–58; O. Sonnentag, K. Hufkens, C. Teshera-Sterne, A.M. Young, M. Friedl, B.H. Braswell, and A.D. Richardson, "Digital Repeat Photography for Phenological Research in Forest Ecosystems," *Agricultural and Forest Meteorology* 152 (2012): 159–77.
6 For more on the active use of photography in urban analysis, see E. Arnold, "Photography, Composition, and the Ephemeral City," *Area*, https://doi.org/10.1111/area.12725.
7 P. Marcuse, "From Critical Urban Theory to the Right to the City," *City* 13, nos. 2–3 (2009), 185–97; see also E. Wyly, "Things Pictures Don't Tell Us: In Search of Baltimore," *City* 14 no. 5 (2010): 497–528.

Chapter 1

1 Murray White, "ROM Opens a Digital Window," *Toronto Star*, 25 April 2016, A1 and A4.
2 Jennifer Keesmaat, "It's Time to Reimagine Toronto's Streetcar 'King,'" *Toronto Star*, 15 June 2016, A17. Keesmaat was the chief planner for the City of Toronto, a position she left in September 2017. In August of 2018, she became a candidate in the race to become mayor of Toronto.
3 Bill Freeman, *The New Urban Agenda: The Greater Toronto and Hamilton Area* (Toronto: Dundurn, 2015), 24.
4 The St. George Mansion is discussed in Richard Dennis, *Toronto's First Apartment-House Boom: An Historical Geography, 1900–1920*, Research Paper No. 177 (Toronto: Centre for Urban and Community Studies, University of Toronto, 1989).
5 The 2019 cost was calculated using the Bank of Canada's inflation calculator. http://www.bankofcanada.ca/rates/related/inflation-calculator/.
6 Data on building heights and date of development were compiled using information at http://www.emporis.com/city/100993/toronto-canada.
7 The meaning of the term City of Toronto has changed over time, as has its physical size. After an active period of annexation from 1883 to 1914, Toronto's boundaries remained the same until 1967, when the Villages of Swansea and Forest Hill became part of the City of Toronto. In 1998, the provincial government forced the amalgamation of the City of Toronto with the Cities of Etobicoke, North York, Scarborough, and York, and the Borough of East York, with the new municipality known as the City of Toronto. In this book, we employ the terms old, former, and then to refer to the City of Toronto as it existed in the 1960s, and the terms new and current to describe its post-1997 form.

8 Dominion Bureau of Statistics. *Population and Housing Characteristics by Census Tract: Toronto* (Ottawa: Minister of Trade and Commerce, 1963).
9 On the removal of the Sunday stops see Eric Andrew-Gee, "Sunday Streetcar Stops near Churches to Be Shuttered in June," *Toronto Star*, 7 May 2015, and Toronto Transit Commission, "TTC Removing Sunday-Only Bus Stops by June 19," TTC notice, 15 June 2016, https://www.ttc.ca/News/2016/June/06-15-16NR.jsp.
10 Metropolitan Toronto Planning Board, *Shopping Centres and Strip Retail Distribution, Metropolitan Toronto Planning Area, 1966* (Toronto: Research Division, Metropolitan Toronto Planning Board, June 1967), Table 2. The Metropolitan Toronto planning area included Metropolitan Toronto plus Mississauga, Toronto Gore Township, Vaughan, Markham, and Pickering.
11 Toronto Transit Commission, *Eighth Annual Report to the Municipality of Metropolitan Toronto* (Toronto: Toronto Transit Commission, 1961).
12 Ridership figures were taken from Toronto Transit Commission, *Annual Reports*, various years.
13 The decline probably was due to two factors, both of which resulted in suburbanization. First, by 1981 members of the baby boom generation had formed families in greater numbers. Second, Toronto witnessed the undoubling of families as the economic circumstances of recent immigrants improved. To help with mortgage payments, immigrant families often shared their homes with other individuals or families. Those people moved to their own dwellings as their incomes and savings permitted.
14 Statistics Canada, *Toronto: Census Tracts*, Catalogue 95-977 (Ottawa: Minister of Supply and Services Canada, 1983). Also consulted for this section were publications with the following catalogue numbers: 93-930, 93-966, and 95-988.
15 See L.S. Bourne, "Market, Location, and Site Selection in Apartment Construction," *Canadian Geographer* 12 no. 4 (1968): 211–26.
16 Metropolitan Toronto Planning Department, *Retailing in Metropolitan Toronto and Surrounding Regions* (Toronto: Research Division, Metropolitan Toronto Planning Department, July 1987), 5.1 and 5.9.
17 "Pro-Streetcar Lobby Taking Appeal to TTC," *Toronto Star*, 1 December 1973, A7; Margaret Daly, "Fate of Our Streetcars Goes before TTC Tomorrow," *Toronto Star*, 6 November 1972, 8; "TTC Votes to Save the Streetcar System, New Designs Sought," *Toronto Star*, 7 November 1972; and William Bragg, "New Toronto Streetcars Could Move at 70 M.P.H.," *Toronto Star*, 1 February 1975, A1. The TTC ordered almost 200 new streetcars at a cost of about $470,000 each. They began service in 1977. See Stan Fischler, "With Luck, TTC's New Trolleys Will Stay on Track," *Toronto Star*, 20 August 1977, B4.
18 Toronto Transit Commission. *Annual Report for the Year 1981* (Toronto: Toronto Transit Commission, 1981).
19 Michael's grandson (Brian's nephew), Felix, was born at Mount Sinai Hospital in June 2016. That hospital lies between the 506 and 505 streetcar lines. Before the age of two, Felix could identify a streetcar by type. Grandson Emmett was born in the same hospital in January 2019. One of his first words was "streetcar."
20 Statistics on the operations of the TTC will be taken from its 2019 Annual Report, the most recent that was available at the time of writing. Census data will be taken from the 2011 and 2016 files, as appropriate. Building height information will be taken from 2020 information, while employment data will pertain to 2016.
21 On the forced amalgamation see Frances Frisken, *The Public Metropolis: The Political Dynamics of Urban Expansion in the Toronto Region, 1924–2003* (Toronto: Canadian Scholar's Press, 2007), 239–58.
22 Two small extensions to the Bloor-Danforth subway line – extending service to Kennedy Road in the east and Kipling Avenue in the west – did open in November 1980. These added just 4.3 kilometres and two new stations to the system. The Scarborough RT measured 7.2 kilometres in length. Going into the 1980s, Toronto's subway system was 52.6 kilometres long and serviced 57 stations. See Peter Howell, "Why 'The Better Way' Is Now the Bitter Way," *Toronto Star*, 10 March 1990, A1 and A17. Howell's series appeared between 10 March and 14 March, 1990. Toronto's first subway opened in 1954, Montreal's in 1966. "Spinning Its Wheels" aired as a segment of the CBC television program *Monitor* in November 1990. For more on the declining reputation of the TTC, especially during Rob Ford's term as mayor, see Taras Grescoe, *Straphangers: Saving Our Cities and Ourselves from the Automobile* (Toronto: HarperCollins, 2012), 288–305.
23 Toronto Transit Commission, *Russell Hill Subway Train Accident, August 11, 1995: Final Report* (Toronto: Toronto Transit Commission, December 1995).
24 The last trolley coaches operated in Toronto in July 1993, with PCC streetcars last used in revenue service in December 1995. Two PCCs (4500 and 4549) were kept by the TTC for use in parades and charter services. Similarly, one Peter Witt streetcar (2766) was retained. In June 2009, the TTC signed a contract with Bombardier to supply 204 low-floor streetcars at a cost of $1.25 billion. It also committed to spend about $500 million on a new streetcar storage and maintenance facility, now known as the Leslie Barns, to provide for the new vehicles. See http://stevemunro.ca/2013/04/13/the-saga-of-leslie-barns/.
25 Now incorporated into the 504 King route, which has been split between two sections.
26 The 514 Cherry route was eliminated in October 2018, with the service it had provided being replaced by two

branches of the 504 line. One (504B) travelled from Broadview station to Dufferin Gate loop at the western entrance to the CNE grounds, with the other (504A) going from the Dundas West station to the Distillery loop.

27 For a discussion of the trials and tribulations of the St. Clair project see http://transit.toronto.on.ca/streetcar/4126.shtml.

28 According to one study by European-based moving company Movinga, the cost of a monthly transit pass in Toronto was the fifth-highest in the world out of 89 cities surveyed. For details see http://www.cbc.ca/news/canada/toronto/toronto-expensive-public-transportation-1.3963426.

29 TTC statistics for 2018 were taken from http://www.ttc.ca/About_the_TTC/Operating_Statistics/2018/Section_One.jsp and http://www.ttc.ca/PDF/About_the_TTC/Annual_Reports/2018_TTC_Annual_Report.pdf. As of 1 March 2015, children aged 12 and under were able to ride the TTC for free.

30 Information on the history of recessions in Canada was found at https://www.thecanadianencyclopedia.ca/en/article/recession.

31 Tess Kalinowski and Rob Ferguson, "TTC Workers' Union Fights for a Return to the Right to Strike," *Toronto Star,* 26 October 2015. Information on the TTC's strike history was taken from the Canadian Public Transit Discussion Board: https://cptdp.ca/topic/3296-transit-strike-history.

32 Toronto Transit Commission, *Annual Report, 2003* (Toronto: Toronto Transit Commission, 2004), 3.

33 On SARS and Toronto see Center for Infectious Disease Research and Policy, "SARS in Toronto considered contained," *CIDRAP News*, 2 July 2003, https://cidrap.umn.edu/news-perspective/2003/07/sars_toronto_considered_contained. See also Donald E. Low, "SARS: Lessons from Toronto," https://www.ncbi.nlm.nih.gov/books/NBK92467/. Globally, there were 8,098 cases of SARS, with 774 deaths, as reported in Centers for Disease Control, "Fact Sheet: Basic Information about SARS," 13 January 2004, https://www.cdc.gov/sars/about/fs-sars.html.

34 Toronto Transit Commission, *Annual Report, 2003*, 3

35 Statistics on COVID-19 cases in Toronto were accessed from https://www.toronto.ca/home/covid-19/covid-19-latest-city-of-toronto-news/covid-19-status-of-cases-in-toronto/. Information on the progress of the disease was found at Lauren Vogel, "COVID-19: A Timeline of Canada's First-Wave Response," *CMAJ*, 12 June 2020, https://cmajnews.com/2020/06/12/coronavirus-1095847/.

36 On COVID-19 and the TTC see Toronto Transit Commission, "Chief Executive Officer's Report – July 2020 Update," https://www.ttc.ca/About_the_TTC/Commission_reports_and_information/Commission_meetings/2020/July_14/Reports/1_Chief_Executive_Officer_Report_July_2020_Update.pdf; Ben Spurr, "Who's Still Crowding TTC buses amid the Pandemic? Evidence Suggests Many Are Toronto's Working Poor," *Toronto Star*, 7 April 2020; Ben Spurr, "What to Expect from Historic TTC Service Cuts Coming Next Week over the COVID-19 Crisis," *Toronto Star*, 5 May 2020; Ben Spurr, "Bus Riders Will Find It Tough to Physically Distance as Toronto Opens Up, the TTC Has Warned. In fact, for Many It's Already Impossible," *Toronto Star*, 11 June 2020; Ben Spurr, "COVID-19 and Transit: What We Think We Know May Be Wrong," *Toronto Star*, 18 June 2020; Matt Elliott, "There's Been a Torrent of Data Released during the Pandemic. Here's What It's Telling Us about Toronto," *Toronto Star*, 21 July 2020; and Lisa Power, "TTC Ridership on Buses Is Starting to Bounce Back but Still Way Off Pre-pandemic Levels," https://www.blogto.com/city/2020/06/ttc-ridership-buses-pandemic-levels/. On possible relief to help with revenue shortfalls see Alex Boutilier, Jennifer Pagliaro, and Ben Spurr, "Ontario May Get $1B in Federal Transit Funding," *Toronto Star*, 23 July 2020; and Ben Spurr, "TTC to Get $400M Bailout from Province," *Toronto Star*, 13 August 2020. On the installation of the Personal Protection Equipment vending machines see Kevin Jiang, "Vending Machines Selling PPE Coming Soon to 10 TTC stations," *Toronto Star*, 14 November 2020, A22. For a fuller discussion of COVID-19 and the TTC, see Michael Doucet, "COVID-19 and TTC Ridership: Putting the Pandemic in Historical Perspective," 27 April 2021, spacing.ca.

37 Building data were compiled on 17 January 2020 from http://www.emporis.com/city/100993/toronto-canada/status/existing/1. According to one 2017 report, Toronto ranked eighth globally in terms of the number of skyscrapers (buildings of at least 100 metres in height) it contained. See https://www.blogto.com/city/2017/04/toronto-ranked-city-8th-most-skyscrapers-world/. The 2020 skyscraper study by the Council on Tall Buildings and Urban Habitat was Reported in CBC, "Toronto on Track to Have More Skyscrapers than Chicago, But Will Quality Match Quantity?" https://www.cbc.ca/news/canada/toronto/toronto-skyscrapers-chicago-1.5429816 on 26 January 2020. On Aura, see Patricia McHugh and Alex Bozikovic, *Toronto Architecture: A City Guide*, 3rd ed. (Toronto: McClelland & Stewart, 2017), 70.

38 The area that contained the old City of Toronto included census tracts 001 to 142 as defined by Statistics Canada.

39 The 905 area refers to the telephone area code for this part of Ontario, to which the 289 code recently was added, while the new City of Toronto is within the 416 and 647 area codes.

40 By 2016, the Toronto CMA had become much less useful as a determinant of Toronto's commuter shed. There no longer was any room for the Toronto CMA to expand because it had become encircled by other CMAs – Hamilton to the west, Guelph to the northwest, Barrie

to the north, Peterborough to the northeast, and Oshawa to the east.

41 Unfortunately, it was not possible to calculate average house values and the average contract rent for the old City of Toronto from the 2016 census data.

42 Hans Ibelings and PARTISANS, *Rise and Sprawl: The Condominiumization of Toronto* (Montreal and Amsterdam: The Architecture Observer, 2016), 9–10.

43 A. Kramer, "Inside and Outside: A Meditation on the Yellowbelt," in Alex Bozikovic, Cheryll Case, John Lorinc, and Annabel Vaughan (eds.), *House Divided: How the Missing Middle Can Solve Toronto's Affordability Crisis* (Toronto: Coach House Press, 2019), 142–55.

44 David Hulchanski, *The Three Cities within Toronto: Income Polarization among Toronto's Neighbourhoods, 1970–2005* (Toronto: Cities Centre, University of Toronto, 2010). http://www.urbancentre.utoronto.ca/pdfs/curp/tnrn/Three-Cities-Within-Toronto-2010-Final.pdf

45 Maurice Yeates, Paul Du, and Tansel Erguden, *Charting the GTA* (Toronto: Centre for the Study of Commercial Activity, Ryerson University, 2011), 46.

46 Office space statistics were derived from City of Toronto, *Office Space Statistics* (Toronto: City of Toronto, Development Department, Research and Information Division, 1973), 1–3; and Deloitte, *Yonge-Bloor Concentration: Office Market Analysis* (Toronto: Deloitte, 2013), https://www1.toronto.ca/City%20Of%20Toronto/City%20Planning/Community%20Planning/Files/pdf/8/80bloor_office-report.pdf. Figures for 2017 were taken from Avison Young, *Greater Toronto Area: Office Market Report, Fourth Quarter 2017* (Toronto: Avison Young, 2018), https://www.avisonyoung.ca/documents/95732/4519304/GTA+Office+Market+Report+(Q4+2017)/a5587866-57a5-4228-a61e-a4f3130ce8a0?t=526972673.

47 On the history of parking in Toronto, see Deryck W. Holdsworth, *The Parking Authority of Toronto, 1952–1987* (Toronto: Parking Authority of Toronto, 1987).

48 Richard Longley, "Facadism: Is It an Architectural Plague or Preservation," *Now*, 19 May 2016, 10–17.

49 Robert Allsopp, "Killing Yonge Street," *Now*, 7 July 2016, 12.

50 Toronto Real Estate Board, *Market Watch*, December 2019, 1.

51 Mike Filey, "Electric Streetcars 124 Years in the Making," *Toronto Sun*, 1 May 2016, 26.

52 National Geographic, *Journeys of a Lifetime: 500 of the World's Greatest Trips* (Washington: National Geographic, 2007), 126.

53 There were numerous delivery problems with the Bombardier LRVs, the last of which arrived in January 2020. See John Thompson, "Last Flexity Arrives," *Transfer Points* 46 (March 2020): 3–4. On the purchase of the 60 additional streetcars, see Ben Spurr, "TTC to Add 60 Streetcars to Its Fleet," *Toronto Star*, 13 May 2021, A12.

54 On the King Street Pilot see Jaren Kerr, "Streetcars Now Rule the Road along King St.," *Toronto Star*, 11 November 2017, GT1 and GT6; Edward Keenan, "A Better Way?," *Toronto Star*, 14 November 2017, A1 and A13; and Ben Spurr, "King Pilot Boosts [Morning Rush Hour] Ridership 25%," *Toronto Star*, 12 January 2018, GT1 and GT6. On the possibility of replicating the King Street Pilot on other streets see Ben Spurr, "City Mulls New Transit-Priority Measures," *Toronto Star*, 11 November 2020, A15.

55 Ken Greenberg, *Toronto Reborn: Design Successes and Challenges* (Toronto: Dundurn Press, 2019), 275.

56 On these LRT projects see http://www.metrolinx.com/en/projectsandprograms/transitexpansionprojects/toronto_lrt.aspx. See also Edward J. Levy, *Rapid Transit in Toronto: A Century of Plans, Projects, Politics, and Paralysis* (Toronto: BA Consulting Group, 2015).

Chapter 2

1 On the interplay of global economic trends and the development of transportation infrastructure in Montreal, Toronto, and Vancouver see Anthony Perl, Matt Hern, and Jeffrey R. Kenworthy, *Big Moves: Global Agendas, Local Aspirations and Urban Mobility in Canada* (Montreal and Kingston: McGill-Queen's University Press, 2020).

2 Robert Fulford, *Accidental City: The Transformation of Toronto* (Toronto: Mcfarlane Walter & Ross, 1995).

3 James Lemon, *Toronto: An Illustrated History since 1918* (Toronto: Lorimer, 1985), 186.

4 *The Rise and Fall of English Montreal*, National Film Board of Canada, http://onf-nfb.gc.ca/en/our-collection/?idfilm=29198.

5 David Ley and Peter Murphy, "Immigration in Gateway Cities: Sydney and Vancouver in Comparative Perspective," *Progress in Planning* 55 no. 3 (2001): 119–94.

6 Ibid., 122.

7 Lemon, *Toronto*, 183–4.

8 The Inglis plant lasted for 108 years on Strachan Avenue. It closed in 1989. See David Sobel and Susan Meurer, *Working at Inglis: The Life and Death of a Canadian Factory* (Toronto: James Lorimer, 1994).

9 On the history of the stockyards and the meat packing industry in Toronto see D.R. McDonald, *The Stockyard Story (1803–1985)* (Toronto: NC Press, 1985); and Michael Bliss, *A Canadian Millionaire: The Life and Business Times of Sir Joseph Flavelle, Bart., 1858–1939* (Toronto: University of Toronto Press, 1978).

10 Alan Walks, "Stopping the 'War on the Car': Neoliberalism, Fordism, and the Politics of Automobility in Toronto," *Mobilities* 10 no. 3 (2015): 402–22.

11 Gillad Rosen and Alan Walks. "Castles in Toronto's Sky: Condo-ism as Urban Transformation." *Journal of Urban Affairs* 37, no. 3 (2015): 289–310.

12 Julie-Anne Boudreau, Roger Keil, and Douglas Young, *Changing Toronto: Governing Urban Neoliberalism* (Toronto: University of Toronto Press, 2009).
13 Ibid. See also Jason Hackworth, "Why There Is No Detroit in Canada," *Urban Geography* 37, no. 2 (2016): 163.
14 See Paul Knox "World Cities in a World-System," in Paul Knox and Peter Taylor (eds.), *World Cities in a World System* (Cambridge: Cambridge University Press, 1995); John Friedman "Where We Stand: A Decade of World-City Research," in Knox and Taylor, *World Cities*.
15 Peter Eisinger, "The Politics of Bread and Circuses: Building the City for the Visitor Class," *Urban Affairs Review* 35, no. 3 (2000): 316–33; Neil Brenner, "Urban Governance and the Production of New State Spaces in Western Europe, 1960–2000," *Review of International Political Economy* 11, no. 3 (2004): 447–88; Alain Lipietz, *Mirages and Miracles: The Crisis of Global Fordism* (London: Verso, 1989).
16 Knox, "World Cities."
17 Hackworth, "Why There Is No Detroit in Canada," 272–95.
18 George Galster, *Driving Detroit: The Quest for Respect in the Motor City* (Philadelphia: Penn Press, 2012).
19 John Friedmann and Goetz Wolff, "World City Formation: An Agenda for Research and Action," *International Journal of Urban and Regional Research* 6, no. 3 (1982): 309–44; John Friedmann, "The World City Hypothesis," *Development and Change* 17, no. 1 (1986): 70.
20 Friedmann, "The World City Hypothesis," 70.
21 *The World According to GaWC, 2016*, http://www.lboro.ac.uk/gawc/world2016t.html.
22 Stephen P. Meyer, "Finance, Insurance and Real Estate Firms and the Nature of Agglomeration Advantage across Canada and within Metropolitan Toronto," *Canadian Journal of Urban Research* 16, no. 2 (2007): 149–81; Graham Todd, "'Going Global' in the Semi-periphery," in Knox and Taylor, *World Cities*.
23 Todd, "'Going Global.'"
24 F. Hou and L.S. Bourne, "The Migration-Immigration Link in Canada's Gateway Cities: A Comparative Study of Toronto, Montreal, and Vancouver," *Environment and Planning A* 38 (2006): 1505–25; David Ley, "Countervailing Immigration and Domestic Migration in Gateway Cities: Australian and Canadian Variations on an American theme," *Economic Geography* 83 (2007): 231–54.
25 Jason Hickel, "Is Global Inequality Getting Better or Worse? A Critique of the World Bank's Convergence Narrative," *Third World Quarterly* 38, no. 10 (2017): 2208–22.
26 Kate Allen, Jennifer Yang, Rachel Mendleson, and Andrew Bailey, "Lockdown Worked for the Rich, but Not for the Poor: The Untold Story of How COVID-19 Spread across Toronto, in 7 Graphics," *Toronto Star*, 2 August 2020, https://www.thestar.com/news/gta/2020/08/02/lockdown-worked-for-the-rich-but-not-for-the-poor-the-untold-story-of-how-covid-19-spread-across-toronto-in-7-graphics.html; Jay Pitter, "Urban Density: Confronting the Distance between Desire and Disparity," *Azure Magazine*, 17 April 2020, https://www.azuremagazine.com/article/urban-density-confronting-the-distance-between-desire-and-disparity/; Jessica Cheung, "Black people and other people of colour make up 83% of reported COVID-19 cases in Toronto," *CBC Toronto*, 30 July 2020, https://www.cbc.ca/news/canada/toronto/toronto-covid-19-data-1.5669091; Brian Doucet, Rianne van Melik, and Pierre Filion (eds.), *Global Reflections on COVID-19 and Urban Inequalities* (Bristol: Bristol University Press, 2021).
27 Saskia Sassen *The Global City: New York, London, Tokyo* (Princeton: Princeton University Press, 2013).
28 Peter Marcuse, "'Dual City': A Muddy Metaphor for a Quartered City," *International Journal of Urban and Regional Research* 13, no. 4 (1989): 699.
29 Ute Lehrer and Jennefer Laidley, "Old Mega-projects Newly Packaged? Waterfront Redevelopment in Toronto," *International Journal of Urban and Regional Research* 32, no. 4 (2008): 799.
30 Similar trends can be found along the waterfront in Rotterdam, where early redevelopments of industrial spaces in the 1980s focused on social housing. By the 1990s, urban policy shifted towards creating a new city centre, with high-end offices, cultural facilities, and housing. See Brian Doucet, *Rich Cities with Poor People: Waterfront Regeneration in the Netherlands and Scotland* (Utrecht: Koninklijk Nederlands Aardrijkskundig Genootschap, 2010).
31 Peter Taylor, *World City Network: A Global Urban Analysis* (London: Routledge, 2004); John Friedman, "Where We Stand: A Decade of World-City Research," in Knox and Taylor, *World Cities in a World System*.
32 David Bassens and Michiel van Meeteren. "World Cities under Conditions of Financialized Globalization: Towards an Augmented World City Hypothesis," *Progress in Human Geography* 39, no. 6 (2015): 752–75.
33 Sassen, *The Global City*, 554–62.
34 Saskia Sassen, "Global Financial Centers," *Foreign Affairs* 78, no. 1 (1999): 75–87.
35 Domenic Viteillo, *The Philadelphia Stock Exchange and the City That Made It* (Philadelphia: Penn Press, 2010).
36 Scotiabank was founded in Halifax and is now officially headquartered in Toronto. The Bank of Montreal has its headquarters in both Toronto and Montreal.
37 Saskia Sassen, "The Impact of the New Technologies and Globalization on Cities," in Arie Graafland and Deborah Hauptmann (eds.), *Cities in Transition* (Rotterdam: 010 Publishers, 2001), as cited in Richard T. LeGates and Frederic Stout (eds.), *The Global Cities Reader*, 5th edition (London: Routledge, 2001), 560.

38 Todd, "'Going Global.'"
39 Ibid., 199.
40 Ute Lehrer and Thorben Wieditz, "Condominium Development and Gentrification: The Relationship between Policies, Building Activities and Socio-economic Development in Toronto," *Canadian Journal of Urban Research* 18, no. 1 (2009): 82–103.
41 Robert Williamson, "1,650 Ministry Jobs to Quit Metro for Kingston, Oshawa," *Globe and Mail*, 6 April 1977, 1–2.
42 "Peterson Moves 145 Lottery Jobs to Sault Area," *Toronto Star*, 9 July 1986, A3; Sandro Contenta, "Mines Ministry, 290 Jobs to Move to Sudbury Area: Just Beginning of Attempts to Aid Northern Ontario: Premier," *Toronto Star*, 31 July 1986, A2; and Alan Christie, "Premier Moving 230 Ministry Jobs to Thunder Bay," *Toronto Star*, 14 February 1987, A3.
43 Neil Brenner, "Urban Governance and the Production of New State Spaces in Western Europe, 1960–2000," *Review of International Political Economy* 11, no. 3 (2004): 447–88; Peter Eisinger, "The Politics of Bread and Circuses: Building the City for the Visitor Class," *Urban Affairs Review* 35, no. 3 (2000): 316–33.
44 Tim Hall and Peter Hubbard, "The Entrepreneurial City: New Urban Politics, New Urban Geographies?" *Progress in Human Geography* 20, no. 2 (1996): 155.
45 David Harvey, "From Managerialism to Entrepreneurialism: The Transformation in Urban Governance in Late Capitalism," *Geografiska Annaler*, Series B. *Human Geography* 71, no. 1 (1989): 3–17.
46 George Galster, "Detroit's Bankruptcy: Treating the Symptom, Not the Cause," in Brian Doucet (ed.), *Why Detroit Matters: Decline, Renewal and Hope in a Divided City* (Bristol: Policy Press, 2017).
47 David Harvey, *Spaces of Hope* (Berkeley: University of California Press, 2000). Brian Doucet, Ronald van Kempen, and Jan van Weesep, "'We're a Rich City with Poor People': Municipal Strategies of New-Build Gentrification in Rotterdam and Glasgow," *Environment and Planning A* 43, no. 6 (2011): 1438–54.
48 Ute Lehrer, Roger Keil, and Stefan Kipfer, "Reurbanization in Toronto: Condominium Boom and Social Housing Revitalization," *disP - The Planning Review* 46, no. 180 (2010): 81–90.
49 Boudreau, Keil, and Young, *Changing Toronto*, 29.
50 Ibid.
51 Stefan Kipfer and Roger Keil, "Toronto Inc? Planning the Competitive City in the New Toronto," *Antipode* 34, no. 2 (2002): 227–64; see also Roger Keil, "The Urban Politics of Roll-with-It Neoliberalization," *City* 13, nos. 2–3 (2009): 230–45; Boudreau, Keil, and Young, *Changing Toronto*; Hackworth, "Why There Is No Detroit in Canada."
52 See Neil Smith, *The New Urban Frontier: Gentrification and the Revanchist City* (New York: Routledge, 2005); for a Toronto example of revanchism, see Roger Keil, "'Common-Sense' Neoliberalism: Progressive Conservative Urbanism in Toronto, Canada," *Antipode* 34, no. 3 (2002): 578–601.
53 Ibid. See also "Ontario tries wiping streets clean of 'squeegee kids,'" *CBC News*, 31 January 2000, https://www.cbc.ca/news/canada/ontario-tries-wiping-streets-clean-of-squeegee-kids-1.251291; Muriel Draaisma, "New Ontario rule banning carding by police takes effect," *CBC News*, 1 January, 2017, https://www.cbc.ca/news/canada/toronto/carding-ontario-police-government-ban-1.3918134.
54 Lehrer and Laidley, "Old Mega-projects," 786–803.
55 Ken Greenberg, "Toronto: The Urban Waterfront as a 'Terrain of Availability,'" in Patrick Malone (ed.), *City, Capital and Water* (New York: Routledge, 1996); see also Lehrer and Laidley, "Old Mega-projects," 786–803.
56 Lehrer and Laidley, "Old Mega-projects," 786–803.
57 See https://waterfrontoronto.ca/.
58 Lehrer and Laidley, "Old Mega-projects," 786–803. See also Richard Florida, *The Rise of the Creative Class and How It's Transforming Work, Leisure, Community and Everyday Life* (New York: Basic Books: 2004).
59 Florida, *The Rise of the Creative Class*.
60 Boudreau, Keil, and Young, *Changing Toronto*; Hackworth, "Why There Is No Detroit in Canada."
61 Lehrer and Laidley, "Old Mega-projects," 786–803.
62 Ibid. For literature on POPS see Jeremy Németh, "Defining a Public: The Management of Privately Owned Public Space," *Urban Studies* 46, no. 11 (2009): 2463–90; Jeremy Németh and Stephen Schmidt, "The Privatization of Public Space: Modeling and Measuring Publicness," *Environment and Planning B: Planning and Design* 38, no. 1 (2011): 5–23; Langstraat Florian and Rianne van Melik, "Challenging the 'End of Public Space': A Comparative Analysis of Publicness in British and Dutch Urban Spaces," *Journal of Urban Design* 18, no. 3 (2013).
63 Lehrer and Laidley, "Old Mega-projects," 786–803.
64 Ibid.
65 Ibid., 800.
66 Penny Bernstock, *Olympic Housing: A Critical Review of London 2012's Legacy* (London: Routledge, 2014).
67 As quoted in Jim Byers, "Toronto 2008 Olympic Bid Attracts Major Sponsor," *Toronto Star*, 1 November 2000, E9.
68 Bruce Kidd, "The Toronto Olympic Commitment: Towards a Social Contract for the Olympic Games," *Olympika: The International Journal of Olympic Studies* 1 (1992): 156.
69 David Harvey, *The Condition of Post-modernity: An Enquiry into the Origins of Cultural Change* (Oxford: Wiley-Blackwell, 1989).
70 Gordon Waitt, "The Olympic Spirit and Civic Boosterism: The Sydney 2000 Olympics," *Tourism Geographies* 3, no. 3 (2001): 254.

71 Monica Nickelsberg, "Urbanist Richard Florida on what splitting Amazon HQ2 means for cities: 'It never was about a second headquarters,'" *Geekwire*, 7 November 2018, https://www.geekwire.com/2018/richard-florida-splitting-amazon-hq2-means-cities-never-second-headquarters/.
72 Phillip Longman, "Boom and Bust," *Washington Monthly*, November–December 2015, https://washingtonmonthly.com/magazine/novdec-2015/bloom-and-bust/#.W9srGIoxtsw.twitter.
73 Leticia Miranda, Nicole Nguyen, and Ryan Mac, "Here Are the Most Outrageous Incentives Cities Offered Amazon in Their HQ2 Bids," *Buzzfeed News*, 14 November 2018.
74 Nick Boisvert "'We Are Now and Tomorrow': Toronto Makes Its Pitch for Amazon's HQ2," *CBC News*, 19 October 2017.
75 James Strashin, "Why Calgary Passed on the 2026 Olympics – and What's Next for the Games Nobody Seems to want," *CBC News*, 14 November 2018
76 Andy Valvur, "6 Cities That Rejected the Olympics," *DW*, 14 November 2018.
77 C. Ian Kyer, *A Thirty Years' War: The Failed Public/Private Partnership That Spurred the Creation of the Toronto Transit Commission, 1891–1921* (Toronto: Irwin Law, 2015); and Michael Doucet, "Politics, Space, and Trolleys: Mass Transit in Early Twentieth-Century Toronto," in Gilbert A. Stelter and Alan F. J. Artibise (eds.), *Shaping the Urban Landscape: Aspects of the Canadian City-Building Process*, Carleton Library Series No. 125 (Ottawa: Carleton University Press, 1982), 356–81.
78 On the mandate for Metrolinx see http://www.metrolinx.com/en/aboutus/metrolinxoverview/metrolinx_overview.aspx.
79 Royson James, "Let the Eglinton Crosstown LRT Be a Lesson: Beware Public-Private Partnerships," *Toronto Star*, 4 January 2020.
80 Ben Spurr, "Companies Building Eglinton Crosstown LRT Sue Metrolinx over COVID-19 Costs, Delays," *Toronto Star*, 8 October 2020; Ben Spurr. "Companies Building Eglinton Crosstown LRT Sue Metrolinx for Breach of Contract," *Toronto Star*, 11 July 2018.
81 Matti Siemiatycki, "Urban Transportation Public–Private Partnerships: Drivers of Uneven Development?," *Environment and Planning A* 43, no. 7 (2011): 1707–22.
82 Stephen Graham and Simon Marvin, *Splintering Urbanism: Networked Infrastructures, Technological Mobilities and the Urban Condition* (London: Routledge, 2001).
83 Ibid., p. 9.
84 Ibid., Introduction.
85 As quoted in Chris Bateman, "How Toronto Turned an Airport Rail Failure into a Commuter Asset," *CityLab*, 19 January 2018.
86 "Union Pearson Express Launches Reduced Prices Today," *CBC News*, 9 March 2016.
87 Bateman, "How Toronto Turned an Airport Rail Failure."

Chapter 3

1 John Sewell, *The Shape of the City: Toronto Struggles with Modern Planning* (Toronto: University of Toronto Press, 1993), xiii–xiv.
2 R.A. Walks, "The Boundaries of Suburban Discontent? Urban Definitions and Neighbourhood Political Effects," *The Canadian Geographer/Le Géographe canadien* 51, no. 2 (2007): 160–85; R.A. Walks, "The Causes of City-Suburban Political Polarization? A Canadian Case Study," *Annals of the Association of American Geographers* 96, no. 2 (2006): 390–414; A. Walks, "Stopping the 'War on the Car': Neoliberalism, Fordism, and the Politics of Automobility in Toronto," *Mobilities* 10, no. 3 (2015): 402–22. See also John Sewell, "Old and New City," in K. Gerecke (ed.), *The Canadian City* (Montreal: Black Rose Books, 1991); David Ley, "Artists, Aestheticisation and the Field of Gentrification," *Urban studies* 40, no. 12 (2003): 2527–44; Jon Caulfield, *City Form and Everyday Life: Toronto's Gentrification and Critical Social Practice* (Toronto: University of Toronto Press, 1994); Pierre Filion, "Enduring Features of the North American Suburb: Built Form, Automobile Orientation, Suburban Culture and Political Mobilization," *Urban Planning* 3, no. 4 (2018): 4–14.
3 Sam Bass Warner, *Streetcar Suburbs: The Process of Growth in Boston (1870–1900)* (Cambridge: Harvard University Press, 1962).
4 Edward Relph, *Toronto: Transformations in a City and Its Region* (Philadelphia: Penn Press, 2014).
5 Stephen M. Wheeler, "The Evolution of Urban Form in Portland and Toronto: Implications for Sustainability Planning," *Local Environment* 8, no. 3 (2003): 317–36.
6 Dylan Reid, "Down on the Corner: Spotting Former Neighbourhood Commercial Buildings, *Spacing*, no. 54 (2020): 26–7.
7 Patrick Sisson, "How the '15-Minute City' Could Help Post-Pandemic Recovery," *CityLab*, 15 July 2020, https://www.bloomberg.com/news/articles/2020-07-15/mayors-tout-the-15-minute-city-as-covid-recovery; "Want to Build a Better City? It Only Takes 15 Minutes," *Globe and Mail*, 2 August 2020, https://www.theglobeandmail.com/opinion/editorials/article-want-to-build-a-better-city-it-only-takes-15-minutes/; Fergus O'Sullivan and Laura Bliss, "The 15-Minute City – No Cars Required – Is Urban Planning's New Utopia," *Bloomberg Business Week*, 12 November 2020, https://www.bloomberg.com/news/features/2020-11-12/paris-s-15-minute-city-could-be-coming-to-an-urban-area-near-you; Sharee Hochman, "Organizing Accessibility and Intersectionality through 15-Minute Cities," *Monitor* 28, no. 1 (May–June 2021): 27–9.
8 Marshall McLuhan, *Understanding Media* (Toronto: Signet Press, 1964); see also Edward Relph, *Toronto: Transformations in a City and Its Region* (Philadelphia: Penn Press, 2014).

9 Relph, *Toronto: Transformations*. On self-building in Toronto see Richard Harris, *Unplanned Suburbs: Toronto's American Tragedy, 1900 to 1950* (Baltimore: Johns Hopkins University Press, 1996).

10 On the development of Lawrence Park see Karina A. C. Bordessa, "A Corporate Suburb of Toronto: Lawrence Park, 1905–1930," MA thesis, York University, 1980. On Kingsway Park see Ross H. Paterson, "Kingsway Park, Etobicoke: An Analysis of the Development of an Early Twentieth Century Suburb," MA thesis, York University, 1982.

11 On the development of Leaside see Jane Pittfield (ed.), *Leaside* (Toronto: Natural Heritage Books, 1999).

12 Sewell, *The Shape of the City*, 22–42.

13 On the development of Regent Park, Canada's first public housing project, see Albert Rose, *Regent Park: A Study in Slum Clearance* (Toronto: University of Toronto Press, 1958). On public housing in Toronto see also Michael McMahon, *Metro's Housing Company: The First 35 Years* (Toronto: Metropolitan Toronto Housing Company Limited, 1990).

14 Sewell, *The Shape of the City*, 201.

15 Pierre Filion and Trudi Bunting, "Urban Transitions: The History and Future of Canadian Urban Development," in Pierre Filion, Markus Moos, Tara Vinodrai, and Ryan Walker (eds.), *Canadian Cities in Transition*, 5th edition (Don Mills: Oxford University Press Canada, 2015).

16 Jonathan English, *The Better Way: Transit Service and Demand in Metropolitan Toronto, 1953–1990* (PhD thesis, Columbia University, 2020).

17 Stephen M. Wheeler, "The Evolution of Urban Form in Portland and Toronto: Implications for Sustainability Planning," *Local Environment* 8, no. 3 (2003): 317–36; see also Pierre Filion, "Enduring Features of the North American Suburb: Built form, Automobile Orientation, Suburban Culture and Political Mobilization," *Urban Planning* 3, no. 4 (2018): 4–14. For further reading on the different periods of urban spatial development in Canada, see Pierre Filion and Trudi Bunting "Urban Transitions: The History and Future of Canadian Urban Development," in Filion and Bunting (eds.), *Canadian Cities in Transition: Perspectives for an Urban Age* (Don Mills, ON: Oxford University Press, 2015).

18 Sewell, *The Shape of the City*, 80–96.

19 Ibid., 88.

20 Jill Grant, "Shaped by Planning: The Canadian City through Time," in Bunting and Filion, *Canadian Cities in Transition*.

21 ERA Architects, Planning Alliance, and Cities Centre at the University of Toronto, *Tower Neighbourhood Renewal in the Greater Golden Horseshoe*, November 2010, http://towerrenewal.com/tower-neighbourhood-renewal-in-the-greater-golden-horseshoe/.

22 Filion, "Enduring Features," 4–14.

23 In addition to streetcars, there were a number of rural radial lines around Toronto: the Toronto and York Radial Railway lines up Yonge Street to Lake Simcoe, along Kingston Road to West Hill, and west to Mimico and Port Credit and Toronto Suburban Railways lines on Weston Road to Woodbridge and from Toronto to Guelph. For defining the Streetcar City, we only included the parts of these radial railways that were upgraded to streetcar lines by the TTC in the 1920s. Generally this involved double tracking the route, increasing frequency, and running streetcars. Former radial lines included in the Streetcar City are the Long Branch line, the Weston line and the Kingston Road line to Birchmount.

24 Rosedale was constructed in the early twentieth century, with some of the principles of modernist planning. However, it was an anomaly, and was designed as an elite neighbourhood and not something to be replicated across the city. See Sewell, *The Shape of the City*, 21; and Bess Hillery Crawford, *Rosedale* (Erin, ON: Boston Mills Press, 2000).

25 Due to the availability of data, this was calculated using the 2001 census, which has more variables readily accessible at the census tract level.

26 In three CTs in Etobicoke around Bloor Street and the Kingsway the majority of housing stock was constructed before 1945. There is a very small gap between them and the CTs on the other side of the Humber River, where the Bloor line ended. This gap is due to an oddly shaped CT to the north, and the presence of a cemetery (established in 1892) on the south side of Bloor. In both these CTs, the majority of the area is far away from Bloor Street. Therefore, it has been decided to include these Etobicoke CTs in the tally for the Streetcar City, as they also share many characteristics with neighbourhoods such as Leaside.

27 The reliability and availability of data at the census tract level from the 1961 census was not as plentiful as that from 1971, hence the decision to analyse 1971 data on the Streetcar City.

28 On the Jane's Walk movement see http://janeswalk.org/.

29 Jane Jacobs, *The Death and Life of Great American Cities* (New York: Vintage Books, 1961), 34.

30 Ibid., 50.

31 For more on this, see Anthony Flint, *Wrestling with Moses* (New York: Random House, 2011). Although he never mentions Jacobs by name, the most definitive account of Robert Moses is in Robert Caro, *The Power Broker: Robert Moses and the Fall of New York* (New York: Alfred A. Knopf, 1974).

32 This issue will be discussed in more detail later in the book under the context of a history of Toronto's streetcars.

33 James Lemon, *Toronto: An Illustrated History since 1918* (Toronto: Lorimer, 1985).

34 Sewell, *The Shape of the City*, 177–82. Also on the Spadina Expressway see David and Nadine Nowlan, *The Bad*

Trip: The Untold Story of the Spadina Expressway (Toronto: new press/House of Anansi, 1970); and Anthony Perl, Matt Hern, and Jefferey Kenworthy, *Big Moves: Global Agendas, Local Aspirations and Urban Mobility in Canada* (Montreal and Kingston: McGill-Queen's University Press, 2020), 166–74.

35 Julie-Anne Boudreau, Roger Keil, and Douglas Young, *Changing Toronto: Governing Urban Neoliberalism* (Toronto: University of Toronto Press, 2009); Jason Hackworth, "Why There Is No Detroit in Canada," *Urban Geography* 37, no. 2 (2016): 272–95; Ute Lehrer, Roger Keil, and Stefan Kipfer, "Reurbanization in Toronto: Condominium Boom and Social Housing Revitalization," *disP - The Planning Review* 46, no. 180 (2010): 81–90.

36 Lemon, *Toronto*, 168.

37 Seila Rivzik, "How Non-Profit Housing Developers Could Ease Toronto's Affordability Crisis," *The Tyee*, https://thetyee.ca/News/2018/11/20/How-Developers-Ease-Toronto-Affordability-Crisis/.

38 Lemon, *Toronto*.

39 For more on this, see Sewell, *The Shape of the City*.

40 Mark Osbaldeson *Unbuilt Toronto: A History of a City That Might Have Been* (Toronto: Dundurn, 2008), 48.

41 Ibid.

42 Sewell, *The Shape of the City*; John Sewell, *Up against City Hall* (Toronto: James Lorimer, 1972). See also Graham Fraser, *Fighting Back: Urban Renewal in Trefann Court* (Toronto: Hakkert, 1972).

43 Sewell, *The Shape of the City*, 181. On the Quebec/Gothic proposal, see 187–91.

44 Sewell, *The Shape of the City*, 182–5

45 Ibid., 191–8

46 M. Moos, T. Vinodrai, N. Revington, and M. Seasons, "Planning for Mixed Use: Affordable for Whom?," *Journal of the American Planning Association* 84, no. 1 (2018): 7–20.

47 Tess Kalinowski, "Commuting costs eat up house savings," *Waterloo Region Record*, 14 November 2018, B6.

48 See B. Borzykowski, "Goodbye Toronto, Hello Winnipeg: Are Canada's Young Giving Up Their Big City Dreams?" *Globe and Mail*, 7 October 2018; S. Berman, "The New Hamiltonians," *Toronto Life*, 21 June 2017; R. Harris, J. Dunn, and S. Wakefield, *A City on the Cusp: Neighbourhood Change in Hamilton since 1970*, Neighbourhood Change Research Partnership, Research Paper 236 (Toronto: Cities Centre, University of Toronto, 2015); R. Harris, "The Gentrification of Hamilton," *Hamilton Spectator*, 21 April 2018; R. Harris, "Gentrification Poses Challenges to Hamilton, but None That the City Can't Address," *Hamilton Spectator*, 28 April 2018.

49 Mark Davidson and Loretta Lees, "New-build 'Gentrification' and London's Riverside Renaissance," *Environment and Planning A* 37, no. 7 (2005): 1187.

50 Ruth Glass, *Introduction: Aspects of Change*, in Ruth Glass (ed.), *London: Aspects of Change* (London: MacGibbon and Kee, 1964).

51 Neil Smith, "Toward a Theory of Gentrification: A Back to the City Movement by Capital, Not People," *Journal of the American Planning Association* 45, no. 4 (1979): 538–48.

52 David Ley, *The New Middle Class and the Remaking of the Central City* (Oxford: Oxford University Press, 1996).

53 Markus Moos, "From Gentrification to Youthification? The Increasing Importance of Young Age in Delineating High-Density Living," *Urban Studies* (2015), https://doi.org/10.1177/0042098015603292.

54 Adam Gopnik, *Through the Children's Gate: A Home in New York* (New York: Vintage, 2006), 147.

55 Jason Hackworth and Neil Smith, "The Changing State of Gentrification." *Tijdschrift voor Economische en Sociale Geografie* 92 (2001): 464–77.

56 Brian Doucet, "A Process of Change and a Changing Process: Introduction to the Special Issue on Contemporary Gentrification," *Tijdschrift voor economische en sociale geografie* 105, no. 2 (2014): 125–39.

57 Loretta Lees, "Super-gentrification: The Case of Brooklyn Heights, New York City," *Urban Studies* 40 (2003): 2487–509.

58 Cody Hochstenbach and Willem R. Boterman, "Intergenerational Support Shaping Residential Trajectories: Young People Leaving Home in a Gentrifying City," *Urban Studies* (2015): https://doi.org/10.1177/0042098015613254.

59 See, for example, Giuseppe Tolfo and Brian Doucet, "Gentrification in the Media: The Eviction of Critical Class Perspective," *Urban Geography* (2020): 1–22, https://doi.org/10.1080/02723638.2020.1785247.

60 Whitepainting dates from the early 1960s in Toronto. See Harry Bruce, "Glory Be, the Whitepainters Are Coming!" *Maclean's*, 18 April 1964, 25–8.

61 Sewell, *The Shape of the City*, 176.

62 On the conversion and deconversion of housing in Parkdale, see Ana Teresa Portillo and Mercedes Sharpe Zayas, "The Urban Legend: Parkdale, Gentrification and Collective Resistance," in Alex Bozikovic, Cheryll Case, John Lorinc, and Annabel Vaughan (eds.), *House Divided: How the Missing Middle Can Solve Toronto's Affordability Crisis* (Toronto: Coach House Books, 2019), 74–82; on population decreases see Anna Kramer, "Inside and Outside: A Meditation on the Yellowbelt," in Bozikovic et al., *House Divided*, 142–53.

63 Sharon Zukin, "Consuming Authenticity, from Outposts of Difference to Means of Exclusion," *Cultural Studies* 22 (2008): 724–48; Gary Bridge and Robin Dowling, "Microgeographies of Retailing and Gentrification," *Australian Geographer* 32 (2001): 93–107.

64 Katharine N. Rankin and Heather McLean. "Governing the Commercial Streets of the City: New Terrains of

Disinvestment and Gentrification in Toronto's Inner Suburbs," *Antipode* 47, no. 1 (2015): 216–39.

65 See Jerilou Hammett, Kingsley Hammett, and Martha Cooper, *Suburbanization of New York* (New York: Princeton Architectural Press, 2007).

66 Brian Doucet and Enda Duignan, "Experiencing Dublin's Docklands: Perceptions of Employment and Amenity Changes in the Sheriff Street Community," *Irish Geography* 45 (2012): 45–65.

67 Brian Doucet, "Living through Gentrification: Subjective Experiences of Local, Non-gentrifying Residents in Leith, Edinburgh," *Journal of Housing and the Built Environment* 24 (2009): 299–315; Sharon Zukin, "New Retail Capital and Neighborhood Change: Boutiques and Gentrification in New York City," *City and Community* 8 (2009): 47–64.

68 Peter Marcuse, "Gentrification, Abandonment, and Displacement: Connections, Causes, and Policy Responses in New York City," *Journal of Urban and Contemporary Law* 28, no. 1 (1985): 195–240.

69 On "loss of place" see Kate S. Shaw and Iris W. Hagemans, "'Gentrification without Displacement' and the Consequent Loss of Place: The Effects of Class Transition on Low-Income Residents of Secure Housing in Gentrifying Areas," *International Journal of Urban and Regional Research* 39, no. 2 (2015): 323–41; on "slow violence" see Leslie Kern, "Rhythms of Gentrification: Eventfulness and Slow Violence in a Happening Neighbourhood," *Cultural Geographies* 23, no. 3 (2016): 441–57; on "un-homing" see Adam Elliott-Cooper, Phil Hubbard, and Loretta Lees, "Moving beyond Marcuse: Gentrification, Displacement and the Violence of Un-homing," *Progress in Human Geography* (OnlineFirst) (2019): 1–18; on "symbolic displacement" see Rowland Atkinson, "Losing One's Place: Narratives of Neighbourhood Change, Market Injustice and Symbolic Displacement," *Housing, Theory and Society* 32, no. 40 (2015): 373–88; see also Brian Doucet and Daphne Koenders, "'At Least It's Not a Ghetto Anymore': Experiencing Gentrification and 'False Choice Urbanism,'" in Rotterdam's Afrikaanderwijk," *Urban Studies* 55, no. 16 (2018): 3631–49; Brian Doucet, "The 'Hidden' Sides of Transit-Induced Gentrification and Displacement along Waterloo Region's LRT corridor," *Geoforum*, 125 (2021): 34–46.

70 Katharine N. Rankin and Heather McLean, "Governing the Commercial Streets of the City: New Terrains of Disinvestment and Gentrification in Toronto's Inner Suburbs," *Antipode* 47, no. 1 (2015): 216–39.

71 Ibid., 235.

72 Jason Hackworth, "Race and the Production of Extreme Land Abandonment in the American Rust Belt." *International Journal of Urban and Regional Research* 42, no. 1 (2018): 51–73.

73 Gillad Rosen and Alan Walks. "Castles in Toronto's Sky: Condo-ism as Urban Transformation," *Journal of Urban Affairs* 37, no. 3 (2015): 289–310.

74 Davidson and Lees, "New-build 'Gentrification,'" 1165–90.

75 W. Curran, "'From the Frying Pan to the Oven': Gentrification and the Experience of Industrial Displacement in Williamsburg, Brooklyn," *Urban Studies* 44, no. 8 (2007): 1427–40.

76 Leslie Kern, "Reshaping the Boundaries of Public and Private Life: Gender, Condominium Development, and the Neoliberalization of Urban Living," *Urban Geography* 28, no. 7 (2007): 659.

77 Lehrer, Keil, and Kipfer, "Reurbanization in Toronto," 81–90.

78 Rosen and Walks, "Castles in Toronto's Sky," 289–310.

79 Ute Lehrer and Thorben Wieditz, "Condominium Development and Gentrification: The Relationship between Policies, Building Activities and Socio-Economic Development in Toronto," *Canadian Journal of Urban Research* 18, no. 1 (2009): 82–103.

80 Hans Ibelings and PARTISANS, *Rise and Sprawl: The Condominiumization of Toronto* (Montreal and Amsterdam: The Architecure Observer, 2016).

81 Rosen and Walks, "Castles in Toronto's Sky," 299.

82 Ibid.

83 Neil Smith, "New Globalism, New Urbanism: Gentrification as Global Urban Strategy," *Antipode* 34, no. 3 (2002): 427–50; Neil Smith, *The New Urban Frontier: Gentrification and the Revanchist City* (New York: Routledge, 2005), 39.

84 See Smith, *The New Urban Frontier*, 39, and David Ley and Daniel Hiebert, "Immigration Policy as Population Policy," *Canadian Geographer/Le Géographe canadien* 45, no. 1 (2001): 120–5.

85 https://www12.statcan.gc.ca/census-recensement/2016/dp-pd/prof/details/Page.cfm?Lang=E&Geo1=CSD&Code1=3520005&Geo2=PR&Data=Count&B1=All.

86 https://www150.statcan.gc.ca/n1/pub/91-209-x/2018001/article/54958-eng.htm.

87 Lehrer, Keil, and Kipfer, "Reurbanization in Toronto," 81–90.

88 Noah Quastel, Markus Moos, and Nicholas Lynch, "Sustainability-as-Density and the Return of the Social: The Case of Vancouver, British Columbia," *Urban Geography* 33, no. 7 (2012): 1055–84. See also Rosen and Walks, "Castles in Toronto's Sky," 289–310.

89 Boudreau, Keil, and Young, *Changing Toronto*, 105–8.

90 Rosen and Walks, "Castles in Toronto's Sky," 303

91 Lehrer and Wieditz. "Condominium Development and Gentrification," 82–103.

92 Stephen Wickens, "Downtown Toronto Went All In with a Pair of Kings," *Globe and Mail*, 16 February 2016.

93 Lehrer and Wieditz, "Condominium Development and Gentrification," 82–103.

94 Ibid., 92.
95 Rosen and Walks, "Castles in Toronto's Sky," 304.
96 Ibid., 303–4.
97 Martine August and Giuseppe Tolfo, "Inclusionary Zoning: Six Insights from International Experience," *Plan Canada* 58, no. 4 (2018): 6–11; Matt Elliott, "A Plan to Force Developers to Build More Cheap Rentals Is a Game Changer: But Does Toronto Have the Stomach for It?," *Toronto Star*, 2 November 2021; Beth Wilson, Jeremy Withers, and Sean Meagher, "Opportunity Knocks: Toronto Council's Chance to Create Tens of Thousands of Affordable Homes," *Social Planning Toronto*, October 2021.
98 Tess Kalinowski, "Removing Rent Control on New Units Won't Ease Toronto's Housing Crisis, Tenant and Housing Experts Say," *Toronto Star*, 15 November 2018.
99 Moos, "From Gentrification to Youthification?"
100 Lehrer, Keil, and Kipfer, "Reurbanization in Toronto," 81–90.
101 For a discussion on why there is no affordable housing in Toronto's condo developments and what possible benefits inclusionary zoning could bring, see Catherine McIntyre, "Torontoist Explains: What Is Inclusionary Zoning?" *Torontoist*, 4 April 2016, http://torontoist.com/2016/04/torontoist-explains-what-is-inclusionary-zoning/.
102 All studies cited in Rosen and Walks, "Castles in Toronto's Sky," 303.
103 Alan Walks and Martine August, "The Factors Inhibiting Gentrification in Areas with Little Non-Market Housing: Policy Lessons from the Toronto Experience," *Urban Studies*, 45, no. 12 (2008): 2594–625.
104 Elvin K. Wyly and Daniel J. Hammel, "Islands of Decay in Seas of Renewal: Housing Policy and the Resurgence of Gentrification," *Housing Policy Debate* 10, no. 4 (1999): 711–71.
105 Martine August and Alan Walks, "Gentrification, Suburban Decline, and the Financialization of Multi-Family Rental Housing: The Case of Toronto," *Geoforum* 89 (2018): 124–36.
106 Ibid. See also Manuel B. Aalbers, "The Financialization of Home and the Mortgage Market Crisis," in *The Financialization of Housing* (New York: Routledge, 2016), 40–63.
107 Martine August, "The Financialization of Canadian Multi-Family Rental Housing: From Trailer to Tower, *Journal of Urban Affairs*, 42, no. 7: 975–97; August and Walks, "Gentrification, Suburban Decline," 124–36.
108 For a very good account of gentrification in Parkdale, and the role of REITs, see M. Whyte, "My Parkdale Is Gone: How Gentrification Reached the One Place That Seemed Immune," *The Guardian*, 14 January 2020.
109 August and Walks, "Gentrification, Suburban Decline," 124–36
110 Lemon, *Toronto*, 105.
111 Lehrer, Keil, and Kipfer, "Reurbanization in Toronto," 81–90.
112 Sharon Kelly, "The New Normal: The Figure of the Condo Owner in Toronto's Regent Park," *City & Society* 25, no. 2 (2013): 173–94.
113 Ibid. and Lehrer, Keil, and Kipfer, "Reurbanization in Toronto," 81–90
114 See Martine August, "Challenging the Rhetoric of Stigmatization: The Benefits of Concentrated Poverty in Toronto's Regent Park," *Environment and Planning A* 46, no. 6 (2014): 1317–33; Kelly, "The New Normal," 173–94.
115 See G. Bridge, T. Butler, and L. Lees, *Mixed Communities: Gentrification by Stealth* (Bristol: Policy Press 2012), 1.
116 Kelly, "The New Normal," 175–6.
117 August, "Challenging the Rhetoric of Stigmatization," 1318.
118 Ibid., 1321.
119 Kelly, "The New Normal," 173–94.
120 Ibid.
121 Rosen and Walks, "Castles in Toronto's Sky," 289–310.
122 Lehrer, Keil, and Kipfer, "Reurbanization in Toronto," 88.
123 Martine August, "Revitalisation Gone Wrong: Mixed-Income Public Housing Redevelopment in Toronto's Don Mount Court," *Urban Studies* 53, no. 16 (2016): 3405–22.
124 See ibid. and Lehrer, Keil, and Kipfer, "Reurbanization in Toronto," 81–90.
125 Moos, Vinodrai, Revington, Seasons, "Planning for Mixed Use," 7–20; Moos, "From Gentrification to Youthification?"
126 David Hulchanski, *The Three Cities within Toronto: Income Polarization among Toronto's Neighbourhoods, 1970–2005* (Toronto: Cities Centre, University of Toronto, 2010).
127 Ibid., 7.
128 David Hulchanski, *Neighbourhood Socioeconomic Polarization and Segregation in Toronto: Trends and Processes since 1970*, http://neighbourhoodchange.ca/documents/2018/09/hulchanski-2018-toronto-segregation-presentation.pdf; Alan Walks, "Inequality and Neighbourhood Change in the Greater Toronto Region," in Jill L. Grant, Alan Walks, and Howard Ramos (eds.), *Changing Neighbourhoods Social and Spatial Polarization in Canadian Cities* (Vancouver: UBC Press, 2020).
129 Lemon, *Toronto*, 184.
130 Moos, Vinodrai, Revington, and Seasons, "Planning for Mixed Use," 7–20.
131 City of Toronto Social Housing Waiting List Reports, 2018 Q2. https://www.toronto.ca/city-government/data-research-maps/research-reports/housing-and-homelessness-research-and-reports/social-housing-waiting-list-reports/.
132 Donovan Vincent, "Black People 'Segregated,' City Map Shows; Black Torontonians Clustered outside Core, Census Indicates," *Toronto Star*, 9 November 2018.
133 David Hulchanski, "Toronto's Mayoral Election in Four Maps, *Spacing Magazine*, 29 October 2014.

134 https://www.cbc.ca/news/canada/toronto/toronto-covid-19-data-1.5669091.

Chapter 4

1. Sarah Bassnett, *Picturing Toronto: Photography and the Making of a Modern City* (Montreal and Kingston: McGill-Queen's University Press, 2016), 3.
2. See Susan Sontag, *On Photography* (New York: Farrar, Strauss and Giroux, 1977).
3. E. Arnold, "Photography, Composition, and the Ephemeral City," Area. (2021) https://doi.org/10.1111/area.12725. E. Arnold, "Aesthetics of Zero Tolerance," *City* 23, no. 2 (2019): 143–69.
4. Arnold, "Photography, Composition, and the Ephemeral City."
5. E. Wyly, "Things Pictures Don't Tell Us: In Search of Baltimore," *City* 14, no. 5 (2010).
6. Ibid., 503
7. J. Collier, *Visual Anthropology: Photography as a Research Method* (New York: Holt, Rinehart and Winston, 1967), 5, as quoted in Gillian Rose, *Visual Methodologies: An Introduction to Researching with Visual Materials* (Los Angeles: Sage, 2012), 299.
8. K. Lemmons, C. Brannstrom, and D. Hurd, "Exposing Students to Repeat Photography: Increasing Cultural Understanding on a Short-Term Study Abroad," *Journal of Geography in Higher Education* 38, no. 1 (2014): 93.
9. Gillian Rose, *Visual Methodologies: An Introduction to Researching with Visual Materials* (Los Angeles: Sage, 2016), 24–46.
10. Gillian Rose, "Engendering the Slum: Photography in East London in the 1930s," *Gender, Place and Culture: A Journal of Feminist Geography* 4, no. 3 (1997): 277–300.
11. Rose, *Visual Methodologies*, 17.
12. Rose, "Engendering the Slum," 277–300.
13. Wyly, "Things Pictures Don't Tell Us," 497–528.
14. A.S. Metcalfe, "Repeat Photography and Educational Research," in Julianne Moss and Barbara Pini (eds.), *Visual Research Methods in Educational Research* (London: Palgrave Macmillan, 2016), 156.
15. Wyly, "Things Pictures Don't Tell Us," 506.
16. E. Margolis, "Class Pictures: Representations of Race, Gender and Ability in a Century of School Photography," *Visual Studies* 14, no. 1 (1999): 7.
17. P. Marcuse, "From Critical Urban Theory to the Right to the City," *City* 13, nos. 2–3 (2009): 194.
18. E. Arnold, "Aesthetic Practices of Psychogeography and Photography," *Geography Compass* 13, no. 2 (2019); E. Arnold, "Aesthetics of Zero Tolerance," *City* 23, no. 2 (2019): 143–69; C. Suchar, "Grounding Visual Sociology Research in Shooting Scripts," *Qualitative Sociology* 20, no. 1 (1997): 33–55; M.H. Krieger, "Taking Pictures in the City," *Journal of Planning Education and Research* 24, no. 2 (2004): 213–15.
19. For a milestone early example of urban photography see Jacob Riis, *How the Other Half Lives: Studies among the Tenements of New York* (New York: Dover Publications, 1971 [1890]).
20. G. DeVerteuil, "The Changing Landscapes of Southwest Montreal: A Visual Account," *Canadian Geographer* 48, no. 1 (2004): 82.
21. Bassnett, *Picturing Toronto*.
22. J. Maloof, *Vivian Maier: Street Photographer* (Brooklyn: Power House Books, 2011).
23. P. Bannos, *Vivian Maier: A Photographer's Life and Afterlife* (Chicago: University of Chicago Press, 2017).
24. See *Fred Herzog: Modern Color* (Berlin: Hatje Kantz, 2017).
25. Barbara Harrison, "Snap Happy: Toward a Sociology of 'Everyday' Photography," in Christopher J. Pole (ed.), *Seeing Is Believing? Approaches to Visual Research* 7 (Bingley, UK: Emerald Publishing, 2004), 22–39.
26. John Urry, *The Tourist Gaze: Leisure and Travel in Contemporary Societies* (London: Sage Publications, 1990).
27. Ayona Datta, "Where Is the Global City? Visual Narratives of London amongst East European Migrants," *Urban Studies* 49, no. 8 (2012): 1725–40; M. Zebracki, B. Doucet, and T. de Brant, "Beyond Picturesque Decay: Detroit and the Photographic Sites of Confrontation between Media and Residents," *Space and Culture* 22, no. 4 (2018): 489–508; Kevin Lynch, *The Image of the City* (Cambridge, MA: MIT Press, 1960).
28. In Toronto, streetcars long have been viewed as iconic. See, for example, the discussion of the city's Canadian Light Rail Vehicles (CLRVs) in Edward Keenan, "Six Endangered Toronto Icons," *Toronto Star*, 22 June 2016, GT2. The CLRVs were replaced by a fleet of low-floor streetcars, the last being retired in December 2019.
29. Yves Marchant and Romain Meffre, *The Ruins of Detroit* (Gottingen: Steidl, 2011).
30. Dan Austin, *Lost Detroit: Stories behind the Motor City's Majestic Ruins* (Charleston, SC: History Press, 2014), 101.
31. Joshua Akers, "A New Urban Medicine Show: On the Limits of Blight Remediation," in Brian Doucet (ed.), *Why Detroit Matters: Decline, Renewal and Hope in a Divided City* (Bristol: Policy Press, 2017). In 2018, the building was purchased by the Ford Motor Company, but again is part of a wider story of governance, politics, and the economy. See A. Gross, "Ford Paid $90 Million for the Michigan Central Station," *Detroit Free Press*, 24 September 2018.
32. Zebracki, Doucet, and de Brant, "Beyond Picturesque Decay."
33. Sontag, *On Photography*.
34. Marcus Banks, *Visual Methods in Social Research* (London: Sage, 2001), 144, as quoted in Rose, *Visual Methodologies*, 298.

35 J.H. Rieger, "Photographing Social Change," *Visual Studies* 11, no. 1 (1996): 6.
36 https://www.ctvnews.ca/sci-tech/they-change-too-rockies-photo-archive-documents-high-altitude-shifts-1.5016937. See also A. Trant, E. Higgs, and B.M. Starzomski, "A Century of High Elevation Ecosystem Change in the Canadian Rocky Mountains," *Scientific Reports* 10, no. 9698 (2020), https://doi.org/10.1038/s41598-020-66277-2.
37 Mark Klett, Joann Verberg, Gordon Bushaw, and Rick Dingus, *Second View: The Rephotographic Survey Project* (Albuquerque: University of New Mexico Press, 1984).
38 Mark Klett, Kyle Bajakian, William L. Fox, Michael Marshall, Toshi Ueshina, and Byron G. Wolfe, *Third Views, Second Sights: A Rephotographic Survey of the American West* (Santa Fe: Museum of New Mexico Press, 2004).
39 M. Klett, M. Lundgren, P. L. Fradkin, R. Solnit, and K. Breuer, *After the Ruins, 1906 and 2006: Rephotographing the San Francisco Earthquake and Fire* (Berkeley: University of California Press, 2006).
40 A.C. Byers, "An Assessment of Contemporary Glacier Fluctuations in Nepal's Khumbu Himal Using Repeat Photography," *Himalayan Journal of Sciences* 4, no. 6 (2008): 21–6; M.A. Crimmins and T.M. Crimmins, "Monitoring Plant Phenology Using Digital Repeat Photography," *Environmental Management* 41, no. 6 (2008): 949–58; O. Sonnentag, K. Hufkens, C. Teshera-Sterne, A.M. Young, M. Friedl, B.H. Braswell, and A.D. Richardson, "Digital Repeat Photography for Phenological Research in Forest Ecosystems," *Agricultural and Forest Meteorology* 152 (2012), 159–77.
41 Jon H. Rieger, "Rephotography for Documenting Social Change," in Luc Pauwels and Dawn Mannay (eds.), *The SAGE Handbook of Visual Research Methods* (London: Sage, 2011), 132–49; Rieger, "Photographing Social Change," 5–49.
42 Rieger, "Photographing Social Change," 27.
43 DeVerteuil, "The Changing Landscapes of Southwest Montreal," 82.
44 R. Fishman, "Camillo Vergara's Detroit," in J. Vergara, *Detroit Is No Dry Bones* (Ann Arbor: University of Michigan Press, 2016).
45 Camillo José Vergara, *Tracking Time: Documenting America's Post-Industrial Cities* (Bielefeld: Kerber Verlag, 2014).
46 Camillo José Vergara, *Detroit Is No Dry Bones: The Eternal City in the Industrial Age* (Ann Arbor: University of Michigan Press, 2016).
47 Ibid.
48 A. Mountz, A. Bonds, B. Mansfield, J. Loyd, J. Hyndman, M. Walton-Roberts, R. Basu, et al., "For Slow Scholarship: A Feminist Politics of Resistance through Collective Action in the Neoliberal University," *ACME: An International E-Journal for Critical Geographies* 14, no. 4 (2015): 1–25.

49 Rieger, "Photographing Social Change," 5–49.
50 W. Wyckoff, *On the Road Again: Montana's Changing Landscape* (Seattle: University of Washington Press, 2006).
51 Thomas Vale and Geraldine Vale, *U.S. 40 Today: Thirty Years of Landscape Change in America* (Madison: University of Wisconsin Press, 1983).
52 D. D. Arreola and N. Burkhart, "Photographic Postcards and Visual Urban Landscape," *Urban Geography* 31, no. 7 (2010): 885–904.
53 J. Finn, A. Fernandez, L. Sutton, D.D. Arreola, C.D. Allen, and C. Smith, "Puerto Peñasco: Fishing Village to Tourist Mecca," *Geographical Review* 99, no. 4 (2009): 596.
54 A.S. Metcalfe, "Repeat Photography and Educational Research," in Julianne Moss and Barbara Pini (eds.), *Visual Research Methods in Educational Research* (London: Palgrave Macmillan, 2016).
55 J. Hwang and R.J. Sampson, "Divergent Pathways of Gentrification: Racial Inequality and the Social Order of Renewal in Chicago Neighborhoods," *American Sociological Review* 79, no. 4 (2014): 726–51.
56 Evelyn Dawn Ravuri, "A Google Street View Analysis of Gentrification: A Case Study of One Census Tract in Northside, Cincinnati, USA," *GeoJournal* (2021): 1–21; Sarah Christensen, Richard Harris, and Kathleen Kinsella, "A Litmus Test for Neighbourhood Change," *Plan Canada* 58, no. 1 (2018).

Chapter 5

1 Anne Lyden, *Railroad Vision: Photography, Travel and Perception* (Los Angeles: Getty Publications, 2003).
2 Ibid., 8
3 Ibid., 9
4 Ibid., 115
5 H. Roger Grant, *Electric Interurbans and the American People* (Bloomington: Indiana University Press, 2016).
6 See https://hcry.org/.
7 The Toronto Railway Historical Association's website has a detailed history of enthusiast clubs in Canada. http://www.trha.ca/about.html.
8 Gillian Rose, *Visual Methodologies* (London: Sage Publications, 2012).
9 Ibid.
10 Tony Reevy, *The Railroad Photography of Lucius Beebe and Charles Clegg* (Bloomington: University of Indiana Press, 2019).
11 J. De Souza, *Digital Railway Photography: A Practical Guide* (Stroud, UK: Fonthill Media, 2015).
12 Reevy, *The Railroad Photography of Lucius Beebe and Charles Clegg*.
13 Jim Shaughnessy and Kevin P Keefe, *Jim Shaughnessy Essential Witness: Sixty Years of Railroad Photography* (London: Thames and Hudson, 2017).

14 T. Benson, *One Track Mind: Photographic Essays on Western Railroading* (Erin, ON: Boston Mills Press, 2000); L.G. Niemann and J. Jensen, *Railroad Noir: The American West at the End of the Twentieth Century* (Bloomington: University of Indiana Press, 2010); D. Kahler, *The Railroad and the Art of Place* (Madison: Center for Railroad Photography and Art, 2016); T. Reevy, *The Railroad Photography of Jack Delano* (Bloomington: University of Indiana Press, 2015); O. Winston Link and Thomas H. Garver, *The Last Steam Railroad in America* (New York: Harry N. Abrams, 2000); J. Parker Lamb, *Steel Wheels Rolling: A Personal Journey of Railroad Photography* (Erin ON: Boston Mills Press, 2001); Greg McDonnell *Rites of Passage: A Canadian Railway Retrospective* (Erin, ON: Boston Mills Press, 2000); Scott Lothes, Kevin P. Keefe, Wallace W. Abbey, *Wallace W. Abbey: A Life in Railroad Photography* (Bloomingdale: Indiana University Press, 2018).

15 Susan Sontag, *On Photography* (New York: Farrar, Strauss and Giroux, 1977), 3.

16 Ben Spurr, "Track of Dreams: 24-Year-Old Transit Fan Buys Himself an Old TTC Streetcar," *Toronto Star*, 30 July 2020.

17 Morihiro Satow, "Railfan and Photographic Collection: A Way to Possess the World," *International Yearbook of Aesthetics* 18 (2014): 424.

18 Jeremy De Souza. *Digital Railway Photography: A Practical Guide* (Stroud, UK: Fonthill Media, 2015), 9.

19 After the Bloor-Danforth subway opened in 1966, the streetcar network saw considerable contraction and many of the PCCs constructed before 1945 were surplus to the TTC's needs and were sold. The strict running of specific classes of cars on specific routes also largely disappeared and it was common to see different classes running throughout the system.

20 Zoom lenses, which allow a photographer to winnow in on their subject matter, were not common until the 1970s. Prior to that, photographers used so-called prime lenses, especially those with focal lengths of 35mm, 50mm, 55mm, and 135mm. One of our sources of historic photos, John F. Bromley, often recorded photographic data on his slides. Using Kodachrome film, with an ASA of 25, he often used a 50mm lens, with an f-stop number, or aperture value, of 5.6 and a shutter speed of 1/200th of a second.

21 C. Slater, "General Motors and the Demise of Streetcars," *Transportation Quarterly* 51 (1997): 45–66.

22 On Chicago, see J. Wien and D. Sadowski, *Chicago Streetcar Pictorial: The PCC Car Era 1936–1958* (Chicago: Central Electric Railfans Association, 2015); for a variety of cities across the Great Lakes region, see R. Halpern, *Great Lakes Trolleys in Color* (Scotch Plains, NJ: Morning Sun Books, 2004).

23 The components of these PCCs were sold to Brussels, Belgium, after Johnstown abandoned its streetcars in 1960; rebuilt into new PCCs, they ran in Brussels until 2009. These cars were photographed extensively by Brian Doucet during his time living in the Netherlands.

24 Harold A. Smith, *Touring Pittsburgh by Trolley: A Pictorial Review of the Early Sixties* (New York: Quadrant Press, 1992); Blaine Hayes and James Toman, *Transit in the Triangle: A Century Look at Pittsburgh Public Transit* (Chicago: Central Electric Railfans Association, 2012).

25 Thanks to Ken Josephson, Rob Hutchinson, and Rob Lubinski for providing some information on Roberta Hill's life and photography.

26 For more on walking and photography as a research methodology, see E. Arnold, "Aesthetic Practices of Psychogeography and Photography," *Geography Compass* 13, no. 2 (2019).

27 Rose, *Visual Methodologies*, 30–8.

28 https://www.nycsubway.org/wiki/Main_Page; https://transit.toronto.on.ca/index.shtml.

29 H. Roger Grant, *Railroads and the American People* (Bloomington: University of Indiana Press, 2017), 262.

30 Reevy, *The Railroad Photography of Lucius Beebe and Charles Clegg*, 8.

31 Ibid.

32 Ibid., 10.

33 Ibid., 25; Jim Shaughnessy and Kevin P. Keefe, *Jim Shaughnessy Essential Witness: Sixty Years of Railroad Photography* (New York: Thames & Hudson, 2017).

34 http://www.railphoto-art.org/.

35 For a full list of books in the Railroads Past and Present Series from Indiana University Press, see https://iupress.org/. Jack Delano's images depicting railways in Puerto Rico in the 1940s and 1980s can be found in Jack Delano, *From San Juan to Ponce on the Train* (San Juan: University of Puerto Rico, 1990)

36 Brian Solomon, *The Twilight of Steam* (Minneapolis: Voyageur Press, 2014).

37 O. Winston Link, *The Last Steam Railroad in America* (New York: HNA Books, 1995).

38 See Richard Steinheimer and Jeff Brouws, *A Passion for Trains* (New York: WW Norton, 2011); Joel Jensen, John Gruber, and Scott Lothes, *Steam: An Enduring Legacy* (New York: W.W. Norton, 2011); Ted Benson, *One Track Mind: Photographic Essays on Western Railroading* (Erin, ON: Boston Mills Press, 2000). For a good list of popular railway photography books see http://www.dewitzphotography.com.

39 Joel Jensen and Linda Niemann, *Railroad Noir* (Bloomington: Indiana University Press, 2010).

40 Kenneth Springirth, *Toronto Streetcars Serve the City* (Stroud, UK: Fonthill Media, 2014); Kenneth Springirth, *Toronto Transit Commission Streetcars* (Scotch Plains, NJ: Morning Sun Books, 2014).

41 A list of books about Toronto's streetcars includes John F. Bromley, *TTC '28: The Electric Railway Services of the Toronto Transportation Commission in 1928* (Toronto:

Upper Canada Railway Society, 1968); John F. Bromley and Jack May, *Fifty Years of Progressive Transit: A History of the Toronto Transit Commission* (New York: Electric Railroaders' Association, 1973); Mike Filey, *Not a One-Horse Town: 125 Years of Toronto and Its Streetcars* (Toronto: Toronto of Old, 1986); Mike Filey, *The TTC Story: The First Seventy-Five Years*, with photographs from the TTC Archives as selected by Ted Wickson (Toronto: Dundurn Press, 1996); Mike Filey, *From Horse Power to Horsepower: Toronto: 1890–1930* (Toronto: Dundurn Press, 1993); Mike Filey, Richard Howard, and Helmut Weyerstrahs, *"Passengers Must Not Ride on Fenders": A Fond Look at Toronto: Its People, Its Places, Its Streetcars* (Toronto: Green Tree Publishing, 1974); J. William Hood, *The Toronto Civic Railways: An Illustrated History* (Toronto: Upper Canada Railway Society, 1986); Larry Partridge, *The Witts: An Affectionate Look at Toronto's Original Red Rockets* (Erin, ON: Boston Mills Press, 1982); Larry Partridge, *Mind the Doors, Please! The Story of Toronto and Its Streetcars* (Erin, ON: Boston Mills Press, 1983); Louis H. Pursley, *Street Railways of Toronto, 1861–1921: Interurbans*, Special No. 25 (Los Angeles: Electric Railway Publications, 1958); Louis H. Pursley, *The Toronto Trolley Car Story, 1921–1961: Interurbans*, Special No. 29 (Los Angeles: Electric Railway Publications, 1961); James V. Salmon, *Rails from the Junction: The Story of the Toronto Suburban Railway* (Toronto: Lyon Productions, n.d.).

42 To mark the 150th anniversary of trams in The Hague, Maurits van den Toorn wrote *Lijnenspel: 150 jaar trams in de Haagse regio* (Haarlem: Krikke Special Books, 2014). In a conversation with the author, he noted that the book had done very well in general bookshops in and around The Hague and, in many cases, sold out quickly. It did not sell as well at the bookshop of the national rail and tram enthusiasts society, NVBS, largely because it did not cover more traditional aspects featured in a tramway book. After the success of this book, van den Toorn was commissioned by the Rotterdam Elektrische Tram (Rotterdam Electric Tram), the public transport provider in the Netherlands' second-largest city to write *RET: Anderhalve eeuw openbaar vervoer in Rotterdam* (Amsterdam: Boom, 2017).

43 Oscar Israelowitz and Brian Merlin, *Subways of New York City in Vintage Photographs* (New York City: Israelowitz Publishing, 2004).

44 Three books about the PCC streetcar were published by Interurban Press in the 1980s: Stephen P. Carlson and Fred W. Schneider III, *PCC: The Car That Fought Back* (Glendale, CA: Interurban Press, 1980); Fred W. Schneider III and Stephen P. Carlson, *PCC from Coast to Coast* (Glendale, CA: Interurban Press, 1983); Seymour Kashin and Harre Demoro, *PCC: An American Original* (Glendale, CA: Interurban Press, 1986).

45 Colin Garratt, *The Golden Years of British Trams* (Leicester: Milepost, 1995). Collin Garratt also produced another volume on Priestley's railway photography. For more information on Henry Priestley see https://funeral-notices.co.uk/notice/Henry+Priestley/2117518. Robert Halpern, *Great Lakes Trolleys in Color* (Scotch Plains, NJ: Morning Sun Books, 2004).

46 Rose, *Visual Methodologies*, 40.

47 Jamie Bradburn, "The Oldest Known Photographs of Toronto," http://torontoist.com/2013/02/the-oldest-known-photos-of-toronto/. See also Steve MacKinnon, Karen Teeple, and Michele Dale, *Toronto's Visual Legacy: Official City Photography from 1856 to the Present* (Toronto: James Lorimer, 2009).

48 Michael Redhill, *Consolation: A Novel* (Toronto: Doubleday Canada, 2006). The book won the City of Toronto Book Award in 2007. Toronto's archival photographs also had an influence on Michael Ondaatje's award-winning novel *In the Skin of a Lion* (Toronto: McClelland and Stewart, 1997). This novel captured the Trillium Book Award in 1987 and the City of Toronto Book Award in 1988.

49 On James see Vincenzo Pietropaolo, "William James: Toronto's First Photojournalist," in John Lorinc, Michael McClelland, Ellen Scheinberg, and Tatum Taylor (eds.), *The Ward: The Life and Loss of Toronto's First Immigrant Neighbourhood* (Toronto: Coach House Books, 2015), 123–7; and Christopher Hume, *William James' Toronto Views* (Toronto: James Lorimer, 1999).

50 On Goss, see Stephen Bulger, "Arthur Goss: Documenting Hardship," in Lorinc, McClelland, Scheinberg, and Taylor, *The Ward*, 106–11; Michel Lambeth, *Made in Canada: Photographs of Toronto, circa 1910, from the Collection of Michel Lambeth* (Toronto: Editions Grafikos, 1967); and Sarah Bassnett, *Picturing Toronto: Photography and the Making of a Modern City* (Montreal and Kingston: McGill-Queen's University Press, 2016). Many of Goss' photographs of immigrant life in the city were reproduced in Robert Harney and Harold Troper, *Immigrants: A Portrait of the Urban Experience* (Toronto: Van Nostrand Reinhold, 1975).

51 Bassnett, *Picturing Toronto*.

52 Ibid., 7.

53 Ibid.

54 On the creation of the City Archives see Scott James and Victor Russell, "Institutional Memory," in Lorinc, McClelland, Scheinberg, and Taylor, *The Ward*, 301–3.

55 See http://www.myseumoftoronto.com/programming/the-boris-spremo-story/, and Boris Spremo, *Boris Spremo: Twenty Years of Photojournalism* (Toronto: McClelland and Stewart, 1983); Boris Spremo, *Boris Spremo and His Camera Look at Toronto* (Toronto: McGraw-Hill, 1967).

56 James Lemon, *Toronto since 1918: An Illustrated History* (Toronto: Lorimer, 1985).

57 William Dendy, *Lost Toronto* (Toronto: McClelland & Stewart, 1993); Michael Kluckner, *Toronto: The Way It Was* (Toronto: Whitecap Books, 1988); Charles P. de Volpi, *Toronto: A Pictorial Record, 1813–1882* (Montreal: Dev-Sco

Publications, 1965); Eric Arthur, *Toronto No Mean City* (Toronto: University of Toronto Press, 1964); William Dendy and William Kilbourn, *Toronto Observed: Its Architecture, Patrons, and History* (Toronto: Oxford University Press, 1986); Patricia McHugh, *Toronto Architecture: A City Guide* (Toronto: Mercury Books, 1985); Margaret Goodfellow and Philip Goodfellow, *A Guidebook to Contemporary Architecture in Toronto* (Toronto: Douglas & McIntyre, 2010); Mark Osbaldeston, *Unbuilt Toronto: A History of the City That Might Have Been* (Toronto: Dundurn Press, 2008); Mark Osbaldeston, *Unbuilt Toronto 2: More of the City That Might Have Been* (Toronto: Dundurn Press, 2011); Geoffrey Simmins, *Fred Cumberland: Building the Victorian Dream* (Toronto: University of Toronto Press, 1997); Glenn McArthur, *A Progressive Traditionalist: John M. Lyle, Architect* (Toronto: Coach House Books, 2009); Michael McClelland and Graeme Stewart, *Concrete Toronto: A Guidebook to Concrete Architecture from the Fifties to the Seventies* (Toronto: Coach House Books and E.R.A. Architects, 2007); Tim Morawetz, *Art Deco Architecture in Toronto* (Toronto: Tim Morawetz, 2009); Christopher Armstrong, *Making Toronto Modern: Architecture and Design, 1895–1975* (Montreal and Kingston: McGill-Queen's University Press, 2014); Sean Stanwick and Jennifer Flores, *Design City Toronto* (Chichester: John Wiley & Sons, 2007); Shawn Micallef and Patrick Cummins, *Full Frontal T.O.: Exploring Toronto's Architectural Vernacular* (Toronto: Coach House Books, 2012); Doug Taylor, *Lost Toronto* (London: Pavilion Books, 2018); Toronto Harbour Commissioners, *Toronto Harbour: The Passing Years* (Toronto: Toronto Harbour Commissioners, 1985); Ted Wickson, *Reflections of Toronto Harbour: 200 Years of Port Activity and Waterfront Development* (Toronto: Toronto Port Authority, 2002); Katherine Taylor, *Toronto: City of Commerce, 1800-1960: Stories of a City's Factories, Businesses and Storefronts* (Toronto: James Lorimer, 2021).

58 Marjorie Harris, *Toronto: The City of Neighbourhoods* (Toronto: McClelland & Stewart, 1984); George H. Rust-D'Eye, *Cabbagetown Remembered* (Erin, ON: Boston Mills Press, 1984); Penina Coopersmith, *Cabbagetown: The Story of a Victorian Neighbourhood*, photographs by Vincenzo Pietropaolo (Toronto: James Lorimer, 1998); Denis De Klerck and Corrado Paina, *College Street, Little Italy: Toronto's Renaissance Strip* (Toronto: Mansfield Press, 2006); Jack Batten, *The Annex: The Story of a Toronto Neighbourhood* (Erin, ON: Boston Mills Press, 2004); Jean Cochrane, *Kensington*, with photos by Vincenzo Pietropaolo (Erin, ON: Boston Mills Press, 2000); Bess Hillery Crawford, *Rosedale* (Erin, ON: Boston Mills Press, 2000); F.R. Berchem, *The Yonge Street Story, 1793–1860* (Toronto: McGraw-Hill Ryerson, 1977); Tom Cruickshank and John De Visser, *Old Toronto Houses* (Toronto: Firefly Books, 2003); Margo Salnek and Donna Griffith, *Coach Houses of Toronto* (Erin, ON; Boston Mills Press, 2005); and Rosemary Donegan, *Spadina Avenue* (Vancouver and Toronto: Douglas & McIntyre, 1985).

59 For many years, Filey has written a weekly column on Toronto's history in the Sunday edition of the *Toronto Sun*. These columns have been collected into books entitled *Toronto Sketches: "The Way We Were."* The first of these was published in 1992 by Dundurn Press. At the time of writing, there had been 12 volumes released. Other books by Filey include *A Toronto Album: Glimpses of the City That Was* (Toronto: University of Toronto Press, 1970); *Look at Us Now* (Toronto: Mike Filey and The Toronto Telegram, 1971); *Toronto: Reflections of the Past* (Toronto: Nelson, Foster, and Scott, 1972); *Toronto City Life: Old and New* (Toronto: Nelson, Foster, and Scott, 1979); *Toronto: Then and Now* (Ottawa: Magic Light Publishing, 2000); *Toronto: A Photographic Portrait* (Ottawa: Magic Light Publishing, 2005); *Toronto: Spirit of Place* (Toronto: John McQuarrie Photography, 2019). See also Doug Taylor, *Toronto Then and Now* (London, UK: Pavilion Books, 2016). Taylor's book features about 60 sets of before and after photos of Toronto buildings and landmarks. Shawn Micallef and Patrick Cummins, *Full Frontal T.O.: Exploring Toronto's Architectural Vernacular* (Toronto: Coach House Books, 2013).

60 Even today, images of scenes along the Long Branch line itself are rare. Most photos were taken at either the Long Branch or Humber loop at each end of the line, or at a special turn back at Hillside Drive, known to transit enthusiasts as the Hillside or Mimico Wye, which, until it was paved over, allowed streetcars to turn around in an emergency; the turn back was also a popular photo stop on charters.

Chapter 6

1 Edward Relph, *Toronto: Transformations in a City and Its Region* (Philadelphia: University of Pennsylvania Press, 2014), 42.

2 John F. Bromley and Jack May, *Fifty Years of Progressive Transit: A History of the Toronto Transit Commission* (New York: Electric Railroaders' Association, 1973).

3 On this early period of Toronto's transit history see Christopher Armstrong and H.V. Nelles, *The Revenge of the Methodist Bicycle Company: Sunday Streetcars and Municipal Reform in Toronto, 1888–1897* (Toronto: Peter Martin Associates Ltd., 1977); J. William Hood, *The Toronto Civic Railways: An Illustrated History* (Toronto: Upper Canada Railway Society, 1986); Robert M. Stamp, *Riding the Radials: Toronto's Suburban Electric Streetcar Lines* (Erin, ON: Boston Mills Press, 1989); Mike Filey, Richard Howard, and Helmut Weyerstrahs, *"Passengers Must Not Ride on Fenders": A Fond Look at Toronto: Its People, Its Places, Its Streetcars* (Toronto: Green Tree Publishing, 1974); Larry Partridge, *Mind the Doors, Please! The Story of Toronto and Its Streetcars* (Erin, ON: Boston Mills Press, 1983); Louis H. Pursley, *Street Railways of Toronto, 1861–1921*, Interurbans Special No. 25 (Los

Angeles: Electric Railway Publications, 1958); Michael Doucet, "Politics, Space, and Trolleys: Mass Transit in Early Twentieth-Century Toronto," in Gilbert A. Stelter and Alan F.J. Artibise (eds.), *Shaping the Urban Landscape: Aspects of the Canadian City-Building Process*, Carleton Library Series No. 125 (Ottawa: Carleton University Press, 1982), 356–81.

4 On Toronto's Peter Witt streetcars see Larry Partridge, *The Witts: An Affectionate Look at Toronto's Original Red Rockets* (Erin, ON: Boston Mills Press, 1982); Louis H. Pursley, *The Toronto Trolley Car Story, 1921–1961*, Interurbans Special No. 29 (Los Angeles: Electric Railway Publications, 1961). See also Kenneth C. Springirth, *Toronto Transit Commission Streetcars* (Scotch Plains, NJ: Morning Sun Books, 2014).

5 On the TTC's early modernization efforts see Toronto Transportation Commission, *Wheels of Progress: A Story of the Development of Toronto and Its Public Transportation Services* (Toronto: Toronto Transportation Commission, nd); Toronto Transit Commission, *Transit in Toronto* (Toronto: Toronto Transit Commission, 1967); Mike Filey, *The TTC Story: The First Seventy-Five Years* (Toronto: Dundurn Press, 1996); Toronto Transit Commission, *A Century of Moving Toronto: TTC 1921–2021* (Toronto: Toronto Transit Commission, 2021); Anthony Perl, Matt Hern, and Jefferey Kenworthy, *Big Moves: Global Agendas, Local Aspirations and Urban Mobility in Canada* (Montreal and Kingston: McGill-Queen's University Press, 2020), 41–3. Sean Marshall has produced an excellent series of maps depicting Toronto's streetcar network from its earliest years down to the present day. See https://seanmarshall.ca/2017/01/11/mapping-torontos-street-railways-in-the-ttc-era-1921-2016/.

6 On the development of the PCC streetcar see Stephen P. Carlson and Fred W. Schnieder III, *PCC: The Car That Fought Back* (Glendale, CA: Interurban Press, 1980); Fred W. Schneider III and Stephen P. Carlson, *PCC: From Coast to Coast* (Glendale, CA: Interurban Press, 1983); Seymour Kashin and Harre Demoro, *The PCC Car: An American Original* (Glendale, CA: Interurban Press, 1986). While developed in North America, PCCs also could be found in European cities. See, for example, Geoffrey Skelsey and Yves-Laurant Hansart, *PCCs of Western Europe, 1950–2010: The Tram That Belgium Made* (Welling, UK: Light Rail Transit Association, 2011). On the initial TTC PCC purchase see "T.T.C. Buys 140 Street Cars for $3,000,000: Fast, Silent Trolleys Bought with Reserves – In Operation by Fall," *Toronto Star*, 8 April 1938, A1 and A19. The cost for each car was almost $21,429 ($385,158 in 2019 dollars).

7 See for example, Tom Sugrue, *The Origins of the Urban Crisis: Race and Inequality in Postwar Detroit* (Princeton: Princeton University Press, 1996). For a direct comparison with Canada, see Jason Hackworth, "Why There Is No Detroit in Canada," *Urban Geography* 37, no. 2 (2016): 272–95.

8 On the disappearance of streetcars in US cities see Cliff Slater, "General Motors and the Demise of Streetcars," *Transportation Quarterly* 51, no. 3 (Summer 1997): 45–66.

9 As quoted in Schneider and Carlson, *PCC: From Coast to Coast*, 108. On the TTC's PCC fleet see the Commission's June 1982 brochure "The PCC Car." Each purchase of PCC cars by the TTC, whether new or second-hand, was given a classification by the TTC. There were fourteen classes of TTC PPCs, A1 through A14.

10 The classic work on North American streetcar suburbs is Sam Bass Warner Jr., *Streetcar Suburbs: The Process of Growth in Boston, 1870–1900* (Cambridge, MA: Harvard University Press, 1962). On streetcar suburbs in Toronto see Richard Harris, *Unplanned Suburbs: Toronto's American Tragedy, 1900 to 1950* (Baltimore: Johns Hopkins University Press, 1996); and Edward Relph, *Toronto: Transformations in a City and Its Region* (Philadelphia: University of Pennsylvania Press, 2014). On development along Yonge Street after 1954, see James H. Kearns, "The Economic Impact of the Yonge Street Subway," address to the 83rd Annual Meeting of the American Transit Association, New York City, September 1964. According to Kearns, between 1952 and 1962, the assessed value of land along the Yonge line increased by 58 per cent, compared to 25 per cent elsewhere in the City.

11 Jonathan English, *The Better Way: Transit Service and Demand in Metropolitan Toronto, 1953–1990*, PhD thesis, Columbia University, 2020.

12 Only two of the TTC's current bus routes do not intersect with at least one rapid transit station. The two non-feeder routes (99 Arrow Road and 171 Mt. Dennis) exist primarily to take TTC personnel to and from bus garages.

13 Jonathan English, "Toronto's Secret Success: Suburban Buses," *Globe and Mail*, 25 October 2019.

14 Paul Mees, *Transport for Suburbia: Beyond the Automobile Age* (London: Earthscan, 2009), 167.

15 On the development of rapid transit in Canadian cities see J.W. Borse Jr., *Rapid Transit in Canada* (Philadelphia: Almo Press, 1968); and James W. Kerr, *Illustrated Treasury of Rail Rapid Transit Systems and Cars of North America* (Montreal: DPA-LTA Enterprises, 1983).

16 The majority of streetcar slides along the Rogers Road line were taken in the final months of its service before abandonment on 19 July 1974. Being peripheral in the network and operating with lower frequencies than the downtown lines meant it was less popular with streetcar photographers. It was only after the TTC's announcement that the line would be abandoned that the route received much attention from enthusiasts.

17 On the makeup and activities of the Streetcars for Toronto group see "Streetcars for Toronto – 35th Anniversary," http://spacing.ca/toronto/2007/12/01/streetcars-for-toronto-35th-anniversary/.

18 For a comparison of how the CLRV fared in relation to other streetcars developed during this era, see Brian Doucet, "A Comparative Look at Toronto's CLRVs," *Spacing Blog*, 18 December 2019.
19 The 19 PCCs that were rebuilt for the Harbourfront line were completed between 1986 and 1992 and were classified as A15 cars by the TTC.
20 On the St. Clair project see James Bow, "The Battle of St. Clair," http://transit.toronto.on.ca/streetcar/4126.shtml.
21 On Toronto's eleven streetcar lines see Kenneth C. Springirth, *Toronto Streetcars Serve the City* (Stroud, UK: Fonthill Media, 2014).
22 On delivery problems for the Bombardier streetcars see Ben Spurr, "A Long, Bumpy Road for New Streetcars," *Toronto Star*, 27 January 2020, A13. The final unit in the 204-car order did not arrive until 14 January 2020, three weeks after the last CLRVs were withdrawn from service.
23 Delivery of the new Bombardier cars was frustratingly slow. On this matter see http://projects.thestar.com/bombardier-ttc/
24 On the iconic nature of the CLRVs see J.P. Larocque, "A Lost Piece of Toronto Iconography," *Toronto Star*, 25 January 2020, L1 and L5.
25 Ben Spurr, "Track of Dreams: 24-Year-Old Transit Fan Buys Himself an Old TTC Streetcar," *Toronto Star*, 30 July 2020.
26 See N. Boisvert, "The TTC Is Cracking Down on Fare Evasion: But Why Is Its Fine So Hefty?" *CBC News*, 20 February 2020.
27 Ben Spurr, "TTC Officers Have Collected More than 40,000 Records on Riders Who Weren't Charged with an Offence," *Toronto Star*, 11 March 2020.
28 K.N. Rankin and H. McLean, "Governing the Commercial Streets of the City: New Terrains of Disinvestment and Gentrification in Toronto's Inner Suburbs," *Antipode* 47, no. 1 (2015), 216–39.

Chapter 7

1 https://www.youtube.com/watch?v=dW9BsOuROno.
2 Derek Flack, "A Visual Ode to the TTC streetcar," *BlogTO*, 19 November 2010, https://www.blogto.com/arts/2010/11/a_visual_ode_to_the_ttc_streetcar/; and "The 35 Most Iconic Photos You Can Take in Toronto," *BlogTO*, 2018, https://www.blogto.com/slideshows/iconic-photos-toronto/12225.
3 Kate McGillivray, "Now 40 Years Old, the CLRV Streetcar Is a Piece of Toronto History That Almost Never Came to Be," *CBC News*, 30 December 2017.
4 Perhaps the best example of a novel in which streetcars play a role is Graham Jackson's *The Jane Loop* (Toronto: Cormorant Books, 2016). Other examples of Toronto fiction in which streetcars play a role include two poetry collections – Chris Faiers, *College Streetcar Runs All Night* (Toronto: Unfinished Monument Press, 1979) and Ted Plantos, *She Wore a Streetcar to the Wedding* (Toronto: Missing Link Press, 1973) – and two works for children – Michael Bedard, *Redwork* (Toronto: Lester & Orpen Dennys, 1990) and Helen Huyk, *The Baby Streetcar* (Toronto: Three Trees Press, 1978).
5 P. Marcuse, "From Critical Urban Theory to the Right to the City," *City* 13, nos. 2–3 (2009): 185–197; see also E. Wyly, "Things Pictures Don't Tell Us: In Search of Baltimore," *City* 14, no. 5 (2010): 497–528.
6 Jason Hackworth, *The Neoliberal City* (Ithaca: Cornell University Press, 2007); see also Ute Lehrer and Thorben Wieditz, "Condominium Development and Gentrification: The Relationship between Policies, Building Activities and Socio-Economic Development in Toronto," *Canadian Journal of Urban Research* 18, no. 1 (2009): 82–103.
7 For an excellent overview of how different modes of transport shape potential exposure to diversity and the types of interactions one has with their community, see M. Te Brömmelstroet, A. Nikolaeva, M. Glaser, M.S. Nicolaisen, and C. Chan, "Travelling Together Alone and Alone Together: Mobility and Potential Exposure to Diversity," *Applied Mobilities* 2, no. 1 (2017): 1–15.
8 Jonathan English, *The Better Way: Transit Service and Demand in Metropolitan Toronto, 1953–1990*, PhD thesis, Columbia University, 2020.
9 *2016 Census: Education, Labour, Journey to Work, Language of Work, Mobility and Migration*, City of Toronto, 5 December 2017, https://www.toronto.ca/wp-content/uploads/2017/12/94ce-2016-Census-Backgrounder-Education-Labour-Journey-to-work-Language-Mobility-Migration.pdf.
10 Julian Agyeman, "Poor and Black 'Invisible Cyclists' Need to Be Part of Post-pandemic Transport Planning Too," *The Conversation*, 27 May 2020. See also Jay Pitter, "Urban Density: Confronting the Distance between Desire and Disparity," *Azure Magazine*, 17 April 2020; Melody L. Hoffmann, *Bike Lanes Are White Lanes* (Lincoln: University of Nebraska Press, 2016); Adonia E. Lugo *Bicycle/Race: Transport, Culture and Resistance* (Portland, OR: Microcosm Publishing, 2018).
11 See https://www.toronto.ca/home/covid-19/covid-19-protect-yourself-others/covid-19-reduce-virus-spread/covid-19-activeto/covid-19-activeto-expanding-the-cycling-network/.
12 Alan Walks, "Stopping the 'War on the Car': Neoliberalism, Fordism, and the Politics of Automobility in Toronto," *Mobilities* 10, no. 3 (2015): 402–22.
13 Kate Allen, Jennifer Yang, Rachel Mendleson, and Andrew Bailey, "Lockdown Worked for the Rich, but Not for the Poor: The Untold Story of How COVID-19 Spread across Toronto, in 7 Graphics," *Toronto Star*, 2 August 2020.

14 Jessica Cheung, "Black People and Other People of Colour Make Up 83% of Reported COVID-19 Cases in Toronto," *CBC News*, 30 July 2020.
15 Allen, Yang, Mendleson, and Bailey, "Lockdown Worked for the Rich."
16 For rates of COVID-19 infection in each census tract, see https://www.toronto.ca/home/covid-19/.
17 Cheung, "Black People and Other People of Colour."
18 S. Biglieri, L. De Vidovich, and R. Keil, "City as the Core of Contagion? Repositioning COVID-19 at the Social and Spatial Periphery of Urban Society," *Cities & Health*, 1–3; Pitter, "Urban Density"; A. Yasin and D. Ferguson, "Pandemic Patios and 'Flat-White' Urbanism," *Plan Canada* 60, no. 4. (2021): 21–6.
19 See also Brian Doucet, Rianne van Melik, and Pierre Filion (eds.), *Global Reflections on COVID-19 and Urban Inequalities: Volume 1, Community and Society* (Bristol: Bristol University Press, 2021).
20 John Sewell, "Old and New City," in K. Gerecke (ed.), *The Canadian City* (Montreal: Black Rose Books, 1991). See also R.A. Walks, "The Boundaries of Suburban Discontent? Urban Definitions and Neighbourhood Political Effects," *Canadian Geographer/Le Géographe canadien* 51, no. 2 (2007): 160–85.
21 Ibid. See also R.A. Walks, "Place of Residence, Party Preferences, and Political Attitudes in Canadian Cities and Suburbs," *Journal of Urban Affairs* 26, no. 3 (2004), 269–95.
22 Walks, "The Boundaries of Suburban Discontent?" 183.
23 Ibid.
24 R.A. Walks, "The Causes of City-Suburban Political Polarization? A Canadian Case Study," *Annals of the Association of American Geographers* 96, no. 2 (2006): 390–414; see also Justin van der Merwe, *Spillover Gentrification? Mid-Sized Cities within Commuter Sheds of Global Cities* (master's thesis, University of Waterloo School of Planning, 2021).
25 Marcus Gee, "From Downtown to Rob Ford's House," *Globe and Mail*, 8 June 2013.
26 Tess Kalinowski and David Rider, "'War on the Car Is Over': Ford Scraps Transit City; Move Could Leave Toronto on Hook for Millions in Penalties, Wasted Work," *Toronto Star*, 2 December 2010.
27 Alan Walks, "Stopping the 'War on the Car.'"
28 Walks, "Stopping the 'War on the Car.'" On Rob Ford's election and time as mayor see Robyn Doolittle, *Crazy Town: The Rob Ford Story* (Toronto: Viking Press, 2014).
29 Walks, "Stopping the 'War on the Car.'"
30 A. Clark, "Doug Ford and the Revival of the 'War on the Car,'" *Globe and Mail*, 23 March 2018.
31 Samantha Craggs, "Mulroney Says She Told Hamilton about LRT Overruns in September 'in Good Faith,'" *CBC Hamilton*, 18 December 2019.
32 While a small sliver of pre–World War II areas exist along the southern flank of the Etobicoke-Lakeshore riding, much of this suburban riding is auto-dependent. However, many of the polling stations along Lake Shore Boulevard, where the 501 Queen streetcar runs, voted for the NDP. For a map of the results per polling station, see https://globalnews.ca/news/4260146/ontario-election-poll-level-results-2/.
33 See previous endnote.
34 "2018 Ontario Election Results Map," *Toronto Star*, 7 June 2018, https://www.therecord.com/news-story/8658115-2018-ontario-election-results-map/.
35 Zach Taylor, "Suburbanization and Politics," *School of Public Policy Publications: SPP Briefings* 11, no. 23 (2018).
36 P. Filion, T. Bunting, K. McSpurren, and A. Tse, "Canada-US Metropolitan Density Patterns: Zonal Convergence and Divergence," *Urban Geography* 25, no. 1 (2004): 42–65.
37 Pierre Filion, "Enduring Features of the North American Suburb: Built Form, Automobile Orientation, Suburban Culture and Political Mobilization," *Urban Planning* 3, no. 4 (2018): 14.
38 "Could TTC Buses Permanently Replace Queen Street Streetcars?" *Global News*, 8 June 2017.
39 Brian Doucet, "I Clog Toronto's Highways Because There Is No Train Service," *Toronto Star*, 22 February 2019.
40 This comment was part of Roger Keil's presentation "Infra-Structuring Suburbia: Challenging Fixities and Mobilities in Toronto's (Post) Suburban Expanse," during the Cities and Infrastructure one-day workshop organized by the City Institute at York University, 8 November 2018; see also Jean-Paul Addie and Roger Keil, "Real Existing Regionalism: The Region between Talk, Territory and Technology," *International Journal of Urban and Regional Research* 39, no. 2 (2015): 407–17.
41 Paul Hess and Jane Farrow, *Walkability in Toronto's Highrise Neighbourhoods* (Toronto: Cities Centre, University of Toronto, 2010); See also Paul M. Hess, "Avenues or Arterials: The Struggle to Change Street Building Practices in Toronto, Canada," *Journal of Urban Design* 14, no. 1 (2009): 1–28.
42 See also A. Kramer, "Inside and Outside: A Meditation on the Yellowbelt," in Alex Bozikovic, Cheryll Case, John Lorinc, and Annabel Vaughan (eds.), *House Divided* (Toronto: Coach House Press, 2019), 142–55; Tess Kalinowski, "Why It's So Hard to Get Housing into Toronto's 'Yellowbelt' Neighbourhoods – and How Experts Say It Can Be Done," *Toronto Star*, 16 March 2019.
43 C. Bateman, "Margaret Atwood Is Mad Online bout a Condo Building," *CityLab*, 5 September 2017.
44 K. Karamali, "Toronto Neighbourhood Residents Upset about Plans to Turn Parking Lot into Affordable Housing," *Global News*, 26 February 2021, https://globalnews.ca/news/7666729/east-york-cedarvale-avenue-affordable-modular-housing-conflict/.

45 Gil Meslin, "A City of Houses," in Bozikovic, Case, Lorinc, and Vaughan, *House Divided*.
46 Alex Bozikovic, "Toronto Has Lots of Room to Grow: It's Time to Let That Happen," *Globe and Mail*, 26 July 2018; Marcus Gee, "Toronto's Missing Middle," *Globe and Mail*, 18 August 2017.
47 "Proposal for a New Plan for Toronto," Toronto Planning Board, 1966, as quoted in Meslin, "A City of Houses," in Bozikovic, Case, Lorinc, and Vaughan, *House Divided*, 92.
48 As quoted in Meslin "A City of Houses," in Bozikovic, Case, Lorinc, and Vaughan, *House Divided*, 92.
49 Blair Scorgie, "Dissecting Official Plan Amendment 320," in Bozikovic, Case, Lorinc, and Vaughan, *House Divided*.
50 Kramer, "Inside and Outside"; Alex Bozikovic, 'Why Density Makes Great Places,'; both in Bozikovic, Case, Lorinc, and Vaughan, *House Divided*.
51 Don Pittis, "From Real Estate to Businesses: Signs the Pandemic Is Boosting Wealth Concentration: Don Pittis," *CBC News*, 10 August 2020.

Chapter 8

1 T. Butler, "LiveMove Speaker Series: The Intersection of Racism and Transportation," 31 October, 2020, https://www.tamikabutler.com/media/2020/10/31/livemove-speaker-series-the-intersection-of-racism-and-transportation; T. Butler, "Why We Must Talk about Race When We Talk about Bikes," bicycling.com, 9 June 2020, https://www.bicycling.com/culture/a32783551/cycling-talk-fight-racism/.
2 John Sewell, *The Shape of the Suburbs: Understanding Toronto's Sprawl* (Toronto: University of Toronto Press, 2009), 85.
3 Toronto was the only Canadian city to retain its streetcars. New LRT systems in Canada include Edmonton (1978), Calgary (1981), Ottawa (2019), Waterloo Region (2019). Vancouver's Skytrain (now Translink) opened in 1985, but has more characteristics of a metro than a light rail system.
4 See https://www.toronto.ca/city-government/planning-development/planning-studies-initiatives/king-street-pilot/king-street-transit-pilot-overview/.
5 This paradox lies at the heart of Samuel Stein's book *Capital City: Gentrification and the Real Estate State* (New York: Verso, 2019).
6 See K.N. Rankin and H. McLean, "Governing the Commercial Streets of the City: New Terrains of Disinvestment and Gentrification in Toronto's Inner Suburbs," *Antipode* 47, no. 1 (2015): 216–39; D. Immergluck and T. Balan, "Sustainable for Whom? Green Urban Development, Environmental Gentrification, and the Atlanta Beltline," *Urban Geography* 39, no. 4 (2018): 546–62; Julian Agyeman, "Poor and Black 'Invisible Cyclists' Need to Be Part of Post-pandemic Transport Planning Too," *The Conversation*, 27 May 2020.
7 T. Butler, "Why We Must Talk about Race"; D. Thomas, "'Safe Streets' Are Not Safe for Black Lives," *Bloomberg CityLab*, 8 June 2020.
8 Sewell, *The Shape of the Suburbs*, 216.
9 Jonathan English, "Toronto's Secret Success: Suburban Buses," *Globe and Mail*, 25 October 2019; Paul Mees, *Transport for Suburbia: Beyond the Automobile Age* (London: Earthscan, 2009).
10 Jarett Walker, "Frequency Is Freedom," in *Human Transit: How Clearer Thinking about Public Transit Can Enrich Our Communities and Our Lives* (Washington, DC: Island Press, 2012).
11 Oliver Moore, "GO Transit Calls Time on Free Parking," *Globe and Mail*, 6 April 2018, https://www.theglobeandmail.com/canada/toronto/article-go-transit-calls-time-on-free-parking/.
12 Jonathan English, "Integration with Local Transit Is Key to GO's Success," *Globe and Mail*, 1 February 2020; see also Sean Marshall, "Disappearing GO-TTC Fare Discount a Major Blow to Regional Transit in Toronto," *Marshall's Musings* (blog), 22 January 2020, https://seanmarshall.ca/2020/01/22/disappearing-go-ttc-fare-discount-a-major-blow-to-regional-transit-in-toronto/.
13 In York Region to the north of the City of Toronto, bus rapid transit (BRT) facilities, with dedicated lanes, have been constructed along parts of Highway 7 and are under construction on Yonge Street. However, the frequency of service on this BRT is far lower than most suburban TTC bus routes.
14 Gregg Culver, "Mobility and the Making of the Neoliberal 'Creative City': The Streetcar as a Creative City Project?" *Journal of Transport Geography* 58 (2017): 22–30; K. Olesen, "Infrastructure Imaginaries: The Politics of Light Rail Projects in the Age of Neoliberalism, *Urban Studies* 57, no. 9 (2020): 1811–26; Brian Doucet, "The 'Hidden' Sides of Transit-Induced Gentrification and Displacement along Waterloo Region's LRT corridor," *Geoforum* 125 (2021): 37–46; https://doi.org/10.1016/j.geoforum.2021.06.013; M. Ellis-Young and B. Doucet, "From 'Big Small Town' to 'Small Big City': Resident Experiences of Gentrification along Waterloo Region's LRT Corridor," *Journal of Planning Education and Research*, https://doi.org/10.1177/0739456X21993914.
15 On the relationship between transit improvements and gentrification, see E. Delmelle and I. Nilsson, "New Rail Transit Stations and the Out-Migration of Low-Income Residents," *Urban Studies* 57, no. 1 (2020): 134–51; Anne E. Brown, "Rubber Tires for Residents," *Transportation Research Record: Journal of the Transportation Research*

Board 2539 (2016): 1–10; I. Nilsson and E. Delmelle, "Transit Investments and Neighborhood Change: On the Likelihood of Change," *Journal of Transport Geography* 66 (2018): 167–179; A. Loukaitou-Sideris, S. Gonzalez, and P. Ong, "Triangulating Neighborhood Knowledge to Understand Neighborhood Change: Methods to Study Gentrification," *Journal of Planning Education and Research* 39, no. 2 (2019): 227–42; Dwayne M. Baker and Lee Bumsoo, "How Does Light Rail Transit (LRT) Impact Gentrification? Evidence from Fourteen US Urbanized Areas." *Journal of Planning Education and Research* 39, no. 1 (2019): 35–49; Culver, "Mobility and the Making of the Neoliberal 'Creative City,'" 22–30; Annelise Grube-Cavers and Zachary Patterson, "Urban Rapid Rail Transit and Gentrification in Canadian Urban Centres: A Survival Analysis Approach," *Urban Studies* 52, no. 1 (2015): 178–94; Craig E. Jones and David Ley, "Transit-Oriented Development and Gentrification along Metro Vancouver's Low-Income Skytrain Corridor," *Canadian Geographer* 60, no. 1 (2016): 9–22; Doucet, "The 'Hidden' Sides of Transit-Induced Gentrification."

16 Brian Doucet and Robin Mazumder, "COVID-19 Cyclists: Expanding Bike Lane Network Can Lead to More Inclusive Cities," *The Conversation*, 22 November 2020.

17 A. El-Geneidy, D. van Lierop, E. Grisé, G. Boisjoly, D. Swallow, L. Fordham, and T. Herrmann, "Get on Board: Assessing an All-Door Boarding Pilot Project in Montreal, Canada," *Transportation Research Part A: Policy and Practice* 99 (2017): 114–24.

18 Ben Spurr, "TTC Officers Have Collected More than 40,000 Records on Riders Who Weren't Charged with an Offence," *Toronto Star*, 11 March 2020.

19 Tricia Wood, "Covid Recovery: Timid Bus Priority Plan in Toronto Needs to be Bigger and Bolder," *Spacing Magazine*, 21 July 2020.

20 Sean Marshall, "Mapping TTC Crowding during a Pandemic," *Spacing Magazine*, 1 April 2020.

21 Tricia Wood, "Post-pandemic Economic Recovery Requires Strong Public Transit," *Spacing Magazine*, 5 June 2020; Wood, "Covid Recovery."

22 "RapidTO Bus Lanes Improve Transit Reliability and Capacity in Scarborough," City of Toronto media release, 28 July 2021, https://www.toronto.ca/news/rapidto-bus-lanes-improve-transit-reliability-and-capacity-in-scarborough/.

23 Ben Spurr, "In Jane-Finch, Leaders Fume at 'Terrible Betrayal' after Metrolinx Goes Back on Plan to Donate Land for Community Centre," *Toronto Star*, 23 July 2020; Angelyn Francis, "The Numbers behind the Need: Why Metrolinx's Cancelled Community Hub Plan Is Such a Blow to Jane and Finch," *Toronto Star*, 23 July 2020.

24 Spurr, "In Jane-Finch, Leaders Fume." See also Royson James, "Don't Underestimate Power in Community," *Toronto Star*, 9 September 2020, A12.

25 Rankin and McLean, "Governing the Commercial Streets of the City."

26 Romain Baker, Dane Gardener-Williams, Anyika Mark, Elizabeth Antczak, Mona Dai, Samuel Ganton, and Tura Wilson, *Report: A Black Business Conversation on Planning for the Future of Black Businesses and Residents on Eglinton Ave. W.* (Toronto: Black Urbanism Toronto, 2020), http://www.mediafire.com/file/rsh7es62c7x6fjb/BUSINESS_CONVERSATIONS_REPORT__2020-07-22.pdf/file. See also Eric Strober, "Little Jamaica Needs Protection Now: Councillors Team Up to Protect Area from Gentrification," *Village Post*, September 2020, 9; Barbara Saba, "Little Jamaica Businesses Are 'Barely Hanging On' Due to LRT Construction and COVID-19: Councillors Say They Have a Plan," *Toronto Star*, 28 September 2020.

27 On the TTC's accessibility plans see 2019–2023 TTC Multi-Year Accessibility Plan.

28 The rollout of the Presto fare system was neither easy nor inexpensive. For some sense of the problems encountered see https://transittoronto.ca/spare/0021.shtml.

29 The Fair Pass requires possession of a Presto card and provides discounts of 33 per cent on single rides and 20 per cent on the cost of a monthly pass. By April of 2018, some 80,000 people had taken advantage of Fair Passes. On this program see https://www.toronto.ca/community-people/employment-social-support/support-for-people-in-financial-need/assistance-through-ontario-works/transit-discount/ and https://www.ttc.ca/Coupler/Editorial/News/020_December__citynr_fair_pass_discount.jsp.

30 On the two-hour transfer program see https://www.ttc.ca/News/2018/August/08_23_18NR_2_hour_transfer.jsp.

31 Ibid.

32 Lindsey Vodarek and Swathi Meenakshi Sadagopan, "Rise of Ghost Hotels Casts Pall over Toronto Rental Market," *Toronto Star*, 22 April 2019.

33 Irelyne Lavery, "Nearly 20 Per Cent of GTA Homeowners under 35 Own More than One Property, Survey Finds," *Toronto Star*, 28 July 2021.

34 E. Wyly, "Things Pictures Don't Tell Us: In Search of Baltimore," *City* 14, no. 5 (2010): 497–528; J.H. Rieger, "Photographing Social Change, *Visual Studies*, 11, no. 1 (1996): 5–49; Gillian Rose, *Visual Methodologies: An Introduction to Researching with Visual Materials* (Los Angeles: Sage, 2016).

35 John Lorinc, "Introduction: The Stability Trap," in Alex Bozikovic, Cheryll Case, John Lorinc, and Annabel Vaughan (eds.), *House Divided: How the Missing Middle*

36 J. Rieti, "Jennifer Keesmaat Puts Ambitious Affordable Rental Housing Target at Heart of Mayoral Campaign," *CBC Toronto*, 7 August 2018, https://www.cbc.ca/news/canada/toronto/keesmaat-housing-announcement-1.4776396.

37 Annabel Vaughan, "Radical Typologies," in Bozikovic, Case, Lorinc, and Vaughan, *House Divided*, 157–64.

38 See https://mikelayton.to/2020/07/23/new-affordable-housing-in-kensington-market/.

39 Philip Zigman and Martine August, *Above Guideline Rent Increases in the Age of Financialization*. Report by Renovictions-TO, 2021, https://www.renovictionsto.com/agi-report.

40 S. Boynton, "New Westminster's Anti-Renoviction Bylaw Upheld by B.C. Supreme Court," *Global News*, 12 February 2020; I. Olson, "City of Montreal to Exercise Right of First Refusal to Turn Properties for Sale into Social Housing," *CBC News*, 17 February 2020.

41 Oliver Wainwright, "The Radical Model Fighting the Housing Crisis: Property Prices Based on Income," *The Guardian*, 16 January 2017; see also http://www.londonclt.org/about-us/.

42 Justin McElroy, "Whistler's Affordable Housing Model Is Below-Market and Free of Speculation: Why Isn't It Used Elsewhere?" *CBC News*, 24 January 2019.

43 Ana Teresa Portillo and Mercedes Sharpe Zayas, "The Urban Legend: Parkdale, Gentrification and Collective Resistance," in Bozikovic, Case, Lorinc, and Vaughan, *House Divided*, 74–82; see also http://www.pnlt.ca/about/.

44 Alex Bozikovic, "Conclusion," in Bozikovic, Case, Lorinc, and Vaughan, *House Divided*, 164.

45 Julian Agyeman, "Urban Planning as a Tool of White Supremacy – The Other Lesson from Minneapolis," *The Conversation*, 27 July 2020.

46 Marcus Gee, "Toronto's Missing Middle," *Globe and Mail*, 18 August 2017.

47 Patrick Sisson, "Can Minneapolis's Radical Rezoning Be a National Model?" *Curbed*, 27 November 2018.

48 For a critique of the Minneapolis upzoning policy, see J. Russell, "Minneapolis and the End of the American Dream House," *Architectural Record*, 2 October 2019, https://www.architecturalrecord.com/articles/14266-minneapolis-and-the-end-of-the-american-dream-house. For research on the impact of upzoning policies on the creation of new affordable housing, see Andres Rodriguez-Pose and Michael Storper, "Housing, Urban Growth and Inequalities: The Limits to Deregulation and Upzoning in Reducing Economic and Spatial Inequality," *Urban Studies* 57, no. 2 (2020): 223–48; Yonah Freemark, "Upzoning Chicago: Impacts of a Zoning Reform on Property Values and Housing Construction," *Urban Affairs Review* 56, no. 3 (2020): 758–89. For the impact of transit-oriented developments, see Karen Chapple and Anastasia Loukaitou-Sideris, *Transit-Oriented Displacement or Community Dividends?* (Cambridge, MA: The MIT Press, 2019).

49 See http://lanewayhousingadvisors.com/faq-financial-considerations/. For median income information see *Backgrounder: 2016 Census, City of Toronto*.

50 For more on the Right to the City, see Henri Lefebvre, "The Right to the City," in E. Kofman and E. Lebas (eds.), *Writing on Cities*, 63-184 (London: Blackwell, 1996); David Harvey, "The Right to the City," in *The Urban Sociology Reader*, 443-46 (Routledge, 2012); Peter Marcuse, "From Critical Urban Theory to the Right to the City," *City* 13, no. 2-3 (2009): 185-97. See also: David Madden and Peter Marcuse, *In Defense of Housing* (London: Verso, 2016), Brian Doucet, "Housing is both a human right and a profitable asset, and that's the problem," *The Conversation*, 14 December 2021.

51 Brian Doucet, "National Housing Strategy Fails to Go Deep Enough on Root of Ills," *Toronto Star*, 3 December 2017; David Hulchanski, "No, Ottawa Has Not Put Forth a National Housing Strategy," *Globe and Mail*, 4 December 2017; Daniel Hertz, "Housing Can't Be Both Affordable and a Good Investment," *CityLab*, 19 November 2018.

52 Martine August, "The Coronavirus Exposes the Perils of Profit in Seniors' Housing," *The Conversation*, 26 July 2020.

53 Wyly, "Things Pictures Don't Tell Us."

54 Annabel Vaughan, "Radical Typologies," in Bozikovic, Case, Lorinc, and Vaughan, *House Divided*, 164.

55 For a good personal example of this type of housing anxiety, see Tatum Taylor Chaubal "Our Own Front Door," in Bozikovic, Case, Lorinc, and Vaughan, *House Divided*, 83–7; see also E. Mathieu, "Toronto Saw a 77 Per Cent Rise in Personal-Use Eviction Applications over Four Years, Data Shows," *Toronto Star*, 12 February 2020.

56 H. Aldridge and R. Tranjan, "Rents Keep Going Up, Pandemic or Not, *Canadian Centre for Policy Alternatives*, 9 February 2021, https://behindthenumbers.ca/2021/02/09/rents-keep-going-up-pandemic-or-not/.

57 M. Moffatt, "We Need to Pay Attention to Migration Patterns and 'Drive until You Qualify': Here's Why," *Ontarians on the Move*, 16 February 2021.

58 Julie-Anne Boudreau, Roger Keil, and Douglas Young, *Changing Toronto: Governing Urban Neoliberalism* (Toronto: University of Toronto Press, 2009), 140.

59 Jay Pitter, "Urban Density: Confronting the Distance between Desire and Disparity," *Azure Magazine*, 17 April 2020.

60 On the intersection between COVID-19, race/ethnicity, and density/overcrowding, see K. Grant, "Data Shows Poverty, Overcrowded Housing Connected to COVID-19 Rates among Racial Minorities in Toronto," *Globe and Mail*, 2 July 2020; D. Carrington, "Covid-19 Impact on Ethnic Minorities Linked to Housing and Air Pollution," *The Guardian*, 19 July 2020; J. Agyeman, "Poor and Black 'Invisible Cyclists' Need to be Part of Post-pandemic

Transport Planning Too," *The Conversation*, 27 May 2020; J. Cheung, "Black People and Other People of Colour Make Up 83% of Reported COVID-19 Cases in Toronto," *CBC News*, 30 July 2020; D. Carrington, "'Compelling' Evidence Air Pollution Worsens Coronavirus – Study," *The Guardian*, 13 July 2020; Brian Doucet and Justin Van der Merwe, "Housing Challenges, Mid-Sized Cities and the COVID-19 Pandemic," *Canadian Planning and Policy/Aménagement et politique au Canada* (2021): 70–90. See also Markus Moos, Amanda McCulley, and Tara Vinodrai, *COVID-19 and Urban Density: Evaluating the Arguments* (Toronto: Innovation Policy Lab of the Munk School of Global Affairs and Public Policy, 2020), https://munkschool.utoronto.ca/ipl/publication/covid-19-and-urban-density-evaluating-the-arguments/; Brian Doucet, Pierre Filion, and Rianne van Melik (eds.), *Global Reflections on COVID-19 and Urban Inequalities: Volume 2, Housing and Home* (Bristol: Bristol University Press, 2021).

INDEX

Please note: Page numbers in italics refer to photographs, illustrations, and maps. Subentries for "portfolios of photographs" appear in page order rather than alphabetical order.

Abbey, Wallace W., 99, 114
ActiveTO, 234, 250
Adel, Aaron, 111
aerial views: from CN Tower (1985/86 and 2020), *28*; from Commerce Court, *14*, *23*; from helicopter above Toronto (1972), *26–33*
affordable housing: community land trusts and, 80, 259; density and, 68, 189, 241–2, 244, 260–1; inclusionary zoning and, 68, 78, 257, 262; reform politics and, 65–7, 259–60; rent controls and, 78, 258, 261, 262–3; tenure/ownership and, 40, 67, 257–60, 261, 262, 263. *See also entry below*; density; tenure/ownership; zoning
affordable housing, obstacles/ threats to: condo construction, 22, 73–9, 247, 257, 262, 263; financialized properties, 79–80, 257–9, 262–3; gentrification, 68–82; government policies, 75, 77–8; opposition to increased density, 68, 189, 242, 244, 260; Toronto as world city/competitive city, 39–40, 42–8. *See also* condominiums; gentrification
Agyeman, Julian, 234, 250
Airbnb, 79, 257, 261
air-electric PCC streetcars, 121, 124, *125*, 127
Akers, Joshua, 89
Allsopp, Robert, 22
ALRVs, 129; in TTC fleet, 15, 129–30
Alstom (French LRV manufacturer), 23, 130

amalgamation, of Metro Toronto, 15, 42, 43–4, 75, 77, 265n7
Amazon, 37, 40; second headquarters of (HQ2), 39, 47
Annex, The, 52, 71, 262; Jacobs as resident of, 60, 64, 65, 68; NIMBYism in, 242, 244
Armstrong, Beere, and Hime (civil engineering firm), 117
Arreola, Daniel, and Nick Burkhart, 96
Art Gallery of Ontario (AGO), 47, *153*
Articulated Light Rail Vehicles. *See* ALRVs
Artscape Weston Common, 73, 250
Artscape Wychwood Barns, *224–5*, 250. *See also* Wychwood carhouse/yard
attractions, civic, 43–5, *46*, 47; bidding for, 47; in Davis era, 65–6; landmarks threatened by construction of, 66; and neoliberalism, 44–5; policing/exclusion at, 45; and "urban competitiveness," 42–8; and "world city" concept, 38–42. *See also* events and spectacles; *specific attractions*
Atwood, Margaret, 242, 244
August, Martine, 81, 82, 261; and Giuseppe Tolfo, 78; and Alan Walks, 80
Aura condominium, 19, 24
Austin, Dan, 88
Automobile City (suburban Toronto): apartment buildings in, 59, 84, 241; bus routes in, 126, 233–4, 252–3, 255; characteristics of (1971/2016), 60, 62, 63; as constructed after World War II, 7, 9, 51, 58, 63–4, 83, 84, 238; as contrasted with Streetcar City, 51–9, 62–3, 231–45, 254; curved/non-grid street pattern of, 58, 59; cycling in, 84, 234, 236, 241, 250, 252, 253; density in, 233, 236, 238, 241; economy/industry of, 36–7, 234–5; house designs of, 58; modernist ideas/visions and, 58–9, 63–4; pandemic issues in, 234, 235–6, *237*; politics of, 236, 238–40; strict zoning/separation of functions in, 58–9; walking in, 84, 234, 241, 252, 253
automotive industry, 36, 37, 235

Baltimore, 43, 47, 103, 127, 248
Bannos, Pamela, 87
Baranowski, Damian, 116
Barthes, Roland, 85
Bassnett, Sarah: *Picturing Toronto*, 117
Bathurst Street: at Bloor Street, 79, *91*; at College Street, *207*; at/near Fleet Street, 36, *90*, *168–9*; at Fort York Boulevard, *76*, *101*; at King Street, 104, *180–1*; at Queen Street, *94–5*, *195*; streetcar line on, 7, 8, 14, 104, 129, 130, *131*, 250
Bathurst Street Bridge (now Sir Isaac Brock Bridge), *182*
Bay Street, 129, *143*, *152*, *160–1*
Beebe, Lucius, 99, 114; *Highball: A Pageant of Trains*, 114; *High Iron: A Book of Trains*, 114; *Mixed Train Daily*, 114

Bell, Daniel, 70
Benson, Ted, 99; *One Track Mind*, 115
bike lanes, 132, 233; in Amsterdam, *249*; along Bloor-Danforth line, *217*, 234; dearth of, in suburbs, 84, 234, 236, 250, 252, 253; design flaws of, 255; Fords' opposition to, 238, 239; and gentrification, 250, 254
Blemiller, Andrew, 14
Bloor-Danforth streetcar line, 5, 68, 102, 110, 125, 127, *157*, *217*, 219, 278n19
Bloor-Danforth subway line, 11, 68, 102, 118, 127; affluent neighbourhoods south of, 83; bike lanes along, *217*, 234; extensions of, 127, 266n22
Bloor Street, 21, 65, 79, *91*, *157*, *185*, *216–17*, 242. *See also* Danforth Avenue
Boeing-Vertol: LRVs made by, 128, 129
Bombardier. *See* Flexity Outlook LRVs
Boudreau, Julie-Anne, Roger Keil, and Douglas Young, 43–4
Bow, James, 111, 116
Bozikovic, Alex, 245
Broadview Avenue, *53*, 68, 130; at Gerrard Street, 104, *202–3*
Bromley, John F., 116, 118–19, *119*, 278n20; *Fifty Years of Progressive Transit* (with Jack May), 115, 116, 118, 123; slide mount of, 118, *119*; *TTC '28*, 118
Bromley, Margaret, 119
Butler, Tamika, 247; and Destiny Thomas, 250

Calgary: streetcars/LRT in, 126, 127
Canadian Car and Foundry (Montreal), 124
Canadian Light Rail Vehicles. *See* CLRVs
Canadian Railway Historical Association (CRHA), 98
Canadian Railway Museum (Exporail), 98
Carlton Street, *156*, *190–1*; streetcar line on, 11, 14, 15, 104, 129. *See also* College Street
Center for Railroad Photography and Art, 114
Central Electric Railfans Association (CERA), 98
Cherry Street streetcar line, 15, 130, 266n26
Chow, Olivia, 239
churches: as converted into condos, 22, *207*; Sunday transit stops at, 9, *10*
Cincinnati, 96; PCC streetcars purchased from, 125, *126*, 127, *127*, *162*
CityPlace, *76*, *171*, 242
Clegg, Charles, 99, 114. *See also* Beebe, Lucius
Cleveland, 22, 37, 103, 123; PCC streetcars and, 125, 128
CLRVs, 14, *17*, *91*, *133*, *159*, *182*; development of, 128–9; farewell to, *131*; preservation of, 97, 100, *101*, 130; prototype of, *128*, 129; as symbol of Toronto, 231; in TTC fleet, 13–14, 15, 129, 130
CN Tower, 12, 18, 24, 66, 88, 231, 241; aerial views from (1985/6 and 2020), *28*
College Street, 7, 21, 68, 242, 260; streetcar line on, *159*, *207*, *214*, 250. *See also* Carlton Street
Commerce Court (now Commerce Court North), 12, 18, 24, 67, *138*, *161*; aerial views from, *14*, *23*
community land trusts (CLTs), 80, 259
competitive cities, 42–5; characteristics of, 44. *See also* Toronto, as competitive city
condominiums, 8, 13; architecture of, 19, 21; bike parking spaces at, 234, *235*; in converted churches, 22, *207*; in converted factories, 22, 36, 40, 52, *57*, 71–2, 74, 77, *215*; and deindustrialization, 36, 40, 42, 71–7, 234–5; density of, 77–8, 264; and gentrification, 39, 69, 71–9, 254; and housing unaffordability, 22, 73–9, 247, 257, 262, 263; legislation governing, 10; with preserved façades, 137, 143, *174*; and public housing, 67, 81–2; as rental properties, 79, 257, 261; small-scale, 68, 189; tenure of, 74; Toronto boom in, 20–1, 73–9, 233, 245; and Toronto skyline, 24, 74, *163*, 242. *See also* deindustrialization; gentrification
construction industry, 36
COVID-19 pandemic, 16–18, 48, 261; cycling during, 52, 132, 234, 236, 250, 252; and demand for office space, 22; and impact on housing/house prices, 69, 79, 245, 257, 261, 263–4; and impact on transit, 17–18, 49, 132–3, 234, 255, *255*; overcrowded buses during, 18, 234, 264; photographs of Toronto during, *142*, *147*, *149*, *199*; socioeconomic disparities revealed by, 18, 39, 84, 234, 235–6, *237*, 250, 252, 255, 256, 263–4
Crombie, David, 44, 66, 67
Curran, Winifred, 74, 77
Currelly, Charles Trick, 7
cycling: in Automobile City, 234, 241; in Streetcar City, 234, 248, 255
cycling infrastructure: ActiveTO and, 234, 250; in Automobile City, 84, 234, 236, 250, 252, 253; in Streetcar City, 73, *217*, 234, 250, *251*, 252. *See also* bike lanes

Dagnino, Michelle, 255–6
Danforth Avenue, 21, 242, 260; Main Street at, *210*. *See also* Bloor-Danforth streetcar line; Bloor-Danforth subway line; Bloor Street
Darkes, Maureen Kempston, 46
Datta, Ayona, 88
Davidson, Mark, and Loretta Lees, 74
Davis, William, 42–3; and Spadina Expressway, 13, 65
deindustrialization, 22, 35–7, 52, 66, 102, 103; and condo construction, 36, 40, 42, 75, *76*, 77, 165, *168–9*, *171*, *181–2*, 235; and condo/loft conversions, 22, 36, 40, 52, *57*, 71–2, 74, 77, *167*, *215*, 234–5; and free trade agreements, 13, 36, 83–4; and gentrification, 73–5; and globalization, 37–42; and shift to financial services sector, 37–42, 234–5; as shown in repeat photography, 77, *92–3*, 165, *166–87*; "terrain of availability" created by, 44, 67, *170–1*, 219. *See also* factories; manufacturing, inner city loss of
Delano, Jack, 99; *From San Juan to Ponce on the Train*, 114, 116

density: in Automobile City, 233, 236, 238, 241; in exchange for public amenities, 77–8; in "forgotten" neighbourhoods, 263–4; government policies on, 75, 77–8; Jacobs' advocacy of, 64, 68; Minneapolis plan to increase, 260–1; "missing middle," 260–1, 262; of New Urbanism, 59; opposition to, 68, 189, 242, 244, 260; politics and, 239–40; in Streetcar City, 52, 58, 67, 68, 241–2, 244, 260; in Toronto (1961–2016), 9, 12, 19, 21; near transit, 78, 125, 234; in unaffordable housing, 68, 257, 261; young adults/young families and, 70. *See also* zoning

De Sousa, Jeremy: *Digital Railway Photography: A Practical Guide*, 99, 101

Detroit, 22, 37–8, 64, 65, 69, 165; repeat photography of, 88–9, 93; streetcars in, 103, 125

DeVerteuil, Geoffrey, 92–3, 116

displacement: by financial services sector, 41; gentrification and, 21, 47, 68, 69, 71–3, 74, 79–81, 132, 229, 245, 250, 254, 256, 261–2; neighbourhood redevelopment and, 250, 261–2; policy/zoning changes and, 75, 77; retail, 72, 73, 79, 242, 244; of tenants, 79–80, 257–9, 262–3; urban renewal and, 66

Dominion Land Survey, 89

Don Mills, 68–9

Doucet, Brian, 3, *4*, 8, 11–12, 36, 116, 236; Toronto of (1981), 12–14; with Zebracki and de Brant, 89

Doucet, Michael, 3, *4*, 7–8, 236; Toronto of (1961), 8–10, 21; and Toronto of grandsons' generation (2016 and beyond), 15–21

downtown: repeat photography portfolio of, 137, *138–63*; as streetcar photography location, 103–4, *105*

Dundas Street, *57*, 68, 72, *152–3, 185, 215*; at Dufferin Street, 212, *213*, 238; at Parliament Street, *204*; at Spadina Avenue, *198–201*; streetcar line on, 7, 14, 15, 102, 129, 143, 238, 250; at/near Yonge Street, *148–51*

East London Community Land Trust (UK), 259

Eaton Centre, 13, 66; initial plans for, 45; as superblock project, 137, *143, 145, 152*

Eaton's department store, 7, 10, 13, 21–2, 56, *143*

Edmonton: streetcars/LRT in, 126, 127

Eglinton Avenue, 7–8, 83, 238, 252; dedicated bus lanes on, 254, 255, *255*; as expressway terminus, 13, 65; at Oakwood Avenue, *229*, 256

Eglinton Crosstown LRT, 25, 255; and fears of gentrification/displacement, 73, 132, 229, 254, 256; and Metrolinx lawsuit, 48

Eglinton subway station, 7, 83, 126

Electric Railroaders' Association (ERA), 98, 115

Electric Railway Presidents' Conference Committee (PCC), 124. *See also* PCC streetcars

employment, in Toronto: census data on (1961), 10; census data on (1981), 13; and largest employers (1983/2019), 13, 15; structure of (1983/2003/2019), 13, 14; survey of (1983), 13; survey of (2019), 21. *See also* deindustrialization; factories; manufacturing, inner city loss of

English, Jonathan, 60, 252–3, 254

events and spectacles, 45–8; bidding for, 47–8; as distraction from societal problems, 46–7; and "urban competitiveness," 42–8; and "world city" concept, 38–42. *See also* attractions, civic; *specific events*

Exporail (Canadian Railway Museum), 98

factories: closure of, 36–7, 52, 63, 75, 83–4; conversion of, into condos/lofts, 22, 36, 40, 52, *57*, 71–2, 74, 77, *215*; as requiring on-site work during pandemic, 235, *237*, 255; in suburban/regional locations, 36–7, 235, 239, 255; survival of, in inner city, 52, 56. *See also* deindustrialization; manufacturing, inner city loss of; *specific factories*

Filey, Mike: *Toronto Sketches: The Way We Were*, 118; *Toronto Then and Now*, 118

Filion, Pierre, 240, 241

financialized properties: long-term care homes as, 251; rental buildings as, 79–80, 257–9, 262–73

financial services sector, 37–42, 234–5; globalization of, 41, 42

Finch West LRT, 25, 132, 255–6

First Canadian Place, 12, 18, 24, 138

Fleet Street: at/near Bathurst Street, 36, *90*, *168–9*

Flexity Outlook LRVs: as built by Bombardier, 23, 130, 132, 266n24; delivery problems with, 268n53, 282nn22–3; on King Street route, 132–3; Leslie Barns storage built for, 23, 132, *133*, 266n24; shortage of, for all-route coverage, 133; as similar to those in Brussels, 130, *130*; on Spadina Avenue route, *158*; in TTC fleet, 15

Florida, Richard, 45, 47, 77

Floyd, George, 256

Ford, Doug, 77–8, 239–40

Ford, Rob, 23, 238–9, 240

Freeman, Bill, 7

free trade agreements, 13, 36, 83–4, 124

Friedmann, John, 38–40; and Goetz Wolff, 38. *See also* "world city"

Fulford, Robert, 35

Garden City movement, 80

Gardiner Expressway, 65, 66, 233, 241, 257; construction of, *160*

Garratt, Colin: *The Golden Years of British Trams*, 116

Gee, Marcus, 238

gentrification: affordable housing and, 68–82; attractions/events and, 44, 46, 47; condo boom and, 39, 69, 71–9, 254; as defined by Glass, 69; displacement and, 21, 47, 68, 69, 71–3, 74, 79–81, 132, 229, 245, 250, 254, 256, 261–2; early Toronto example of, *172*; financialized rental buildings and, 79–80, 257–9; historic preservation and, 44, 68; of inner city, 39, 68–9, 102, 233, 238, 239; Jacobs and, 68–9; new-build developments as, 74, 79; NIMBYism and, 68, 189, 242, 244, 260; of public

gentrification (*continued*)
  housing, 80–2; resistance to, 80; retail, 72, *72*, *73*, 79, *197*, 242, 244; socioeconomic disparities and, 82–4; as tracked using Google Street View, 96; traditional explanations of, 69–73; transit and, 21, 49, 130, 132, 233, 254, 256. *See also entries below*; affordable housing, *and entry following*
gentrification, characteristics of: cafés/bistros, 22, 72, *73*, 189, *193*, *197*, *213*, *217*; loft/condo conversion, 22, 36, 40, 52, *57*, 71–2, 74, 77, *215*; replacement of original homes with larger ones, 242, 260; retail conversion, 72, *72*; rooming house de-conversion, 21, 71; "whitepainting," 71, 242
gentrification, frontiers of, 69, 79–82, 189; Lansdowne Avenue, *217*; Little Jamaica, *229*, 254, 256; Mount Dennis, 73, 132, 256; Parkdale, 80, *192–93*, 259; Rogers Road, *226–28*, 242, *243*
gentrification, in specific neighbourhoods: Annex, 262; Cabbagetown, *190–91*; Danforth, 21, 262; Leslieville, *197*; Queen Street West, *194*; "Two Kings," 75, 77, *172*; "Upper Beaches," *209*, 262
Geological Survey of Canada, 89
Gerrard Street, 68, 189, *208–9*, 242, 260; at Broadview Avenue, 104, *202–3*; retail gentrification on, *72*, *73*
Glass, Ruth, 69
Glista, Alex, 100, 130; CLRV streetcar acquired by, 100, *101*, 130, 132
Globalisation and World Cities (GaWC) Institute (UK): classification of cities by, 41–2
Goodyear tire factory, 36, 121, *186*
Google Street View, 96
Gopnik, Adam: *Through the Children's Gate*, 70
Goss, Arthur, 87, 117
GO Transit, 11, 13, 241; and poor fare integration with TTC, 253; ridership of, compared to streetcars, 130, 248
Goulah, Jesse, 116
Graham, Stephen, and Simon Marvin: on "splintering urbanism," 48–9, 77

Greater Golden Horseshoe, 37, 59, 69, 75, 231, 241
Greater Toronto Area (GTA), 1, 22, 69, 231, 233; as car-dependent, 241; growth of, 75; immigration to, 35; manufacturing in, 35–7, 235, 239; property ownership in, 257
Greater Toronto and Hamilton Area (GTHA), 7, 48, 231, 233
Greenberg, Ken, 25, 44
Grossman, Irving, 71

Hackworth, Jason, 37, 233; and Neil Smith, 70–1
Halpern, Robert: *Great Lakes Trolleys in Color*, 116
Halton County Radial Railway (HCRR) museum: TTC streetcars at, 97–8, *129*, 129–30, *182*
Hamilton, 20, 37, 48, 96, 165, 231, 240; house prices in, 69, 263; LRT of, 25, 133, 239
Hancock, Macklin, 59
Harbord Street, 7, *131*, *158*; streetcar line on, 7, 127, 143, 200, *211*, 219
Harbourfront Centre, 20, 44
Harbourfront LRT and extension, 11, 12, 15, 25, 130, 247; rebuilt PCCs for, 129, 282n19
Harris, Mike, 15, 44, 258
Harris, Richard, 56
Harvey, David, 43, 44, 47, 48, 49, 67
Henry, Robert Selph: *Trains*, 114
Herzog, Fred, 87–8
Hess, Paul, 241
Hidalgo, Anne, 52
Highway 401, 10, 65, 245
Hill, Roberta, 110; photographs by, *110*
Hine, Lewis, 87, 93
historic preservation, 78; of façades, 22, 137, *145*, *174*; and gentrification, 44, 68; and loft/condo conversions, 75, 77; of Toronto landmarks, 66, 67
Hofsommer, Don L., 114, 115
homeless people, 236, 244; policing of, 44, 45, 49
Honest Ed's: demolition/replacement of, 79, *91*
Hood, J. William: *The Toronto Civic Railways: An Illustrated History*, 115
Houser, Charles, 116

housing. *See* affordable housing, *and entry following*; condominiums; public housing; *see also* density; tenure/ownership; zoning
Howard, Ebenezer, 58, 80
Howell, Peter, 15
Hulchanski, David, 21; *The Three Cities within Toronto*, 82, 83, 84, 238, 239
Hume, Christopher, 1
Hurontario LRT (Mississauga), 25
Hutchinson, Rob, 116
Hwang, Jackelyn, and Robert Sampson, 96

Ibelings, Hans, 20–1, 75
Inglis appliance factory, 36, 268n8
Israelowitz, Oscar, and Brian Merlis: *Subways of New York City in Vintage Photographs*, 115

Jacobs, Jane, 60, 63–9; *The Death and Life of Great American Cities*, 60, 63; and influence on Toronto politics/planning, 60, 64–8; as opponent of Spadina Expressway, 64–5, *65*; as opponent of urban renewal, 60, 63–4, 66; and St. Lawrence neighbourhood, 67, 259–60; and Streetcar City, 67–9
James, Royson, 48
James, William, 117
Jensen, Joel, 99; *Steam: An Enduring Legacy*, 115. *See also* Niemann, Linda Grant
Jessell, Joe, 116
Johnstown, PA: trolley system in, 103, *104*
Josephson, Ken, 110

Kahler, David, 99
Keefe, Kevin P., 99
Keesmaat, Jennifer, 7, 258
Keil, Roger, 241; with Boudreau and Young, 43–4; Stefan Kipfer and, 44; with Lehrer and Kipfer, 81
Kern, Leslie, 74
Kidd, Bruce, 46
Kim's Convenience (store in television series), 231, *232*
King Street, 7, 72, 121, 165, 233; at Bathurst Street, 104, *180–1*; at Church Street, *146–7*; during pandemic, 236, *237*; in Parkdale,

192–3; at/near Parliament Street, *172–5*; at Shaw Street, *183–4*; at Simcoe Street, *140*; at Spadina Avenue, *176–9*; at York Street, *138–9*. *See also entries below*; "Two Kings," mixed-use development in

King streetcar line, 14, *17*, 23, 25, 129, 130; as busiest in Toronto, 15, 20, 132, 248; at Queen–Roncesvalles intersection, 104, *106–7*; "walking the line" on, *109*

King Street Pilot project, 25, *177*, 239, *251*; success of, 132–3, 248, 250, 253, 254

Kipfer, Stefan: and Roger Keil, 44; with Lehrer and Keil, 81

Kitchener, 240, 263; GO Transit to/from, 73; industry in, during pandemic, 236, *237*; residential density allowed in, 260

Kitchener-Waterloo, 36, 37; ION LRT line in, 25, 127, *250*

Klett, Mark, 89, 92, 121

Knowles, Jack, 116

Kodachrome film, xv, 87, 88, 102, 119, 278n20

Krambles, George, 116

Kramer, Anna, 245

Kyer, C. Ian, 48

Lakeshore Boulevard, *186*

Lamb, J. Parker, 99

Landau, Gerald, 116

laneways, 56, 59; housing in, 260, *261*

Lansdowne Avenue, 36; at/north of Bloor Street, *217–18*

Lansdowne carhouse, 218, *219*

Lastman, Mel, 17

Le Corbusier, 58, 60, 66, 80

Lehrer, Ute: with Keil and Kupfer, 81; and Jennefer Laidley, 45; and Thorben Weiditz, 75

Lemmons, Kelly, Christian Brannstrom, and Danielle Hurd, 85

Lemon, James, 65–6; *Toronto since 1918: An Illustrated History*, 118

Leslie Barns, 23, 132, 133, 266n24; three generations of streetcars at, *133*

Ley, David: *The New Middle Class and the Remaking of the Central City*, 70

Liberty Village, 23, 36, 234, 242, 248, 257

light rail transit (LRT): Hamilton, 25, 133, 239; Hurontario (Mississauga), 25; ION (Kitchener-Waterloo), 25, 127, *250*; O-Train (Ottawa), 25, 127; Transit City project (Toronto), 238–9, 252, 254. *See also* streetcar/light rail system, Toronto; *specific Toronto LRT routes*

light rail vehicles (LRVs). *See* ALRVs; CLRVs; Flexity Outlook LRVs

Link, O. Winston, 99, 115

Little India: retail gentrification in, *72*

Little Jamaica, *229*, 254, 256

lofts: conversion of factories into, 22, 36, 40, 52, *57*, 71–2, 74, 77, *215*. *See also* condominiums; factories

Long Branch, 60; gentrification in, 260; original streetcar line of, 14, 102, 121, *186*, 280n60

Lorde, Audre, 247

low-income people. *See* poor and low-income people

Lubinski, Rob, 116

Lyden, Anne: *Railroad Vision*, 97

Maier, Vivian, 87, 88

Main Street: at Danforth Avenue, *210*

Maloof, John, 87

manufacturing, inner city loss of, 13, 14, 247; exceptions to, 52, *56*; and free trade, 13, 36, 83–4; and globalization, 14, 21, 35–42; and gentrification, 70, 74, 75, 77; and socioeconomic divide, 82–4; to suburbs/regions, 35–7, 235, 239, 255; on waterfront, 43, 44–5. *See also* deindustrialization; factories

Maple Leaf Gardens, *154–6*; redevelopment of, *156*

Marcuse, Peter: on indirect/exclusionary displacement, 72; on three pillars of critical urban planning, 2–3, 86–7, 232; on wealth/poverty polarization, 39

Marshall, Sean, 255

Massey Ferguson factory, 36, 165, *183–4*

May, Jack. *See* Bromley, John F.

McCuaig, Bruce, 49

McDonnell, Greg, 100

McKenzie, Kwame, 236

McLuhan, Marshall, 52

McMann, Robert D., 116, 119, *119*

Mees, Paul, 126, 252

Meslin, Gil, 244

Metcalfe, Amy, 96

Metro Centre project (unbuilt), 66

Metrolinx, 49, 239, 253; and Jane-Finch community hub, 255–6; lawsuit against, 48; and planned Ontario Line, 132

Metropolitan Toronto, Municipality of: amalgamation of, 15, 43–4; creation of, 8; as employer, 15; employment structure of (1983), 13; office space in, 22; population of, 12; shopping malls in, 13; TTC's two-zone system in, 10

Micallef, Shawn, and Patrick Cummins: *Full Frontal T.O.*, 118

Michigan: central railway station of (Detroit), 88–9; Rieger's repeat photography of, 92

Middleton, William D., 114; *The Interurban Era*, 114; *The Time of Trolley*, 114

Mies van der Rohe, Ludwig, 64

Mihevc, Joe, *225*

Miller, David, 238

Minneapolis, 103, 125, 256; rezoning initiative in, 260–1

Montreal, 8, 12, 15, 36, 42, 52, 70, 98; deindustrialization in, 92–3; Expo in, 66; housing policies in, 257, 259; and loss of financial services firms to Toronto, 35, 37, 39, 41, *139*; PCC assembly in, 124; streetcars in, 102, 103, 124, 126; subway in, 127, 266n22

Moos, Markus, 70, 78; with Quastel and Lynch, 75

Moscoe, Howard, 16

Moses, Robert, 60, 63, 64, 66

Mountain Legacy Project, 89; images from, 89, *92*

Mount Dennis, 73, 132, 256

National Housing Strategy, 261

Naylor, Harvey R., 116, 121

neighbourhood change. *See* Streetcar City, neighbourhood change in

neighbourhoods: repeat photography portfolio of, 189, *190–229*. *See also specific neighbourhoods*

neoliberalism, 42–5; and amalgamation of Toronto, 42, 43–4; and civic attractions, 43–5, *46*, 47;

neoliberalism (*continued*)
and events/spectacles, 45–8; and public-private partnerships, 45, 48–9, 80–2; and policing/exclusion of certain groups, 44, 45, 49; and "urban competitiveness," 42–8; and "world city" concept, 38–42

"new international division of labour" (NIDL), 38

New Urbanism, 59

Niagara Suspension Bridge, 97

Niemann, Linda Grant, and Joel Jensen: *Railroad Noir*, 115

NIMBYism, 68, 169, 242, 244, 260

North Toronto, 36, 56, 83, 123, 126, 236

Oakwood Avenue streetcar line, *229*, 256

Official Plan, of Toronto, 68, 75, 212, 242, 244, 245

Old City Hall, *143*; clock tower of, 8, 24; preservation of, 66, 143

Olympic Games, 44, 46, 47–8

Ontario Electric Railway Historical Association (OERHA), 97

Ontario Line (planned), 25, 132

Ontario Municipal Board (OMB), 65, 66

Ontario Place, 25, 65, 132

Ontario Science Centre, 25, 65, 132

Osbaldeston, Mark: *Unbuilt Toronto*, 66

O'Sullivan, Timothy, and William Henry Jackson, 89

Ottawa, 42, 117; streetcars/LRT in, 25, 126, 127, 240

Pan American Games (Toronto, 2015), 25, 46

Pape Avenue, *110*, *211*, 242

Parkdale, 52, 71, 79, 80, *192–3*, 234

Parkdale Community Land Trust (PCLT), 80, 259

parking lots, surface, *23*, *140*, *144*, *155*, 174; as created by demolition of older buildings, 75, *146*; demise of, 22, 137, *140*, *145*; NIMBYist protection of, 244; as rezoned for construction, 77, 258

Partridge, Larry: *Mind the Doors, Please! The Story of Toronto and Its Streetcars*, 115; *The Witts: An Affectionate Look at Toronto's Original Red Rockets*, 115

PCC streetcars, 12, 36, 102, 124–30, *140*; air-electric, 121, 124, *125*, 127; books on, 115, 279n44; Bromley's photos of, 118, 119; classes of, 102, 118, 278n19, 281n9, 282n19; preservation of, *129*, 129–30; rebuilding of, 128, *129*; retirement/scrapping of, 36, 127, 129, *129*, 266n24; second-hand, as bought/resold, 125–8, *126–8*, *162*, 220; as symbol of Toronto, 231; in TTC fleet, 10, 13; windows of, 124–5, *126*; at Wychwood carhouse, *223*, *224*. *See also entries below*

PCC streetcars, around the world: Brussels, 119, 278n23; Chicago, 125; El Paso, TX, 103; The Hague, 119; Johnstown, PA, 103; Montreal, 124; Pittsburgh, 103, 125; Vancouver, 124

PCC streetcars, second-hand, as purchased by TTC: from Birmingham, AL, 102, 125–6, 128, *128*; from Cincinnati, 125, *126*, 127, *127*, *162*; from Cleveland, 125; from Kansas City, 125–6, 128

PCC streetcars, second-hand, as sold by TTC: to Alexandria, Egypt, 127, *162*; to Philadelphia, 128, *128*; to San Francisco, 128; to Shaker Heights, OH, 128; to Tampico, Mexico, 127, *127*

Perry, Otto, 115

Peterson, David, 43

Peter Witt streetcars, 121, 123–4, *124*; preservation of, 130, 266n24; retirement/scrapping of, 126, 127; in TTC fleet, 10, 13

photography: cameras used in, 87, 98–9, 102, 104, 109, 110, 118, 119, 278n20; digital, 111, 114, 119, 121; editing of, xv, 119; film used in, xv, 87, 88, 98–9, 102, 103, 119, 121, 278n20; power/limitations of, 85–7; repeat, 89–96; of Toronto, 117–18; urban, 85, 87–9. *See also* railway vehicle photography; repeat photography; streetcar photography; urban photography

Pitter, Jay, 263–4

Pittsburgh, 47, 100, 110; Bromley's photos in, 118; LRT network of, 103; PCC streetcars in, 125, 127

politics: of Automobile and Streetcar Cities, 236, 238–40; in reform era, 65–7, 259–60

Polson Pier, view of Toronto from, *162–3*

poor and low-income people: bus use by, 18, 234, 255; as cyclists, 234; displacement/exclusion of, 69–70, 72–3, 74, 79–80, 257–9, 260, 262–3; early photography of, 117; in inner suburbs, 73, 79–80, 82–4, 241–2, 254, 256; Metrolinx and, 255–6; pandemic and, 39, 234, 235–6, 250, 252, 255, 256, 263–4; in polarized city, 1, 39–40, 47, 82–4; policing of, 44, 45, 49; redevelopment/gentrification and, 72–3, 79–82, *204*, 256; retail gentrification and, 72, *192–3*; TTC as unaffordable for, 256–7. *See also* affordable housing, *and entry following*; public housing; racialized people

portfolios of repeat photography: map of locations, *135–6*; downtown, 137, *138–63*; (de)industrialization, 165, *166–87*; neighbourhoods, 189, *190–229*

Priestley, Henry, 116

public housing: condos and, 67, 81–2; redevelopment/gentrification of, 80–2, *204*; "slab" design of, 63; superblocks created by, 63, 66, *204*. *See also* Regent Park

public-private partnerships (P3s): infrastructure/city services and, 48–9; "public" spaces and, 44, 45, 49; Regent Park redevelopment and, 80–2; TTC/Metrolinx and, 48, 49

Pullman Standard Company (Worcester, MA), 124, 125

Quastel, Noah, Markus Moos, and Nicholas Lynch, 75

Queen Street, *54–5*, 68, *110*, *120*, 127, 130, 189, 260; Bathurst Street at, *94–5*, *195*; in Leslieville, *196–7*; at McCaul Street, *194*; in Trefann Court, *205–6*; at Victoria Street, *144–5*; at/near York Street, *105*, *142*

Queen streetcar line, 14, 15, 60, 129, 250; Neville Park terminus of, 104, *108*; as one of world's top ten, 23;

as seen on television, 231, *232*; "walking the line" on, 119. *See also* King streetcar line

Queens Quay Terminal, 36, 44, *166–7*

Queens Quay West, 233; Harbourfront LRT line on, 25, 129, 130, *167, 171*; planned Waterfront East line on, 25, 132; streetcars on, *166, 170*

Queensway, The, *125*, 127

racialized people: and cycling/walking infrastructure, 234, 250, 252; displacement of, 73, 81, 132, 229, 254, 256; Metrolinx and, 255–6; pandemic and, 39, 84, 234, 235–6, *237*, 250, 252, 255, 256, 263–4; in public housing, 80; as segregated in inner/outer suburbs, 68, 73, 83, 84, 254–6; and streetcar fare enforcement, 132; in US cities, 65, 125, 126, 256. *See also* poor and low-income people

railway and streetcar enthusiasts: associations of, 98, 104, 118, 119; excursions of, 104, 109; generations of, *4*, 116; and historic preservation/museums, 97–8, 116; photography by, 98, 99–104, 109–11; photography acquisition/cataloguing by, 100–1, 118–19; as predominantly men, 1, 98, 100, 110; purchase of vehicles by, 100, *101*, 130, 132; websites for, 111, 114, 116, 119. *See also entry below*; streetcar photography

railway vehicle photography, 97–101; books of, 114–15; cameras/film used in, 98–9; compositions/angles of, 99–100; early genres of, 97. *See also* streetcar photography

Rankin, Katharine, and Heather McLean, 73

real estate investment trusts (REITs), 79–80, 193, 259, 261

Redhill, Michael: *Consolation*, 117

Reevy, Tony, 114

Regent Park, 58, 63, 66, 189, *204*; redevelopment/gentrification of, 80–2, *204*

Reid, H., 115

Relph, Edward, 52, 56

rental buildings, financialized, 79–80, 257–9, 262–3; and "renovictions," 79, 258–9, 263

rent controls, 78, 258, 261, 262–3

rent gap, 69–70, 71, 79, 262

repeat photography, 89–96; in books, 118; early scientific uses of, 89, 92; on Google Street View, 96; urban analysis using, *90–1, 92–3, 94–5*, 96; as used by authors to analyse Toronto's changing geographies, 1–3, 118–21. *See also* portfolios of repeat photography

Rephotographic Survey Project, 89, 92

retail: changes in (1961–2016), 10, 13, 21–2; gentrification of, 72, *72, 73*, 79, *197*, 242, 244; grocery, 72, *156, 192, 204, 220–1*; ground floor, on main streets, 52, *54*, 58, 59, 64, 68, 73, 126, 137, *184*, 189, *213, 215*; pandemic and, 235; as redevelopment projects, *156, 199*; at shopping malls/plazas, 10, 13, 36, 59, 102

Richardson, Frederick, and F. Nelson Blount: *Along the Iron Trail*, 114

Rieger, Jon, 92, 93, 116

Riis, Jacob, 87, 93; *How the Other Half Lives*, 87

Roebling, John A., 97

Rogers Road: streetcar line on, 5, 8, 110, 121, 128, *226–8*, 242, *243*, 281n16

Roncesvalles Avenue, 68, 245; streetcars on, *100*, 130, 132; Sunday stop on, *10*. *See also* King streetcar line

Roncesvalles carhouse, 104, 106, 125

Rose, Gillian: on interpretation of visual imagery, 85–6; and site of the audience, 86, 111–17; and site of circulation, 86, 111–17, 121; and site of the image, 86, 98–102; and site of production, 86, 102–11

Rosedale, 60, 65, 272n24

Rosen, Gillad: and Alan Walks, 75, 77–8

"roster" photo shots, 99, 114, 115

Royal Ontario Museum (ROM), 7, 47

Royal York Hotel, 8, 24, *161–3*

Russell, Andrew J.: "last spike" photograph by, 99

Russell carhouse, *19*

Sassen, Saskia, 40, 41, 68; *The Global City: New York, London, Tokyo*, 40

Scarborough Rapid Transit (RT) line, 11, 12, 15, 127, 128, 266n22

Seashore Trolley Museum (Kennebunkport, ME), 97

Severe Acute Respiratory Syndrome (SARS), 16, 17

Sewell, John, 66–7, 71; on suburban planning/transportation, 51, 58, 236, 252; and Trefann Court battle, 66–7, 205

Shaughnessy, Jim, 99, 114

shopping malls and plazas, 10, 13, 36, 59, 102

Siemiatycki, Matti, 48

Simpsons department store, 10, 13, 15, 21–2

Smith, Neil, 69–71, 75; Jason Hackworth and, 70–1; on "rent gap," 69–70, 71, 79; on "revanchist city," 44, 45

Smitherman, George, 238, 239

socioeconomic divide, in Toronto, 21, 82–4, 233–6, 247–64; early photography, as used to document, 117; Hulchanski's study of, 82, 83, 84, 238, 239. *See also* poor and low-income people; racialized people; Toronto, divisions in

Sontag, Susan, 85, 86, 89, 100

Spadina Avenue, 64; College Street at, *159*; Dundas Street at, *198–201*; Harbord Street at, *158*; King Street at, 36, *176–9*; Queens Quay West at, *170–1*

Spadina Expressway, 13, 64–6, 67, 128; and subway line, 65

Spadina streetcar line and LRT, 7, 11, 15, *158, 200–1*

Spremo, Boris, 117

Springirth, Kenneth: *Toronto Streetcars Serve the City*, 115; *Toronto Transit Commission Streetcars*, 115

"stable residential neighbourhoods," *209*, 242, 244–5; population loss in, 21, 189, 245

St. Clair Avenue, 123, 250, 260; stockyards/meat packing plants on, 36, 165, *187*. *See also entry below*

St. Clair streetcar line, 8, 14, 124, *187, 220*; right-of-way for, 15, 130, *187, 221*, 247–8
Steinheimer, Richard, 115; *A Passion for Trains*, 115
Stewart, George R., 96
St. George Mansions, 1, 24
St. Lawrence neighbourhood, 40, 67, 259–60
St. Louis Car Company, 124
stockyards/meat packing industry, 36, 165, *187*
Streetcar City (inner city/inner suburbs), 51–84; characteristics of (1971), 60, 63; characteristics of (2016), 60, 62, 233–4; commercial streets in, 52, *54–5*; as constructed before World War II, 3, 5, 51, 102, 231, 238; as contrasted with Automobile City, 51–9, 62–3, 231–45, 254; cycling in, 73, *217*, 234, 248, 250, *251*, 252, 255; definition/parameters of, 59–60, 73, 272n23, 272n26; density in, 52, 58, 67, 68, 241–2, 244, 260; economy of, 37–42, 234–35; grid street pattern in, 58, 59, 67; house designs of, 52; map of, showing streetcar network (1928), 60, *61*; map of, showing streetcar/subway network (2021), 60, *61*; mixed-use neighbourhoods in, 51, 52, *56*, 63–4; pandemic issues in, 234, 235–6, *237*; politics of, 236, 238–40; population loss in, 19, 21, 71, 189, 245, 247, 260; as predominantly white, 68, 84, 247; residential parking in, 56, 58; residential streets in, 52, *53*; suburbs of, 51–2, 56, 58, 71, 126, 242; as type of city advocated by Jacobs, 64, 67–8; walking in, 52, 58, 234, 250, 255. *See also entry below*; Automobile City
Streetcar City, neighbourhood change in: and condo boom, 20–1, 73–79; and gentrification, 69–82; and Jacobs' effect on planning, 60, 64–8; and public housing, 63, 66, 67, 81–2; and reform politics, 65–7, 259–60; and Toronto's socioeconomic divide, 21, 82–4, 233–6, 247–64. *See also* deindustrialization; gentrification; manufacturing, inner city loss of

streetcar enthusiasts. *See* railway and streetcar enthusiasts
streetcar/light rail system, Toronto, 22–3, 25; accessibility of, 132; expansion of, 23, 25, 132; fare enforcement issues on, 132; improvements to, 132; new vehicles ordered/purchased for, 23, 133. *See also* King Street Pilot project; streetcars, Toronto; *specific streetcar types and routes*
streetcar photography, 97–121; acquisition/cataloguing of, 100–1, 118–19; books of, 115–16; cameras/film used in, 102; compositions/angles of, 98–100, *100, 101, 102*; as creating "accidental" record of urban change, 2, 98, 110; of deviations from normal operations, 101–2; excursions for, 104, 109; geography of, 102–11; locations for, 103–4; as male-dominated hobby, 1, 98, 100, 110; production of, 102–4, 109–11; viewing of, 111, 114–17; while "walking the line," 102, *109*, 109–10. *See also entry below*
streetcar photography, locations for, 103–4; downtowns, 103–4, *105*; end of a line, 104, *108*; intersections of different lines, 104, *106–7*
streetcars, Toronto, 123–33; books on, 115; classes of, 102, 116, 118, 278n19, 281n9, 282n19; in fiction, 282n4; railway gauge for, 124; as symbols of Toronto, 116, 231, *232*; as viewed by motorists driving behind them, *232*, 241. *See also specific streetcar types*
streetcars and trams, around the world, 110–11, 248; Amsterdam, 104, *249*; Darby, PA, *112*; Duisberg, Germany, *112*; Paris, *113*; Rome, *113*. *See also* trams; trolleys
streetcar suburbs, 51–2, 58; early lack of zoning in, 52, 56, *56*; and mixed housing types of "yellowbelt," 244, 263; outermost, as frontiers of gentrification, 242; rent gap in, 71; as transformed by subways, 126
suburbs, post-1945. *See* Automobile City

subway, Toronto, 7, 11, 15, 66, 132, 234; accessibility on, 256; as built on managerial model, 48; bus routes feeding, 126, 252–3; events affecting service on, 16–17, *18*; expansion of, 11, 12, 14, 60, 65, 68, 83, 123, 127, 240, 266n22; Ford brothers and, 238–9, 240; gentrification and, 254; vs GO Transit, 241, 253; high-density development and, 59, 67, 126, *216*; map of (1960), *135*; maps of (2021), *61, 136*. *See also specific subway lines*
superblocks: commercial megaprojects as, *142, 151*–2; Eaton Centre as, 137, 143, *145*, 152; urban renewal and, 63, 66, 80, 142, *143, 204*
Swanson, Lewis, 116
Swift meat packing factory, 36, *187*

Taylor, E.P., 59
Taylor, Zach, 240
tenure/ownership, 13, 43; and affordable housing, 257–60, 261, 262, 263; of condos, 74; successful mix of, in St. Lawrence neighbourhood, 40, 67, 260; uneasy mix of, in Regent Park redevelopment, 80–2
"terrain of availability," 44, 67, *170–1, 219*
Testagrose, Joe, 116
"three-quarters" or "wedge" photo shots, 99, *100*, 102, 114, 115
ToastMaster bread factory, *185, 216*
Todd, Graham, 39, 42
Toorn, Maurits van den, 115
Toronto: "condominiumization" of, 20–1, 75, *76*; photographic history of, 117–18; politics of, 65–7, 236, 238–40, 259–60. *See also entries below*; Automobile City; Streetcar City
Toronto (1961–2016), 7–25; authors' experiences of, 7–14; census data on (1961–2016), 8–10, 12–13, 19–20, 21; changes in look of, 22, *23*; changing boundaries of, 15, 265n7; demographics of, 9, 12, 19–20; employment trends in, 10, 13, 14, 15, 21; house prices in, 22, *23*; housing types in, 9–10, 12–13, 18–19, 20–1; office space

in, 22; population of (1911–2016), *9*; public health crises in, 16–18; retail changes in, 10, 13, 21–2; socioeconomic disparities in, 18, 21; tall buildings/skyline of, 8, 12, 18–19, 24, 74, *162–3*, 242; TTC ridership/system in, 10, *10*, 11, 13–18, 22–3, 25. *See also entries below*; Greater Toronto Area; Metropolitan Toronto, Municipality of; TTC

Toronto, as competitive city, 42–8; deindustrialization and, 43; neoliberalism and, 42–5; vs previous government decentralization, 42–3; waterfront development and, 44–5; Yonge-Dundas Square and, 45, *46*

Toronto, divisions in, 231–45; affordable housing and, 241–2, 244–5; cars vs transit and, 241; density/zoning and, 67, 241–42, 244; political, 236, 238–40; socioeconomic, 233–6, *237*. *See also* Automobile City; socioeconomic divide, in Toronto; Streetcar City

Toronto, as global city, 35–49; economic restructuring in, 35–42; events/attractions in, 43–8; public-private partnerships in, 48–9; "urban competitiveness" and, 42–8; "world city" concept and, 38–42. *See also* attractions, civic; events and spectacles; neoliberalism; "world city"

Toronto Census Metropolitan Area (CMA). *See* Toronto (1961–2016)

Toronto Civic Railways, 123; book on, 115; streetcar of, at HCRR museum, 98, *182*; Wychwood carhouse built by, *222*. *See also* TTC

Toronto Feather and Down Company (later Feather Factory Lofts), *57*, *215*

Toronto Railway Company, 48, 123, 126; Lansdowne carhouse built by, *219*; streetcar of, at HCRR museum, 97–8. *See also* TTC

Toronto Railway Historical Association, 98

Toronto Railway Museum, 98

Toronto Transit Commission. *See* TTC

Toronto Transportation Commission, 56, 123. *See also* TTC

Toronto Transportation Society (TTS), 98, 104, 111

Trains magazine, 114

trams, 110–11; in Amsterdam, 104, 248, *249*; books on, 115, 116, 279n42; Bromley's photography of, *118*, *119*; dedicated lanes for, 248; in Paris, *113*; in Rome, *113*. *See also* trolleys

Transit City (planned LRT system), 238–9, 252, 254

Transit Toronto website, 111, 119

Trefann Court, 66–7, *205–6*

trolleys: books on, 98, 114, 115, 116; decline/demise of, 97, 102, 103, 118, 125; in El Paso, 103; in Johnstown, PA, 103, *104*; in Philadelphia, 248; photography of, 100, 103, 111, 114, 116, 118; in Pittsburgh, 103; preservation/museums of, 97; in San Diego, 129; storage of, *219*, *224*, 250; TTC, 7, 10, 13, 23, 110, 126, 128, *218*, 266n24. *See also* trams

TTC: books on, 115, 116, 118, 123; fare collection on, 123–4, 126, 132, 252, 254; fare prices of, 10, 14, 15–16, 252, 256–7, 285n29; poor funding/problems of, 15–16; history of, 48, 123; ridership/system of, 10, *10*, 11, 13–18, 22–3, 25; subway/LRT construction by (1954–2017), 11; Sunday stops of, 9, *10*. *See also entry below*; streetcar/light rail system, Toronto; subway, Toronto

TTC operations/ridership, events affecting: blackout (2003), 16, *19*; blizzard (1999), 16, *18*; COVID-19 pandemic (2020), 16–18; ice storm (2004), 16, *17*; labour disruptions, 16; SARS crisis (2003), 16, 17

"Two Kings," mixed-use development in, 75, 77, 248; King-Parliament area, *172–5*; King-Spadina area, *176–80*. *See also* King Street; King Street Pilot project

Union Pearson (UP) Express, 49, 73, 185, 253, 256

Union Station, 7, 41, 129, 130, 132; Bay Street at, *160–1*; as GO Transit/UP Express hub, 11, 13, 49; preservation of, 66

University of Indiana Press, 114

University of Toronto, 7; and bus lane on Scarborough campus route, 255, *255*; as employer, 15, 21; graduate residence of, *158*

Upper Canada Railway Society (UCRS), 98, 104, 119

urban photography, 87–9; history of, in Toronto, 117–18; Wyly on, 85, 86, 98, 116, 257, 262

urban renewal, 13, 43, 58, 60, 63–4, 66–7; and for-profit housing, 63–4; and public housing, 43, 58, 63; and slum clearance, 58, 60, 63–4; and superblock projects, 63, 66, 80, *142*, *143*, *204*; and Trefann Court battle, 66–7, *205*, *206*. *See also* Jacobs, Jane; Regent Park

Urban Transportation Development Corporation (UTDC), 128. *See also* CLRVs

Urry, John, 88

Vale, Thomas and Geraldine, 96

Vancouver, 35, 36, 39, 42, 70, 77; Herzog's photography of, 87–8; SkyTrain of, 127; streetcars in, 124, 126

Van Dusen, Eugene, 116

Vaughan, Annabel, 262

Vergara, Camillo José, 93, 116; *Tracking Time*, 93

Vible, Richard, 116

Victory Burlesque Theatre, *198*; redevelopment of, *199*

Walker, Jarrett, 253

walking: in Automobile City, 84, 234, 241, 252, 253; in Streetcar City, 52, 58, 234, 250, 255; along streetcar lines, to take photos, *109*, 109–10

Walks, Alan, 51, 235, 236, 238, 239; Martine August and, 80; Gillad Rosen and, 75, 77–8

Walsh, Ivor, 116

Warner, Sam Bass: *Streetcar Suburbs*, 51–2

Washington, DC: Bethesda subway station in, 252

Waterfront East LRT (planned), 25, 132

Waterfront Regeneration Trust, 44

Waterfront Toronto, 44–5

Wheel-Trans service, 256
Whistler Housing Authority (WHA), 259
Wickson, Ted, 116, 119, 121
Witt, Peter, 123. *See also* Peter Witt streetcars
women: streetcar photography by, 110, *110*; urban photography by, 87, 88
Wood, Tricia, 255
"world city": Friedmann's hypothesis on, 38–40; Sassen on, 40–1; Toronto's ranking as, 41–2
Wright, Frank Lloyd, 58
Wychwood carhouse/yard, 36, 129, *129*, 222–4; recreation of, as community hub, *224–5*, 250

Wyckoff, William, 96
Wyly, Elvin, 85, 86, 98, 116, 257, 262; and Daniel Hammel, 79

Yonge-Dundas Square, 45, *150*; evolution of, *46*; policing in, 45
Yonge Street: at Bloor Street, *157*; at Dundas Street, *148–9*, *151*
Yonge streetcar line, 56, 126, 157
Yonge subway line, 7, 8, 11–12, 126; affluent neighbourhoods along, 82, 83; high-density development along, 126, 281n10; overcrowding on, 132; as shut down by blizzard, 16, *18*
York Street, *105*, *138–9*, *141*, *142*, *166–7*

Zebracki, Martin, Brian Doucet, and Toha de Brant, 89
zoning: and affordable housing, 68, 78, 242, 244, 257, 260–3; in Automobile City, 58–59; and condo height/density increases, 77–8; and creation of superblocks, 63; and displacement of small manufacturers, 74; early lack of, in streetcar suburbs, 52, 56, *56*; inclusionary, 68, 78, 257, 262; Minneapolis initiatives in, 260–61; neighbourhood battles over, 67, 242, 244; in "Two Kings," 75, 77. *See also* density